AROMATICITY AND ANTIAROMATICITY

AROMATICITY AND ANTIAROMATICITY

Electronic and Structural Aspects

VLADIMIR I. MINKIN
MIKHAIL N. GLUKHOVTSEV
BORIS YA. SIMKIN

A Wiley-Interscience Publication

JOHN WILEY & SONS, INC.

New York · Chichester · Brisbane · Toronto · Singapore

This text is printed on acid-free paper.

Library of Congress Cataloging in Publication Data:

Minkin, V. I. (Vladimir Isaakovich)
    Aromaticity and antiaromaticity : electronic and structural
aspects / Vladimir   I. Minkin, Mikhail N. Glukhovtsev, and Boris Ya.
Simkin.
        p.   cm.
    "A Wiley-Interscience publication."
    Includes bibliographical references and index.
    ISBN 0-471-59382-6
    1. Aromaticity.   2. Aromatic compounds.   I. Glukhovtsev, Mikhail Nikolaevich
II. Simkin, B. Ya. (Boris Yakovlevich)   III. Title.
QD476.M49   1994
547.6'044—dc20   93-10743

Printed in the United States of America

10 9 8 7 6 5 4 3 2 1

CONTENTS

PREFACE

Aromaticity remains one of the most important concepts in modern organic chemistry. This concept continues to evolve and is currently in a most active phase of extension and assessment. Recent years have seen a virtual explosion in the development of various criteria of aromaticity and in theoretical work aimed at gaining deeper insight into the origins of this phenomenon. Numerous novel types of aromatic systems have been discovered both experimentally and computationally.

Much of the current refinement of the concept of aromaticity (antiaromaticity) is due to computational chemistry, which has become a worthy partner of experimental science. These recent contributions from theoretical and computational chemistry to the development of the concept and its expansion into new areas (e.g., organometallic and cluster compounds, three-dimensional structures) provided the stimulus to write this book. Our book is addressed to specialists and students in the fields of physical, organic, and organometallic chemistry. One purpose is to help chemists to integrate the results of calculations on molecules and ions into the general body of chemical knowledge. Important experimental observations on aromatic structures and aromatic behavior have been included in and integrated into the general context of the aromaticity concept.

The structure and reactivity of aromatic compounds are areas where theory and experiment meet closely and combine symbiotically to provide new insights. We hope this book may stimulate further interest of both experimentalists and computationally oriented chemists in this fascinating and rapidly developing subject.

The translation of the manuscript for this book was done by Konstantin

Blokhin, to whom we express our gratitude. We also are deeply indebted to Professor Paul von R. Schleyer, who encouraged us to write this book. One of the coauthors, Mikhail N. Glukhovtsev, worked as an Alexander von Humboldt Fellow in Erlangen (Institut für Organische Chemie des Universität Erlangen-Nürnberg). He is grateful to Professor Schleyer for useful discussions and critical comments on various aspects of aromaticity. Some recent results obtained in the group of Professor Schleyer have been included in this book. Dr. Vera Glukhovtseva is gratefully acknowledged for help in the preparation of the manuscript. The final stages of the preparation of the manuscript were carried out while M. N. Glukhovtsev was a Visiting Research Fellow at the Australian National University (ANU) and B. Ya. Simkin was a Visiting Scientist at Cornell University. They thank Professor Leo Radom and the Research School of Chemistry of ANU and Professor Roald Hoffmann and Department of Chemistry of Cornell University for their hospitality.

VLADIMIR MINKIN
MIKHAIL GLUKHOVTSEV
BORIS SIMKIN

Rostov State University, Rostov on Don, Russian Federation
Australian National University, Sydney, Australia
Cornell University, Ithaca, USA
November 1993

LIST OF ABBREVIATIONS

A	Structural index of aromaticity, Eqs. (2.54) and (2.55)
	ab initio nonempirical quantum mechanical method of calculation
AM1	Austin model 1
AO	Atomic orbital
AS	Aromatic stabilization index, Eq. (2.48)
CASSCF	Complete-active-space SCF calculation
CCMRE	RE calculated with the use of conjugated circuit model, Eq. (2.26)
CEPA	Coupled electron-pair approximation
CI	Configuration interaction
3 × 3 CI	CI for all three possible singlet configurations in the two-electron two-orbital model
CI-SD	Single and double excitations, single reference CI
CI-SDT	CI-SD plus triple excitations
CI-SD(Q) or QCISD	CI-SD plus quadruple excitation correction
CNDO	Complete neglect of different overlap
CSC	Corrected structural count
D	Index of delocalization, Eq. (2.115)
3D-HMO	Three-dimensional HMO theory
DRE	Dewar resonance energy, Eq. (2.6)
DZ	Double-zeta basis set having two basis functions per Slater atomic orbital: DZP means double-zeta plus polarization

ED	Gas-phase electron diffraction
EHMO	Extended Hückel method
GVB	Generalized valence bond method
HF	Hartree–Fock method
HHSE	Hyperhomodesmotic stabilization energy
HMO	Hückel molecular orbital method
HOMO	Highest occupied molecular orbital
HOMAS	Harmonic oscillator model of aromatic stabilization, Eqs. (2.57)–(2.59)
HOSE	Harmonic oscillator stabilization energy, Eq. (2.60)
HRE	Hückel resonance energy, Eq. (2.5)
HSE	Homodesmotic stabilization energy
HSRE	Hess–Schaad resonance energy, Eq. (2.9)
I	Structural index of aromaticity, Eq. (2.66)
I_A	Unified structural index of aromaticity, Eq. (2.66a)
I_1, I_2	Aromaticity indices based on the anisotropic optical polarizability, Eqs. (2.119) and (2.120)
IBA	Index of bond alternation, Eq. (4.28)
IC	Information index, Eq. (2.118)
IEPA	Independent electron-pair approximation
IGLO	Individual gauge for localized orbitals
INDO	Intermediate neglect of differential overlap
ISE	Isodesmic stabilization energy
LM	Logarithmic model, Eq. (2.30)
LMO	Localized MO
LUMO	Lowest unoccupied molecular orbital
MBPT(n)	nth Order of many-body perturbation theory
MC SCF	Multiconfiguration SCF theory
MERP	Minimal energy reaction path
MINDO	Modified INDO
MNDO	Modified neglect of diatomic overlap
MNDOC	Correlation corrected version of MNDO
MO	Molecular orbital
MPN ($N = 2$–4)	Møller–Plesset perturbation theory of order 2–4
MP2/6-31G*//HF/6-31G*	Example of abridged notation to specify the type of level used: the single point calculation with use of the MP2 theory and of the 6-31G* basis set, at the geometry optimized at HF/6-31G*
MRDCI	Multireference double CI
MW	Microwave spectroscopy
N-31G, etc.	Pople's basis set. Notations of these basis sets like n-ijG or n-ijkG can be encoded as: n, number of gaussian primitives for the inner shells; ij or ijk, number of primitives for contractions in the valence shell. The ij notations correspond to

	basis sets of valence DZ quality and *ijk* notations indicate sets of valence triple zeta (TZ) quality. *n-ij*G*, a basis set augmented with *d* type polarization functions on heavy atoms only; *n-ij*G** (or *n-ij*G(d, p)), *n-ij*G* basis set with *p*-functions on hydrogens
ΔN	structural index of aromaticity, Eq. (2.64)
P(X)	Characteristic polynomial, Eq. (2.13)
PES	Potential energy surface
PMO	Perturbation MO theory
POAV	π-Orbital axis vector (Fig. 3.5)
PPP	Parizer–Parr–Pople method
QMRE	Quantum mechanical resonance energy
R(X)	Acyclic (reference) polynomial, Eq. (2.23)
RCI	Ring current index
RE	Resonance energy
REPE	Resonance energy per π electron
RHF	Restricted HF theory
SC	Structure count
SCF	Self-consistent field
SI	Stability index, Eq. (2.42)
SINDO1	INDO method modified on the basis of symmetrically orthogonalized orbitals and commutator relations
SRTRE	RE scheme based on the structure resonance theory, Eq. (2.27)
STO	Slater type orbital
TCRE	Thermochemical resonance energy
TCSCF	Two-configurational SCF
TRE	Topological resonance energy, Eq. (2.25)
UMNDO	UHF version of MNDO
UHF	Unrestricted HF theory
ZDO	Zero differential overlap
ZPVE (ZPE)	Zero point vibration energy

AROMATICITY AND ANTIAROMATICITY

1

INTRODUCTION

Among the theoretical concepts that constitute the rational basis of modern organic chemistry there are some controversial constructs, but, perhaps, none to such a degree as that of aromaticity. With all its versatility and usefulness for the systematization of various characteristics relating to structure, stability, reactivity, and other chemical concepts, the idea of aromaticity lacks secure physical basis and is ill-defined and vague. Numerous attempts at the canonization of this concept have shown again that it could not be confined within any rigid framework whether it be of speculative or empirical nature [1–9]. No wonder, unending debate has been going on for a considerable time whether the term "aromaticity" may at all be rightfully regarded as legitimate [5]. The following comment by Binsch [7] illustrates the intensity of the debate: "Aromaticity is just a name, and we are at liberty to continuously adapt its meaning to our changing needs for conceptualization.... It is indeed suspicious how often magic rules had and have to serve as an alibi for creating an aura of intellectual respectability for chemical research which is on the verge of turning stale."

This is certainly forceful language. However, whatever guesswork and wrangling there is about this concept, the plain fact remains that it constitutes the basis for very useful classification of organic, inorganic and organometallic cyclic compounds, both qualitative and quantitative, into aromatic, antiaromatic, or nonaromatic classes, with the quantitative degree of aromaticity (antiaromaticity) determined within each class.

Apparently, the thorniest problem associated with the application of this concept stems from the fact that there are quite a large number of properties that are invoked to indicate aromaticity. As a result, a need arises for a choice

1

of not only a quantitative but also qualitative measure of aromaticity. So, having devised a scale of aromaticity based on a particular property, we cannot be sure that this classification will be valid with another characteristic even if its physical content is similar. There are a multitude of various criteria of aromaticity [1–6, 10]. Moreover, there is no simple relationship between these criteria [10, 11]. Criteria of at least two types should be chosen as the main criteria, for example, an energy criterion and a structural criterion [11]. That is, aromaticity is at least a two-dimensional phenomenon [11].

All-inclusiveness of the concept predetermines the vagueness it creates; hence uncertainty about the assignment of a particular compound, and sometimes the correctness of a whole classification scheme, may be in doubt.

An illustrative example featuring the confusion associated with the concept of aromaticity is given by the problem of aromaticity of [n]-para-cyclophanes **1** (for details see Chapter 2). Based on the UV and ^{1}H NMR spectra of [5]-para-cyclophane, Jenneskens and co-authors [12] who had synthesized it concluded that the "benzene ring retains its aromatic character with a remarkable tenacity."

$$n = 5 \quad \phi = 23.7°$$
$$n = 6 \quad \phi = 18.6°$$
$$n = 7 \quad \phi = 14.2°$$

1

However, calculations using MNDO [13] and molecular mechanics [14] methods contradicted that judgment and showed the strain energy of **1** ($n = 5$) to exceed the resonance energy of benzene. In a subsequent series of papers, H. F. Schaefer and co-workers [15–17] demonstrated that this contradiction sprang from the use of criteria of aromaticity.

The criterion opted for by the researcher predetermines his/her viewpoint and, accordingly, the assignment of a given compound. Thus, considering the insignificance of bond alternation (0.025 Å) in the benzene ring of [5]-para-cyclophane, this molecule should be classified as aromatic. But looking at it from a different angle and taking note of the substantial nonplanarity of the benzene ring ($\phi = 23.7°$ [15]), one is inclined to view the para-cyclophane as nonaromatic. However, the close correspondence between the predicted vibrational frequencies of the molecule in question and para-dideuterobenzene indicates similarity with benzene. Hence one returns to the verdict of aromaticity. The calculated resonance energy of [5]-para-cyclophane presents another problem: it is negative ($- 50.1$ kcal/mol at HF/DZ, this value also contains the strain energy, which is hard to separate out), in contrast to $+ 28.1$ kcal/mol for benzene. However, it was shown with the use of the p-orbital axis vector (POAV, see Chapter 3) analysis that the boat-shaped benzenes with deviations from planarity upto 25° retain conjugation and can be considered as "essentially aromatic compounds" [18]. Moreover, while ring puckering disrupts cyclic conjugation in benzene, benzene is not very rigid itself. For the range of small torsional deformations (<15°), benzene and cyclohexane have been found to be equally flexible [19]. What can one now say about the aromaticity of *para*-cyclophane?

One more factor that gives rise to the multitude of definitions and criteria of aromaticity is the large variety of structural types (including nonclassical ones) of the compounds to which this concept is applied. As a result, there are many derivatives of the concept, some of which are shown in Scheme 1.1.

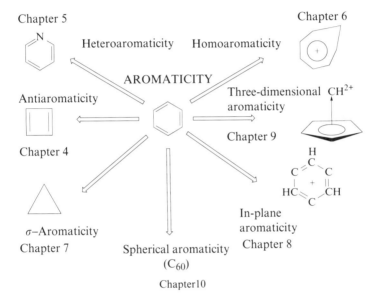

Scheme 1.1

Having grown into such an all-embracing system, the aromaticity has, of necessity, surrendered its original unsophisticated definiteness and has even become a target of jokes (schizoaromaticity) [5].

Unfortunately, in suggesting new types of aromaticity, the authors often fail, as was correctly noted by Marschand [20], to define the prototypical concept, thus putting their constructs on a shaky basis. It was proposed to abandon this term altogether or, at least, to introduce some new ones to describe the principal types of molecular characteristics [5, p. 21; 21] (e.g., "benzenoid" for the description of structure, "meneidic" referring to reactivity, and "hückelian" for ground-state properties [5, p. 85; 21]).

These suggestions, which had mainly to do with semantic distinctions [5, p. 386], did not meet with approval. One of the reasons for this was, apparently, the fact that it is convenient to rationalize the properties, common to compounds of different types, in terms of one unifying concept. Of course, it would be preferable if those diverse properties were related to one and the same physical effect.

The concept of aromaticity plays such an important role in teaching organic chemistry and in conducting research that all proposals to do away with it are simply not realistic [21]. It is indeed hard not to agree with A. T. Balaban when he says: "so instead of trying to turn our thumbs down on a frequently used term, we should better make the use of it" [6].

There is one more source of ambiguities and difficulties in the use of the concept of aromaticity. It is insufficient precision of the terms employed for its interpretation, namely, electron delocalization and conjugation.

It is routinely assumed that electron delocalization is a distinguishing property of the π-conjugated systems, which is most sharply manifested in aromatic compounds. Yet, strictly speaking, all electrons in a molecule, including those of the core shells, are delocalized. Hence electron delocalization is not a feature specific to π-conjugated systems only (for more details see [22–24]).

On the other hand, it is extremely useful to consider the bonding, lone pair, and inner shell electrons to be essentially "localized" in a bond, lone pair, or core region. This assumption underlies the concept of "bond localization," which reflects the fact that many molecular properties, such as the dipole moments, diamagnetic susceptibilities, and heats of formation, can be calculated by means of additive schemes. This means, in effect, that the bond properties are transferable from one molecule to another. This fact, however, does not give any ground for attaching a physical sense to these terms.

Working from the definition of "bond localization," it is easy to formulate the concept of electron (bond) delocalization. This type of delocalization occurs when collective properties of a molecule cannot be represented as a sum of the individual contributions, that is, when nonadditivity effects are operative.

Bond delocalization should not always be related with the concept of "conjugation." Originally, conjugation was used in a topological sense, indicating that each pair of double (multiple) bonds in a conjugated system is separated by just one single bond. Today, the term conjugation simply denotes interactions between single (σ-conjugation) or between multiple bonds (π-conjugation). If the conjugation effects are constant, they do not affect the pattern of "bond localization," since by selecting appropriate increments one may achieve (staying within the framework of the additive scheme), a high degree of accuracy in calculations of collective properties.

Obviously, bond conjugation is far more common than electron (bond) delocalization. Conjugation does not always lead to bond delocalization [23, 24].

Thus aromaticity (antiaromaticity) of cyclic systems will, in the first place, be manifested in the effects of cyclic electron (bond) delocalization and the nonadditivity of the collective, primarily energetic, properties. So the following chapter is devoted to methods for the assessment of the cyclic electron (bond) delocalization effects, which are significant for the various criteria of aromaticity.

REFERENCES

1. G. M. Badger, *Aromatic Character and Aromaticity*. University Press, Cambridge, 1969.
2. D. Lewis and D. Peters, *Facts and Theories of Aromaticity*. Macmillan, London, 1975.
3. D. Lloyd, *Non–benzenoid Conjugated Carbocyclic Compounds*. Elsevier, Amsterdam, 1984.
4. P. J. Garrat, *Aromaticity*. Wiley, New York, 1986.
5. E. D. Bergmann and B. Pullman (Eds.), *Aromaticity, Pseudoaromaticity, Anti-aromaticity*. Israel Academy of Science and Humanities, Jerusalem Symposium on Quantum Chemistry and Biochemistry, Vol. 3, 1971.
6. A. T. Balaban, *Pure Appl. Chem.*, **52**, 1409 (1980).
7. G. Binsch, *Naturwissenschaften,* **60**, 369 (1980).
8. J.-P. Labarre and F. Crasnier, *Topics Curr. Chem.*, **38**, 33 (1971) .
9. M. N. Glukhovtsev, B. Ya. Simkin, and V. I. Minkin, *Russian Chem. Rev.*, **54**, 54 (1985).
10. A. R. Katritzky, P. Barczynski, G. Masummarra, D. Pisano, and M. Szafan, *J. Am. Chem. Soc.*, **111**, 7 (1989).
11. K. Jug and A. M. Köster, *J. Phys. Org. Chem.*, **4**, 163 (1991).
12. L. W. Jenneskens, F. J. J. de Kanter, P. A. Kraakman, L. A. M. Turkenburg, W. E. Koolhaas, W. H. de Woff, F. Bickelhaupt, Y. Tobe, K. Kakiuchi, and Y. Odaira, *J. Am. Chem. Soc.*, **107**, 3716 (1985).
13. H. Schmidt, A. Schweig, and W. Thiel, *Chem. Ber.*, **111**, 1958 (1978).
14. L. Carballeira, J. Casado, E. Gonzalez, and M. A. Rios, *J. Chem. Phys.*, **77**, 5655 (1982).
15. J. E. Rice, T. J. Lee, R. B. Remington, W. D. Allen, D. A. Clabo, and H. F. Schaefer, *J. Am. Chem. Soc.*, **109**, 2902 (1987).
16. T. J. Lee, J. E. Rice, W. D. Allen, R. B. Remington, and H. F. Schaefer, *J. Chem. Phys.*, **123**, 1 (1988).
17. T. J. Lee, J. E. Rice, R. B. Remington, and H. F. Schaefer, *Chem. Phys.*, **150**, 63 (1988).
18. L. W. Jenneskens, E. N. van Eenige, and J. N. Louwen, *New J. Chem.*, **16**, 775 (1992).
19. K. B. Lipkowitz and M. A. Peterson, *J. Comput. Chem.*, **14**, 121 (1993).
20. A. P. Marschand, *Chem. Ind. (London)*, 1760 (1964).
21. D. Lloyd and D. R. Marschall, *Agnew. Chem. Int. Ed. Engl.*, **11**, 404 (1972).
22. M. J. S. Dewar, "Localization and Delocalization," in J. F. Liebman and A. Greenberg (Eds.), *Modern Models of Bonding and Delocalization*, VCH, New York, 1988.
23. D. Cremer, *Tetrahedron*, **44**, 7427 (1988).
24. E. Kraka and D. Cremer, "Chemical Implication of Local Features of the Electron Density Distribution," in Z. B. Zaksič (Ed.), *The Concept of the Chemical Bond*, Part 2, Springer, Berlin, 1990, p. 453.

2

CRITERIA OF AROMATICITY AND ANTIAROMATICITY

2.1 DEMANDS ON THESE CRITERIA

Various criteria for aromaticity are known [1–3]. It is impossible to define aromaticity in a completely exhaustive manner (see Chapter 1). Accordingly, there exists no unambiguous yardstick according to which one might assign a compound to the aromatic or antiaromatic class. The most important requirements that a criterion of aromaticity should meet are as follows: (a) it must be directly related with some known physicochemical effect regarded as a manifestation of aromaticity and this effect must be experimentally quantifiable; and (b) it is essential that the fulfillment of the chosen criterion should indicate the presence of such properties in a given compound as are commonly regarded to be the main attributes of aromaticity.

In determining the aromaticity or antiaromaticity of a compound using different criteria, one should strive to avoid the legendary situation when blind sages attempted to describe the elephant. In some cases not all the main criteria of aromaticity—namely, the energetic, structural and magnetic criteria—are satisfied concurrently [4]. This discordance may be accounted for by the fact that the criteria refer to different, mutually "orthogonal," groups [5].

But then is a coherent system conceivable of interrelated, noncontradictory criteria of aromaticity? The analysis in this chapter of various criteria will suggest answers to this question. According to the definition of aromaticity adopted in Chapter 1, the principal criteria are energetic.

2.2   ENERGETIC CRITERIA

The aromaticity (antiaromaticity) of a compound is associated for the experimentalist with primarily its stability (instability) against valence isomerizations, intra- and intermolecular cyclizations, recyclization reactions, and so on [1–3]. The reactivity of the aromatic compounds is characterized by a so-called regeneration, that is, "the tendency to retain the type" [6]. The original type of the electronic system, lost at a certain reaction stage, is restored in the products. Such regenerative (or meneidic [6]) behavior of aromatic compounds is regarded as a manifestation of their special stability. All this was, apparently, a good reason for assigning the dominant role in determining the aromaticity, and later the antiaromaticity, to the energy criterion, which rests on energy estimates of aromatic stabilization. However, the stability (instability) of a compound characterized by cyclic electron (bond) delocalization may depend not only and even not so much on the aromaticity (antiaromaticity) but rather on various other factors. Therefore, in order to classify a compound as aromatic, antiaromatic, or nonaromatic, it is necessary to single out the stabilization (destabilization) caused by the cyclic electron (bond) delocalization (see Chapter 1).

To determine this contribution, quantum chemical and experimental schemes have been devised for estimating the so-called resonance energy (RE). The content of this term will be examined in the next section; here we merely note that, besides the RE calculation, schemes based on general theoretical models (in certain cases approaches, also resting on the energy criterion, but of a more specific character) may be applied. These models include various other quantitative characteristics concerning the chemical processes attended by the breaking up or, conversely, the formation of an aromatic (antiaromatic) system (see Section 2.5).

The important points that determine the leading role of the energy criteria—the structural and magnetic ones—are covered in Sections 2.3 and 2.4. A quantitative relationship has been established between the RE values and the aromaticity indexes based on other criteria, such as the values of ring currents [7–9]. Also, the dependence has been found on the value of $\Delta G^{\neq}$ an index of the "kinetical stability," for example, for the Diels–Alder reactions of aromatic hydrocarbons with maleic anhydride [10]. Furthermore, of importance is the relationship between energy criteria and electron-count rules, such as the $(4n + 2)$ Hückel rule [11], which represents a convenient tool for qualitative verification of aromaticity of a given compound.

Various schemes of calculation of REs serve as quantitative indexes of the aromaticity (antiaromaticity) within the framework of an energy criterion. Next, we consider these schemes in some detail.

2.2.1   Various Types of Resonance Energies

Before entering upon any particulars concerning the schemes for calculating the resonance energies, it is important to define this term as we shall understand it, because it is often used to describe essentially different characteristics of a conjugated molecule.

In the context of the present book, the resonance energy (RE) will refer to the part of the total energy due to the electron cyclic (bond) delocalization. The definition of the latter term (considered in Chapter 1) implies that in order to find the value of the RE the difference must be calculated between a quantity characterizing the experimentally found energy of a given molecule (such as atomization enthalpy ΔH_a or formation enthalpy ΔH_f) and the same characteristic obtained with the aid of an additive scheme [12]. Thus for benzene, we have

$$RE = \Delta H_a^\circ \text{ (benzene)} - 6E(C-H) - 3E(C-C) - 3E(C=C) \qquad (2.1)$$

The specificity of each scheme for calculating the resonance energy depends on the procedure employed to calculate the bond energies $E(C-H)$, $E(C-C)$ and $E(C=C)$ and the problem comes down to constructing this procedure in such a way as to allow one (after determining with its aid the bond energies) to single out of the total energy of a molecule the contribution coming from the electron cyclic (bond) delocalization. In other words, this procedure should be based on a model reference structure whose energy would differ from that of the cyclic structure precisely by the component corresponding to the delocalization in question.

A solution to the problem of the choice of the reference structure meeting the above requirement was proposed by Dewar and deLlano [13]. They suggested using the energies of acyclic linear polyenes. These energies are contained in equations similar to (2.1), which are employed to calculate the RE of a given compound. Within the framework of this approach, various schemes for the determination of the RE have been devised [13–23]. It should be noted that the REs can be divided into two major classes: the thermochemical type abbreviated to TCRE and the vertical one [24–26]. In the latter case, one has to deal with a variety of the so-called quantum mechanical resonance energy (QMRE) [24, 25]. In the general case, the QMRE corresponds to the energy contribution made by the electron delocalization as a whole but not by the part of it represented by the electron cyclic (bond) delocalization. For this reason, in calculating the general form of the QMRE, the reference structure must have "isolated" (non-interacting) double bonds [24] (in the VB scheme such a calculation takes into account the contribution from one only, the most stable, resonance structure [25, 26]).

On the other hand, to apply the Dewar's approach within the QMRE scheme (e.g., for conjugated cyclic hydrocarbons), it would be necessary to normalize the resonance energies, calculated in terms of the above model, with respect to butadiene in accordance with relationship (2.2) [24].

$$QMRE(Dewar) = QMRE - nQMRE(butadiene) \qquad (2.2)$$

where n is the number of single bonds in the Kekule structure of the molecule.

In calculating the TCRE, the bond energies are determined from the energy of acyclic polyene having equilibrium geometry. The difference between TCRE and QMRE may be visualized by Scheme 2.1 [25], in which $\Delta E_{distort}^{\pi\sigma}$ is the energy required for the distorting of the D_{6h} structure of benzene into the D_{3h} structure.

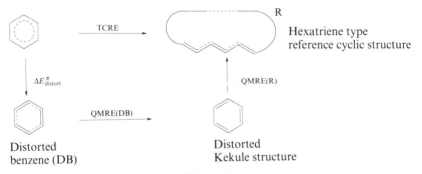

Hexatriene type
reference cyclic structure

Distorted
benzene (DB)

Distorted
Kekule structure

Scheme 2.1

In the following sections some characteristic schemes for calculating the RE will be considered, such as those of Hückel, Dewar, and Hess–Schaad. Particular attention will be given to the currently common schemes based on the isodesmic, homodesmotic, and hyperhomodesmotic reactions (Section 2.2.9), whereby the expressions for the values of bond energies found with a given set of molecules are substituted into the RE so that the problem is reduced to the determination of the RE from the enthalpy of the relevant reaction. For example, when the bond energies are calculated from experimental values of ΔH_a° which are taken for the set of reference molecules CH_4, CH_3CH_3, $CH_2{=}CH_2$, the calculation of the RE comes down to the evaluation of ΔH of the reaction

$$\text{benzene} + 6CH_4 \rightarrow 3CH_3CH_3 + 3CH_2{=}CH_2 \qquad (2.3)$$

in which case $\Delta H = 64.2 \pm 1.7$ kcal/mol [12].

With molecules such as CH_4, $CH_2{=}CH_2$, and *trans*-butadiene ($={C}{-}C{=}$bond), the RE is determined in the form of the enthalpy of reaction (2.4)

$$\text{benzene} + 3CH_2{=}CH_2 \rightarrow 3\text{-}trans\text{-}CH_2{=}CH{-}CH{=}CH_2 \qquad (2.4)$$

with $\Delta H = 21.6 \pm 1.5$ kcal/mol [12].

Reaction (2.3) is isodesmic [27] (equal numbers of formal single and double bonds between the carbon atoms in reactants and products), and reaction (2.4) is classified as homodesmotic [28] (where, unlike (2.3), the number of bonds of every formal type proves equal in reactants and products). In both of them, particularly in (2.4), the energy contribution to ΔH, stemming from the difference between the types of hybridization (bonding states) of the carbon atoms as well as between the types of $C(sp^n){-}H$ bonds in reactions and products, is reduced to a minimum.

Thus the RE determined from the energy of the isodesmic reaction of bond separation is in fact QMRE-like and represents an estimate of various effects of electron delocalization. By contrast, the use of the homodesmotic reaction leads to a Dewar-type RE [28] allowing the evaluation of the contribution by precisely the cyclic electron bond delocalization.

Calculations of the so-called empirical resonance energies, based on ΔH of combustion or hydrogenation reactions, in which the above differences are not minimized, produce a considerable scatter in the RE data [29]. These calculations will not be discussed here; the interested reader may find their critical analysis, for example, in the review by George [29] and in the books by Lewis and Peters [1] and by Garrat [3]. Our analysis starts with the chronologically first scheme, namely, that developed by Hückel.

2.2.2   Hückel Resonance Energy

The determination of resonance energies according to Hückel (HRE) comes down to an evaluation of the QMRE within the framework of the Hückel MO method (HMO), in other words, it is equivalent to the calculation of the delocalization energy [1, 3, 30]:

$$HRE = DE = -(E_\pi - n_{C=C}(2\alpha + 2\beta))\tag{2.5}$$

where n is the number of double bonds in the hypothetical Kekule-type structure (non-resonating double bonds) and E_π is the π-electron energy of the conjugated molecule calculated by the HMO method.[1] For example, the HRE(benzene) $= 2(-\beta)$. The HRE is used to evaluate the energy of the electron delocalization rather than the cyclic electron (bond) delocalization; hence the inadequacy of the HRE (DE) scheme in assessing aromaticity or antiaromaticity is obvious. According to the DE values, all annulenes, except cyclobutadiene (DE = 0), must be aromatic [1, 30] and highly unstable pentalene has an even higher value of DE $(2.46(-\beta))$ than benzene $(2(-\beta))$. The HRE values for benzene and pentalene are closer in per π electron calculations (HREPE) -0.33 and 0.31, respectively. The unsuitability of DE in the assessment of the aromaticity has been well documented in the literature [1, 3]. The use of this scheme may be justified in some special cases (the size of the molecule) for obtaining preliminary estimates only. For example, one may mention the prediction of the stability and aromaticity of the carbon clusters C_n ($n = 60, 120$) based on HRE calculation [31]; even though subjected to just criticism, these calculations started an avalanche of work on this problem [32–34].

---

[1]The minus in Eq. (2.5) is necessary since positive values of the RE are assumed for the aromatic molecules.

2.2.3 Dewar's Resonance Energy

Unlike the Hückel values, the Dewar resonance energy represents solely the contribution coming from the cyclic electron (bond) delocalization. The model reference structure is not a system of isolated π-bonds, but a hypothetical cyclic polyene with the number of the π- and σ-bonds equal to that in a given molecule. Making use of the additivity of bond energies in acyclic polyenes [35, 36], one may calculate the total energy of any acyclic or hypothetical cyclic polyene by summing the energies of all bonds. The Dewar resonance energy (DRE) is found as the difference between the atomization enthalpies of a given conjugated molecule (ΔH_a^M) and of the classical Kekule reference structure (see also Section 2.1):

$$\text{DRE} = (\Delta H_a^M - \Delta H_a^{\text{add}}) \qquad (2.6)$$

where ΔH_a^{add} is the atomization enthalpy calculated for the reference structure.[2] For example, in the case of benzene $\Delta H_a = 57.16$ eV, $\Delta H_a^{\text{add}} = 3E(\text{C}=\text{C}) + 3E(\text{C}-\text{C}) + 6E_{\text{CH}} = 56.29$ eV [13]; hence DRE $= 0.87$ eV (20 kcal/mol).

Usually ΔH_a is employed in view of the possibility of using bond energies in the calculation of ΔH_a^{add}; of course, ΔH_f may also be applied in calculating the DRE [14]. If there are no data on the values of ΔH_a, quantum chemical methods may be employed for their calculation.

The energy additivity of acyclic conjugated compounds has been evidenced by calculations in the π-electron approximation [13, 36]. *Ab initio* calculations [38–40] have confirmed such additivity for linear polyenes, which underlies the DRE scheme. The maximal deviation of the total energy from the additivity amounts to 2.5 kcal/mol (STO-3G) with the optimization of only the C—C bond lengths [40] and comes to a mere 0.08 kcal/mol (STO-3G) for a complete geometry optimization of linear polyene [38]. Calculations of the RE of benzene using the Dewar-type expression (2.7) yield the values of 26 kcal/mol (3-21G) and 23 kcal/mol (6-31G*) [39] (cf. Eqs. (2.6) and (2.7)):

$$\text{RE}(\text{C}_n\text{H}_n) = E(\text{C}_n\text{H}_n) - (n/2)(E_{\text{CH}=\text{CH}} + E_{\text{CH}-\text{CH}}) \qquad (2.7)$$

For cyclobutadiene the values of the RE with allowance made for the strain

---

[2] Here a note is in order regarding the sign before the values of ΔH_a. According to the definition (see [12]) for the hydrocarbons C_mH_n, $\Delta H_a^{\circ} \text{C}_m\text{H}_n) = m \Delta H_f^{\circ}(\text{C})_g + n \Delta H_f^{\circ}(\text{H})_g - \Delta H_f^{\circ}(\text{C}_m\text{H}_n)$, where $\Delta H_f^{\circ}(\text{C})$ and $\Delta H_f^{\circ}(\text{H})$ are the enthalpies of formation of the carbon atom from graphite and the hydrogen atom from H_2. In [13] the opposite sign before ΔH_a is used; that is $\Delta H_a^{\text{exptl}}$ (benzene) $= -57.16$ eV, but this condition is not observed consistently, in Table 2 of [13] $\Delta H_a^{\text{exptl}}$ (benzene) $= -57.16$ eV while on page 794 we read that "the experimental value (Table 2) is 57.16 eV." Moreover, in [13] the value of ΔH_a^{add}(benzene) $= 56.29$ eV is given for expression (16) of [13]. In connection with all this and following the statement that the atomization enthalpies ΔH_a are always positive (in agreement with the thermodynamics sign conversion) we apply definition (2.1). When, however, the negative sign is used for ΔH_a (i.e., ΔH_a (benzene) $= -57.16$ eV), then the minus must be used before the parentheses in Eq. (2.6) [37].

energy comes out at -61.9 kcal/mol (3-21G) and -54.7 kcal/mol (6-31G*). In determining the RE by means of Eq. (2.2), a calculation with the electron correlation included gives RE(benzene) = 26 kcal/mol (DZ basis set) [24]. In the same work, it has been found that vertical *ab initio* resonance energies (see Section 2.2.1) are proportional to HREs (Eq. (2.5)) with the proportionality constant of (20 kcal/mol)·β.

DRE values, some of which are given in Table 2.1, have been calculated for a broad range of systems including the heterocyclic compounds [13–15, 35]. They are well correlated with structural and magnetic criteria of the aromaticity (see Sections 2.3 and 2.4) as well as with the data on the reactivity of given compounds such as the logarithm of the rate constant for the Diels–Alder addition of maleic anhydride to dehydro[*n*]annuleno[*c*]furans (see Section 2.2.10).

2.2.4 Hess–Schaad Resonance Energies

Since the DRE calculation scheme takes into account only two types of CC bonds, its applicability is restricted to such cyclic molecules for which a linear polyene serves as a reference structure. This is explained by the fact that the energy of acyclic polyene depends on its branching. So the π-energy of branching of an acyclic polyene $E_\pi(\text{BP})$ is related to the energy of a linear polyene $E_\pi(\text{LP})$ having the same number of carbon atoms by Eq. (2.8) [41], in β units:

$$E_\pi(\text{BP}) = E_\pi(\text{LP}) - 0.09\ T \qquad (2.8)$$

where T is the number of branching sites in the polyenes. Thus a more detailed differentiation is needed for the values of bond energies that correspond to the different types of bonds. For branching polyenes, the additive energy scheme is also valid. For example, the value of T may be represented as a linear function of the number of carbon–carbon bonds m for a corresponding type of polyene: in the case of linear polyenes $T = 0$, for type (1) $T = 1/2\ (m - 3)$, and so on [41]. Consequently, the π-energy of these molecules may be written as a sum of bond energy terms. Bearing this in mind, Hess and Schaad applied the DRE model for calculating resonance energies (HSRE) within the Hückel MO method [16–19, 42–46]. The successful application of this scheme with even such an unsophisticated method as HMO has highlighted the importance of the correct choice of the reference structure.

Hess and Schaad classified the bonds in acyclic polyenes into eight types [16, 17], depending on the number of attached hydrogens; five comprise CC double bonds and three represent CC single bonds (energies are given in β units):

$$E_{\text{H}_2\text{C}=\text{CH}} = 2.0000 \qquad E_{\text{HC}=\text{CH}} = 2.0699$$
$$E_{\text{H}_2\text{C}=\text{C}} = 2.0000 \qquad E_{\text{HC}=\text{C}} = 2.1083$$
$$E_{\text{C}=\text{C}} = 2.1716 \qquad E_{\text{HC}-\text{CH}} = 0.4660$$
$$E_{\text{HC}=\text{C}} = 0.4362 \qquad E_{\text{C}-\text{C}} = 0.4358$$

In this classification, the branching is implicitly taken into account. The expression for calculating the HSRE has the following form (for the energy in β units):

$$\begin{aligned}
\text{HSRE} = E_\pi \text{ (conjugated molecule)} &- (n_{H_2C=CH} E_{H_2C=CH} + n_{HC=CH} E_{HC=CH} \\
&+ n_{H_2C=C} E_{H_2C=C} + n_{HC=C} E_{HC=C} + n_{C=C} E_{C=C} \\
&+ n_{HC-CH} E_{HC-CH} + n_{HC-C} E_{HC-C} + n_{C-C} E_{C-C}
\end{aligned} \qquad (2.9)$$

In finding the values of π-bond energies, the problem arose of determining eight unknowns from a set of six linear equations [16], which indicates the impossibility of constructing an acyclic conjugated hydrocarbon with more than six types of bonds [16]. Therefore the determination of these values should involve two arbitrary ones, whereby six others might be derived. Hess and Schaad assigned a value of 2.0 to $E_{H_2C=CH}$ and $E_{H_2C=C}$ [16]. As a result, the HSRE of benzene is $8.000 - (3 \cdot 2.0699 + 3 \cdot 0.4660) = 0.392$ (in β units). The value of the HSRE has been calculated for a broad variety of organic compounds including the heterocyclic ones (see Table 2.1) [16–19, 41–46].

The HSRE was extended to cover also radicals and ions [47, 48]. In this case, use was made of a reference structure with alternating long and short bonds analogous to that suggested by Mulliken and Parr [49]. This version of the HSRE calculation consists of the following steps. First, E_π is calculated by the HMO method with all resonance integrals equal to β_0; then the obtained bond orders are used to find bond lengths (Eq. (2.10)) after which new values of β are calculated by means of Eq. (2.11) and the final value of E_π is found. For E_{HC-CH_2} see (2)—the value of 0.6632 was determined (in β_0 units) [48].

$$R = (1.517 - 0.18P) \,(\text{Å}) \qquad (2.10)$$

$$\beta(R) = \beta_0 S(R) / S(1.397 \,\text{Å}) \qquad (2.11)$$

$$E_{HC-CH_2}$$

2

The calculated HSRE values for radicals, cations, anions, dications, and dianions [48, 49] are consistent with the assignment of these species (based on experiment) to the aromatic, antiaromatic, or nonaromatic class. These results will be analyzed in greater detail when examining some representative compounds.

As noted above, Hess and Schaad assumed the value of $E_{H_2C=C} = E_{H_2C=CH} = 2.0000$ [16] for determining the values of π-bond energies for the acyclic polyene reference structure. The choice of another value would have led to a different set of energies. In other words, the values of the π-bond energies obtained by them are purely formal. This is seen, for example, in the relationship between the values of the parametrized bond energies and the corresponding bond orders. According to the parametrization [16], the π-bond energy for the terminal double bond ($E_{H_2C=CH}$ = 2.0000) is less than that for the disubstituted one ($E_{HC=CH}$ = 2.0699), while for the corresponding bond orders this relationship is reversed. In particular, in a linear conjugated polyene, the bond order for a given bond type depends on the position of this bond in the chain—being maximal in the case of the terminal double bond [50, 51]. A solution to this problem was suggested by Moyano and Paniagua [51]; it involves a physically meaningful parametrization of π-bond energies based on the localized molecular π-orbitals (π-LMOs). The π-LMOs and corresponding orbital energies have served as a basis for establishing a natural classification of π-bond types in acyclic conjugated polyenes as well as for the corresponding parametrization of the reference bond energies. In the thus developed classification of bond types, the π-bond energies are largely determined by the topological factors, such as the conjugation and branching. The relevant parametrization (least-square fitting to the π-energies of the same 40 acyclic polyenes as were used for parametrization purposes by Hess and Schaad) does not involve any arbitrary assignment of a bond energy value. Similar to the HSRE scheme, eight bond types have been distinguished, which, however, differ from those of Hess and Schaad (in the brackets are the values of parametrized bond energies as given in β units) [51]:

$H_2C=CH—CH=$	(2.2234)	$HC=CH—C=$	(2.2336)
$=CH—HC=CH—CH=$	(2.5394)	$=CH—HC=CH—C—$	(2.5244)
$=C—HC=CH—C$	(2.4998)	$H_2C=C—$	(2.4320)
$—CH=C—$	(2.7524)	$—C=C—$	(2.9970)

The above values of bond energies are close to the double values of energies of the corresponding LMOs. The values of resonance energies, calculated in this manner, differ slightly from those of the HSRE (e.g., for benzene RE = 0.384 in β units [51]). But it should be emphasized that the scheme in [51] outlines more clearly the essence of the definition of aromaticity (antiaromaticity) based on the effect of the cyclic electron (bond) delocalization. It will be recalled that this effect renders the localized bond model unsuitable for the cyclic conjugated molecules, even though it may be applied in the description of acyclic polyene structures notwithstanding the conjugation present in these (see Chapter 1).

2.2.5 Topological Resonance Energy

The DRE and HSRE schemes have certain shortcomings, such as the following: in attempting to extend them to radicals and ions [48] difficulties arise in regard to their modification and introduction of new parametrizations; empirical parameters have to be used for reference bond energies [13, 16–18] whose number increases considerably in passing to heterocyclic systems [15, 19, 35] and these schemes cannot be applied in the case of excited states. The topological resonance energy (TRE) scheme is free of those shortcomings.

The TRE scheme rests on the formalism of the graph theory. For a conjugated hydrocarbon, the matrix of the Hückel Hamiltonian H and the adjacent matrix $A(G)$ of the corresponding molecular graph $G(A_{rs} = 1$ if v_r and v_s vertices (atoms) are adjacent, otherwise it is zero) are related as follows [52–56]:

$$H = \alpha I + \beta A (G) \tag{2.12}$$

where I is the unit matrix, α and β are the Coulomb and the resonance integrals, respectively. Seeing that the characteristic polynomial $P(G; X)$ of the graph G is the characteristic polynomial of its adjacency matrix A,

$$P(G; X) = \det| XI - A| \tag{2.13}$$

the secular HMO determinant (2.14) may be written as Eq. (2.15) or, to achieve analogy with Eq. (2.13), as Eq. (2.14) [52]:

$$\det|H - EI| \tag{2.14}$$

$$\det\left|\frac{\alpha - E_i}{\beta} I + A\right| \tag{2.15}$$

$$\det\left|\frac{E_i - \alpha}{\beta} I - A\right| \tag{2.16}$$

The comparison between Eqs. (2.13) and (2.16) shows that the Hückel molecular orbital energies E_i are linear functions of the graph eigenvalues X_i (Eq. (2.17)). Making use of the so-called β units – $\alpha = 0$ and $\beta = 1$ – Eq. (2.17) may be written as Eq. (2.18), remembering that β is negative:

$$E_i = \alpha + x_i \beta \tag{2.17}$$

$$E_i = x_i \tag{2.18}$$

The relationship (2.18) shows that graph eigenvalues coincide, in β units, with HMO energies E_i. Hence by determining the roots x_i of the characteristic polynomial $P(G; X)$ of the molecular graph G that corresponds to the conjugated molecule, one may find the total π-electron energy of this molecule [52–56].

Since the matrices H and A commute [52–56] ($HA = AH$), they have identical eigenvectors. In other words, when eigenvectors of the graph of the conjugated molecule are found, the LCAO coefficients of the Hückel molecular π-orbitals of this molecule are also determined. One of the important factors determining the total π-electron energy is the cyclic conjugation; the TRE scheme represents quite an appropriate way for estimating the contribution to this energy coming from the cyclic electron conjugation (or, terminologically more correctly, from the cyclic electron (bond) delocalization; see Chapter 1).

The characteristic polynomial of a conjugated system may be constructed on the basis of the Sachs theorem [52–57], according to which the coefficients a_n of the characteristic polynomial $P(G; X)$ are given by the relationship

$$a_n = \sum_{s \in S_n} (-1)^{c(s)} 2^{r(s)} \qquad 0 \leq n \leq N \qquad (2.19)$$

where N is the number of vertices of the graph G, and $c(s)$ and $r(s)$ denote the total number of components and cycles of the Sachs graphs, respectively. A graph is Sachs in type if its every component represents either a complete graph K_2 or a cycle C_m. [3]S_n is the set of all Sachs graphs with n vertices: in Eq. (2.19) the summation is performed over all Sachs graphs; a_n is the coefficient in $P(G; X)$ with $a_0 = 1$. Thus the polynomial $P(G; X)$ may be written as

$$P(G; X) = \sum_{n=0}^{N} \sum_{s \in S_n} (-1)^{c(s)} 2^{r(s)} x^{N-n} \qquad (2.20)$$

Then the total π-electron energy is

$$E_\pi(\text{conjugated molecule}) = \sum_{i=1}^{N} g_i x_i \qquad (2.21)$$

where x_i ($i = 1, 2,...,N$) are the roots of the characteristic polynomial $P(G; X)$ and g_i is the occupation number of the ith MO.

For calculating the TRE by means of Eq. (2.22), E_π of the reference structure must be determined; this differs from E_π of the conjugated molecule in the absence of the contribution coming from the cyclic electron (bond) delocalization. In terms of the graph theory, a polynomial must be constructed for the reference structure with only the acyclic Sachs graph for the given molecule taken into account [22, 57, 58].

$$\text{TRE} = E_\pi(\text{conjugated molecule}) - E_\pi(\text{reference structure}) \qquad (2.22)$$

The relevant polynomial corresponding to the reference structure is called the acyclic [22, 57] or reference [20, 21] polynomial and, since $r(s) = 0$, it has the form [22, 57]

$$R(G; X) = \sum_{n=0}^{N} \sum_{s \in S_n} (-1)^{c(s)} x^{N-n} \qquad (2.23)$$

The roots x^{ac} of R form the "acyclic spectrum" of the graph G. The construction of R may be simplified by using the recurrence relations [22, 23, 58]. Since

---

[3]K_2 is a complete graph of degree one; the degree (or valency) of a vertex v is the number of vertices adjacent to it. A graph is a complete graph of degree one if it consists of two vertices joined by an edge, that is an isolated bond [55–57].

in most cases R has only the real-valued roots,[4] the energy of the reference structure may be given by

$$E_\pi(\text{reference structure}) = \sum_{i=0}^{N} g_i^{ac} x_i^{ac} \qquad (2.24)$$

Thus the formula for the TRE has the form [22, 23, 57, 58]

$$\text{TRE} = \sum_{i=0}^{N} g_i x_i - g_i^{ac} x_i^{ac} \qquad (2.25)$$

This expression shows that for acyclic polyenes TRE = 0.

Let us consider the construction of $P(X)$ and $R(X)$ for cyclobutadiene
(graph $\mathbf{G}_1$). We obtain for the Sachs graph S_n

$$\mathbf{G}_1$$

Consequently, for the characteristic polynomial $P(G_1; X)$ we have $a_2 = -4$, $a_3 = 0$,
$a_4 = 2 - 2 = 0$ (see Eq. (2.19)), and for the reference polynomial $a_2 = -4$, $a_3 = 0$,
$a_4 = 2$. Thus $P(G_1; X) = x^4 - 4x^2$, $R(G_1; X) = x^4 - 4x^2 + 2$, and TRE = -1.226. By
contrast, in the case of benzene, TRE = 0.273 (in β units).[5] The calculation
of the TRE of cyclobutadiene highlights one important detail: the occupation
numbers g_i, (see Eq. (2.21)), are a concept of the HMO model but are alien to
the graph theory since they are, in principle, not deducible from the molecular

---

[4] This conclusion in [25] was not based on mathematical proof but rather on results of numerous
calculations. Later, a relationship between R and the characteristic polynomial of a certain acyclic
graph with weighted edges was established and it was inferred that the roots of R must be real numbers [59]. However, the use of the TRE for assessing the stabilizing or destabilizing effects of individual cycles in polycyclic conjugated hydrocarbons [60] has shown that in some cases R may have
imaginary roots for these cycles [61, 62]. Since the roots of R are assumed to correspond to the
energy levels (see Eq. (2.25)), the presence of imaginary roots will thwart the determination of the
TRE for individual cycles of polycyclic hydrocarbons.

[5] For [n] annulenes analytical expressions may be used to find the roots x and x_i^{ac} [63, 64].

graph G [63]. This may be illustrated by calculation of the TRE of cyclobutadiene. For this molecule we assume that $g_1 = g_2 = 2$, $g_3 = g_4 = 0$, and $g_1^{ac} = g_2 = 2$, which yield the TRE of -1.226; if $g_1 = 2$, $g_2 = g_3 = 1$, and $g_4 = 0$ are used in accordance with Hund's rule then TRE $= 0.305$, which exceeds that of benzene [64, 65]! The last result is obtained on the assumption that $g_i = g_i^{ac}$ (see Eq. (2.25)) and that only the values of g_i are determined in such a way as to make the obtained value E_π (Eq. (2.21)) the lowest.

However, since for evaluating the total π-electron energy of the acyclic reference structure in the ground state the Aufbau principle may be allied, the correlation between g_i and g_i^{ac} may be dispensed with. To find the reference energy, the π-orbitals of the reference structure should be filled, according to the Aufbau principle, with the same number of electrons as in a conjugated cyclic molecule [9, 66]. This approach may help to avoid the problems connected with the orbital filling.

The results of calculations of the TRE for closed-shell conjugated alternant hydrocarbons are in good agreement with the estimates of the aromaticity or antiaromaticity based on other RE calculation schemes and on other structural and magnetic criteria. This may in part be explained as follows. For the molecules in question, the estimates of relative stability based on the values of E_π calculated by the HMO method do not depend on whether the difference between bond lengths was taken into account (e.g., by means of the "variable β" Hückel method) or not (calculation by the ordinary "topological" HMO method) [67].

For nonalternant hydrocarbons the inadequacy of the ordinary HMO model may be overcome by applying the ω-technique [68]. In this case, the TRE values are occasionally considerably reduced, for example, with azulene it falls from 0.151 to 0.055. A similar modification can also be effected within the HSRE scheme, whereby the energy value drops for azulene from 0.23 to 0.14 (in β units) [68].

A considerable alternation of bond lengths is characteristic of antiaromatic compounds, which should be kept in mind when calculating the TRE. In calculating the TRE of cyclobutadiene by the "variable β" method, its value for the D_{2h} structure is as low as -0.427, while for the D_{4h} structure it amounts to -1.226 [69].

The straightforward and elegant determination of the energy contribution coming from the cyclic electron (bond) delocalization makes the TRE scheme very attractive. However, there are certain problems [63–65, 70]—inevitable for a scheme based on a combination of such heterogeneous elements as the graph theory and the chemical theory of the conjugated molecule structure.

We have already mentioned the imaginary roots that may emerge in determination of a TRE contribution by individual cycles [61, 62]. Another problem is that the TRE values may be manifestly overestimated in calculations on nonclassical structures. Thus, of three isomeric quinodimethanes (**3–5**) the *meta*-isomer has the maximum value of the TRE even though it is a highly reactive biradical [65]. An extremely reactive, unstable, and in some cases, merely hypothetical polyradical species (**6**) turns out to be more aromatic (TRE = 0.209) than its relatively stable singlet ground-state isomer (**7**) (TRE = 0.151) [70].

3	**4**	**5**	**6**	**7**

The failures of the TRE scheme have attracted exaggerated attention for the simple reason that this method has been very extensively exploited for studying compounds of the most diverse types. The plain fact is, however, that in most cases the TRE values are in quite satisfactory agreement with other estimates of aromaticity based on both theoretical and experimental approaches.

The per electron values[6] of the DRE, HSRE, and TRE are given in Table 2.1. A number of correlations have been established between the values of TRE and HSRE [75] as well as TRE and the magnetic susceptibility [8, 76, 77]. A satisfactory assessment of the aromaticity or antiaromacticity can also be made when the TRE scheme is extended to cover heterocyclic molecules [23, 58], radicals and ions [58, 71] (Table 2.1), excited states [78, 79], and organometallic compounds [80], as well as σ-aromatic (σ-antiaromatic) systems [81] and three-dimensional molecules, such as C_{12} [82], C_{60}-footballene [83], and bridged polyenes [84]. This scheme allows also the calculation of the TRE for Möbius systems [78, 79, 84–87].

The foregoing text clearly shows that the TRE scheme is quite an effective energetic criterion for aromaticity and antiaromaticity. There are also basically different approaches to studying the aromaticity (antiaromaticity) of polycyclic molecules, which permit the determination of resonance energies without recourse to quantum chemical calculations. One of these is represented by the conjugated circuits model.

2.2.6 Conjugated Circuits Model

Calculations of resonance energies of polycyclic molecules, such as benzenoid hydrocarbons, help assess the relative stability of molecules. They are particularly important in those cases when experience and intuition cannot be relied upon. It is known that linear acenes are less stable than "kinked" acenes [88], but how

---

[6] Since the total resonance energies of molecules of different sizes cannot be compared, the following specific resonance energies are used to this end: per electron (REPE) [17, 23, 71], per bond (REPB) [72], per atom (REPA) [47], per hundred (%RE), which is the ratio between the RE and the reference energy multiplied by 100 [73], and per face for the polyhedral structures [74]. It is not so easy to make general recommendations in favour of any one of these schemes of normalization.

TABLE 2.1 REPE Calculated by Various Schemes[a]

Compound	DRE, eV [13–15]	HSRE, β [16–19, 42–45]	TRE, β [22, 23, 71]	CCMRE, eV [89–94]	SRTRE, eV [112–114]
Benzene	0.1448	0.065 (−0.014)	0.046[b] (−0.173)	0.145 [−0.028]	0.140
Naphthalene	0.1323	0.055 [0.033]	0.039	0.132 [0.011]	0.135
Anthracene	0.1143	0.047 [0.056]	0.034	0.114 [0.034]	0.114
Cyclobutadiene	−0.193	−0.268 (−0.001)	−0.307[b]	−0.195 [0.145]	−0.162
Cyclooctatetraene	−0.054	−0.060 (0.097)	−0.074 (0.031) [0.019]	[0.025]	
Benzocyclobutadiene	0.054	−0.027 (0.055) [0.034]	−0.049 (−0.007) [−0.006]	0.099 [0.012]	−0.006
Propalene	−0.139	−0.100			
Pentalene	0.001	−0.018 (−0.002) [0.093]	−0.027 (−0.030) [0.046]	[0.107]	−0.033
Heptalene	0.008	−0.004 (0.010) [0.060]	−0.012		
Azulene	0.017	0.023 [0.042]	0.015	[−0.009] 0.0246 [0.021]	0.034
Butalene	−0.046	−0.067 [−0.020]		−0.058	−0.052
Cyclopropenyl cation		0.365	0.268		
Cyclopentadienyl cation		−0.095	−0.153		

Cyclopentadienide anion		0.111	0.053
Pyridine	0.167	0.058	0.038
Pyrimidine	0.146	0.049	0.032
Pyrazine	0.124	0.049	0.022
Azete	-0.168	-0.160	-0.1936
1,3-Diazete		-0.113	-0.1356
Pyrrole	0.038	0.039	0.040
Furan	0.031	0.007	0.007
Thiophene	0.047	0.032	0.033
Quinoline	0.148	0.052	0.036
Isoquinoline		0.051	0.033

[a]Figures in parentheses indicate the REPE for the corresponding dication; figures in square brackets are for the dianion.
[b]For benzene excited state, TREPE = -0.115; for cyclobutadiene excited state, TREPE = 0.076 [79].

can the stability of, for example, benzo[*a*]pyrene (**8**) and benzo[*d*]pyrene (**9**) be compared?

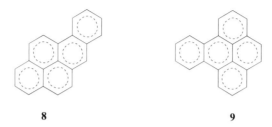

8   **9**

It is true that calculations by the Parizer–Parr–Pople (PPP) method may yield DRE values for these hydrocarbons [16]. However, more attractive would be an approach that could dispense with quantum chemical calculations; this is particularly valuable in the case of large conjugated hydrocarbons. An example of just such an approach is given by the recently developed conjugated circuits model [89–100]. A conjugated circuit is defined as a circuit within an individual Kekule valence structure in which there is a regular alternation of the formal CC single and double bonds [89]. Such circuits necessarily are of even length and either of a $(4n + 2)$ or $(4n)$ type. The former are denoted by R_n, the latter by Q_n. The total number of sets of disjoint conjugated circuits within a single Kekule structure is $k - 1$, where k is the number of Kekule valence structure for a given benzenoid hydrocarbon [89, 91]. Different circuit counts are designated as $\#^{(4n + 2)}$ or $\#^{(4n)}$ and are obtained by the summation of all $(4n + 2)$ or $(4n)$ conjugated circuits.

As noted in Section 2.2.1, the RE corresponds to a difference between the energy of a given molecule and that of the hypothetical reference structure calculated in terms of the bond additivity model. At the same time, the resonance energies themselves may be expressed in the additive form [101, 102]. When the conjugated circuits model (CCM) is applied, the RE of a polycyclic conjugated molecule (CCMRE) may be determined, taking into account that the REs are additive within the model in question [101]:

$$\text{CCMRE} = \frac{1}{k}\sum_n (R_n \#^{(4n + 2)} + Q_n \#^{(4n)}) \tag{2.26}$$

where R_n and Q_n are the parametric values for the conjugated circuits of the $(4n + 2)$ and $(4n)$ type, respectively, containing carbon atoms only.[7] For $n > 3$, the terms R_n and Q_n are neglected since their energy contributions rapidly diminish with the growing circuit size (increase in n). Three types of conjugated circuits of size 14 having different shapes are conceivable, namely, R_3^a, R_3^b, and R_3^c [99] (this detail was not noticed in earlier works [97, 98]).

---

[7] Occasionally, the parametric value is designated by ρ_n [91] to distinguish it from the corresponding R_n circuit.

The numerical values of R_n ($n=1$ to 3) are obtained by means of a parametrization procedure with respect to the DRE values calculated for benzene, naphthalene, and anthracene [95, 101]; the values of Q_n ($n = 1$ to 3) are

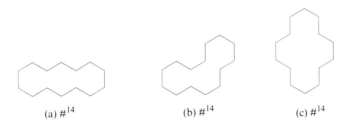

(a) #[14]   (b) #[14]   (c) #[14]

derived from the DRE value of cyclobutadiene (see Table 2.1) making use of the approximations $Q_2 = R_2 Q_1/R_1$ and $Q_3 = R_3 Q_1/R_1$ [94, 95]. The values in question are as follows (in eV): $R_1 = 0.869$, $R_2 = 0.247$, $R_3^a = 0.099$, $R_3^b = -0.006$, $R_3^c = 0.104$, $Q_1 = -0.781$, $Q_2 = -0.222$, $Q_3 = -0.090$. As a result, we have Scheme 2.2 for naphthalene [93].

Kekule structure   Circuit decomposition

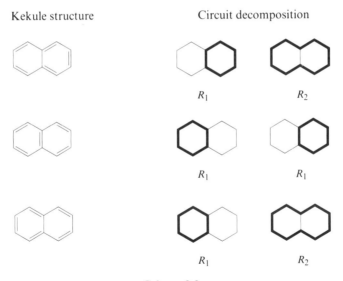

Scheme 2.2

Consequently, benzo[d]pyrene (9) has a stronger aromatic character than benzo[a]pyrene (8) and accordingly, must be more stable (less reactive), which is supported by experimental data [88].

The CCMRE method has in recent years been extended to cover various types of polycylic conjugated hydrocarbons [91] including large molecules [103, 104], benzannelated annulenes [105], annelated [n]annulenes [106], the ions of conjugated hydrocarbons [90], benzenoid hydrocarbon radicals [107], excited states of benzenoid hydrocarbons [95], heterocyclic conjugated molecules [93, 94, 108], Möbius systems [96], and such three-dimensional systems as footballene C_{60} [109, 110] and related elemental carbon cages [33, 110, 111].

In calculating the CCMRE for excited states, one has to take into account excited valence structures involving the "long bond" which is a "double CC bond" without the σ-bond support [95]. In other words, such a "long bond" is formally regarded as a "double bond." As an example, some of the singly excited valence structures of anthracene may be given:

For the CCMRE calculations on heterocyclic molecules, such as the ones containing divalent sulfur, a Kekule-type structure is generated from the structure of polycyclic conjugated hydrocarbon by replacing the —CH=CH fragment with –S–. For example, in the case of 2-thianorbiphenylene [108].

The theoretical considerations on which the conjugated circuits model is based have much in common with the structure-resonance theory advanced by Herndon.

2.2.7 Structure–Resonance Theory and Some Related Models

The resonance energy may be calculated using the structure-resonance theory on which the empirical parametrized valence bond method is based [112–115]:

$$SRTRE = (2/SC)(\sum H_{ij}) \tag{2.27}$$

where SC is the number of Kekule structures that can be drawn for a given molecule, and H_{ij} are the resonance integrals corresponding to the stabilizing exchange energies; they equal γ_1 or γ_2, depending on whether the two resonance structures are interconverted in consequence of the permutation of three pairs of electrons (three-bonds permutation) in a single ring (γ_1) or as a result of the permutation of five pairs of electrons (five bonds) within two annelated rings (γ_2) [113]:

When even numbers of electron pairs are permuted, the corresponding resonance integrals are represented by ω_1 and ω_2.

Thus, in order to calculate the SRTRE, it is necessary to enumerate the Kekule structures and determine the values of H_{ij}. For example, in the case of naphthalene that has the three Kekule structures A, B, and C, SRTRE $= \frac{2}{3}(H_{AB} + H_{AC} + H_{BC})$. Since $H_{AB} = H_{AC} = \gamma_1$ and $H_{BC} = \gamma_2$, SRTRE $= \frac{2}{3}(2\gamma_1 + \gamma_2)$. The

A

B   C

$$\gamma_1 = 0.841 \text{ eV}$$
$$\gamma_2 = 0.336 \text{ eV}$$

$$\omega_1 = -0.65 \text{ eV}$$
$$\omega_2 = -0.26 \text{ eV}$$

numerical value of γ_1 has been ascertained after a comparison with a large number of DREs for aromatic hydrocarbons and the γ_2/γ_1 ratio has been determined from experimental data on energies of electronic transitions of benzene and azulene [111, 113].[8]

The value of $\omega_1 = -0.65$ eV is the arithmetical mean of the DRE value of cyclobutadiene (-0.78 eV) and that of the empirical RE of the same compound (-0.52 eV) derived from the value of the potentials of oxidation leading to compounds containing a cyclobutadiene fragment (see Chapter 4).

The SRTRE values for various compounds correlate fairly well with the DRE and HSRE values [113] (see Table 2.1). Later, the SRTRE scheme was extended to permit calculations of the RE for π-conjugated hydrocarbon radicals [112]. In this case, β_1, the allyl resonance integral, and β_2, the pentadienyl resonance integral, were included in the expression for calculating the SRTRE ($n_1 - n_4$ are the numbers of each type of resonance integral)

$$SRTRE = \frac{2}{SC}(n_1\gamma_1 + n_2\gamma_2 + n_3\beta_1 + n_4\beta_2)$$

A good agreement is also observed between the calculated SRTRE values and the estimates of the aromaticity of benzenoid polycyclic hydrocarbons based on Clar's qualitative sextet concept [116, 117]. This concept singles out within the polyacene molecule fully benzenoid rings designated by a circle (a symbol for the electron sextet) as in **10**, rings that share a migrating sextet of electrons (migration is shown with an arrow), structure **11**, rings with fixed double bonds, as in zethrene (**12**) or completely empty rings, as the central ring in perylene (**13**) [117].

In terms of this qualitative model, a conclusion may be reached that, for example, the RE values for **12** and **13** will be double that of naphthalene (**11**), since in (**12**) a diene system with localized bonds is contained, and in **13** an

[8]Even though the RE for aromatic hydrocarbons may receive contributions also from the conjugated circuits larger than those containing 10 carbon atoms, they may be neglected—similar to R_n in the CCM scheme, for $n > 3$ (see preceding section).

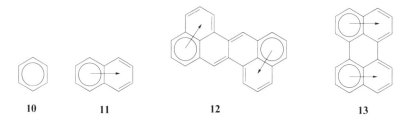

10   **11**   **12**   **13**

empty central ring is present. This conclusion coincides fully with SRTRE calculations [113]. Clar's concept enables various data on properties and the reactivity of polycyclic aromatic hydrocarbons to be successfully interpreted
[113, 115, 116]. A limitation of this concept is that it does not apply to sextet isomers, that is, molecules with the same number of sextets per number of rings.

   A method for calculating the RE involving Clar's concept within the VB
theory has been suggested by Herndon and Hosoya [118]. The Clar structures
serve as a basis set for an empirical VB calculation. In these calculations, the
parameters a, b, and k correspond, respectively, to the cycles with an aromatic
sextet, the cycles with isolated double bonds, and those with a diluted adjacent
sextet (e.g., in the case of naphthalene (**11**) we have $a + 2b + k$). They are determined from the correlation with the DRE values. For linear catacondensed
acenes, the general expression for RE is as follows:

$$RE = a + 2b(n - 1) + 2k \left(\cos \frac{\pi}{n + 1} \right) \qquad (2.28)$$

where n is the number of six-membered rings. The attractiveness of Clar's sextet
concept lies in its simplicity. For the same reason, the method of structure count
(SC) based on the structure-resonance theory, has found a fairly wide acceptance [113, 119–121]. The implementation of this method used to estimate properties of conjugated polycyclic hydrocarbons requires nothing but pencil and
paper. The value of the SC may be found from coefficients of nonbonding MOs
(NBMOs) of conjugated molecules or ions [119, 120]. These coefficients are
determined by means of the Longuet-Higgins rule [112]. For odd alternant systems, the value of the SC equals the sum of the absolute values of unnormalized
coefficients, the least of which is taken to be unity. For example, in the case of σ-
complexes that are formed as a result of aromatic substitution of naphthalene,
the following values are obtained:

SC = 7   SC = 6

As may be seen from the SC values, the substitution into α-position is more favored, which is compatible with experimental results. So to determine the SC values for even systems, one vertex of the molecular graph must be deleted, whereby an odd system is formed (the choice of the vertex is a separate question not treated here). Then the sum is found of the absolute values of unnormalized coefficients at the points adjacent to the deleted vertex [119, 120]:

$$\begin{array}{ccc} \text{SC} = 2 & \text{SC} = 5 & \\ \text{CSC} = 0 & \text{CSC} = 3 & \text{CSC} = 1 \end{array}$$

The term of a corrected structure count (CSC) has been introduced, which excludes the structure that do not make any contribution to the stabilizing res-onance interactions. The CSC is equal to the absolute value of the algebraic sum of the coefficients at the points adjacent to the vertex deleted. Since the CSC for alternant systems equals the square root of the absolute value of the product of all HMO eigenvalues E_i [119, 120], Eq. (2.29), the even systems with CSC = 0 must have two or more nonbonding levels and, consequently, will be unstable:

$$\text{CSC} = (\Pi_i |E_i|)^{1/2} \tag{2.29}$$

The value of the CSC = 1 is characteristic of the structures of acyclic poly-olefins, whereas the values of the CSC > 1 indicates the resonance stabilization of the structure. The determination of the instability of a molecule from the fact that CSC = 0 resembles, in essence, the procedure of ascertaining such possibil-ity by calculating the number of the nonbonding energy levels using the HMO method [122].

An important characteristic of the SC and CSC is the correlation between the SC, CSC, and RE, which may be represented in an analytical form. For exam-ple, for benzenoid hydrocarbons [121], the logarithmic model (LM) gives

$$\text{DRE(eV)} = 1.185 \ln \text{SC} \tag{2.30}$$

The RE values calculated by the structure–resonance method (see Eq. (2.27)) correlate also with the CSC in Eq. (2.31). These values, obtained for π-hydro-carbon–iron tricarbonyl complexes on the basis of this correlation, are in good agreement with experimental data on the stability and reactivity [123]:

$$\text{RE(eV)} = 1.3 \ln \text{CSC} \tag{2.31}$$

In many cases the relation $\ln(\text{SC}_{product} / \text{SC}_{reactant})$ proves a reliable index of reactivity [115]. Thus it correlates with the rate constant for the Diels–Alder reaction of polycyclic aromatic hydrocarbons with maleic anhydride. Based on

Eq. (2.30), the change in the RE, in going from the reactant (R), to the product (P), may be given by the following [116]:

$$\Delta RE(eV) = 1.185 \ln(SC_P/SC_R) \qquad (2.32)$$

In the general case, one should expect the proportionality constant in a relationship like (2.32) to be different, depending on reactive intermediates.

2.2.8 PMO Method of the Dewar Resonance Energy Estimation

Equally simple is the method suggested by Dewar [124, 125], based on the perturbation molecular orbital (PMO) theory. To estimate the resonance energy of an even-alternant hydrocarbon, such as **10**, or even-nonalternant hydrocarbon with two condensed odd-membered cycles, for example, **15**, the conjugated carbocycle is divided into two odd-alternant fragments. If a cyclic and an acyclic system were constructed of such identical odd-alternant fragments, then the first-order π-energy of bonding might be used to compare the energies of the former and the latter. The stabilization energy in the recombination of these two fragments is largely determined by the interaction between nonbonding MOs. It may be expressed as follows [124, 125]:

$$\delta E_\pi = \sum_{r,s} 2a_{or} b_{os} \beta_{rs} \qquad (2.33)$$

where a_{or} and b_{os} are the corresponding coefficients of the NBMO, which may readily be found with the aid of the Longuet-Higgins rule [122]. Thus, for estimating the RE, as the difference between δE_π energies of the carbocyclic of interest and of the corresponding acyclic polyene, e.g. benzene and hexatriene (2.34), the PMO method makes use of the same coefficients, the unnormalized values of which were employed in calculating the SC (see preceding section), which is reflected in the closeness of results [120].

Ordinarily, the fragments are represented by an odd-alternant system containing one less carbon atom than a given cyclic even-conjugated hydrocarbon and by methine, which may be regarded as the simplest "odd-alternant hydrocarbon" whose NBMO is formed merely by one 2p AO or carbon. Hence the corresponding coefficient in Eq. (2.33) will be unity.

For example, in the case of benzene (10) RE $= 2\beta \left[(1\cdot1/\sqrt{3}+1\cdot1/\sqrt{3})\right.$
$\left.-(1\cdot1/\sqrt{3})\right] = 1.156 \ \beta$ and for propalene (15) RE $= -1.4\beta$.

Thus propalene can be assigned to antiaromatic molecules.

This method, the detailed description of which the reader may find in [124, 125], has gained great popularity in organic chemistry and was even included in a number of textbooks [126–128]. It should be noted, however, that some inconsistencies were recently detected in it [129]. When using Eq. (2.33) for estimating the δE_π, it was assumed that the energy of nonfrontier orbitals is not appreciably changed due to the recombination into cyclic or acyclic systems. But the HOMO calculations of changes in the energy of the π-orbital, in the process of recombination of methane with an odd-alternant fragment, have shown that with some molecules, such as bicyclo [3.1.0] hexatriene, the stabilization of the lowest occupied MO during recombination turned out to be greater (1.40β) than that of the $\pi 3$ orbital of pentadienyl (1.22β) [129]. In light of their findings, Durkin and Langler [129] have suggested that the total π-energies of the starting fragments be resorted to as the basis for a scheme for an evaluation of RE. To approximate these energies, the following modification of the above scheme is proposed. The stabilization energy is estimated for each π-MO making use of AO coefficients for the recombination centers as was done earlier in the case of the NBMO (e.g., the coefficients may be found by means of the free-electron method [130]): note that if the stabilization energy for the monocycle is zero (0β), the corresponding energy difference is not taken into account. Finally, the derived difference between the total energies of the recombination stabilizations is multiplied by $4/k$, where k is the number of π-electrons in the annulene under consideration. For example, with octatetraene $\delta E(\pi l) = 0.38\beta$, $\delta E(\pi 2) = 0.70\beta$, $\delta E(\pi 3) = 0.90\beta$, and $\delta E(\pi 4) = \beta$; for cyclooctatetraene $\delta E'(\pi 1) = 0.76\beta$, $\delta E'(\pi 2) = 0$, $\delta E'(\pi 3) = 1.84\beta$, and $\delta E'(\pi 4) = 0$. Hence $(\delta E'(\pi 2) - \delta E'(\pi 2))$ is not included as noted above.

The so obtained estimates of the RE correlate better with the results of their complete calculation in which the energies of recombination stabilizations are found by the HMO method [129] than with those made by use of the original Dewar scheme. Note, however, that in contrast to the latter estimates, they are not in line with the well-known calculational and experimental data on properties of some hydrocarbons. For example, according to the Hückel-calculated DRE, propalene (15) has RE > 0 (in β units) and may thus be classified as aromatic, which contradicts the calculated estimates for this compound [37] that agree with the above-given value RE $= -1.4\beta$.

2.2.9 Estimation of Energies of Aromatic Stabilization and Antiaromatic Destabilization from the Energies of Isodesmic, Homodesmotic, and Hyperhomodesmotic Reactions

As has been noted in Section 2.2.1, depending on the choice of compounds from which the bond energies given in Eq. (2.1) are to be determined, one may obtain

TABLE 2.2 Stabilization Energies (in kcal/mol) for Annulenes and Their Heteroanalogs, Calculated by Means of the Schemes of Isodesmic (ISE), Homodesmotic (HSE), and Hyperhomodesmotic (HHSE) Reactions

Compounds	ISE		HSE		HHSE
	Experimental	Calculated	Experimental	Calculated	
Benzene	64.1 ± 1.7 [139]	58.2 (6-31G*)[a] [137]	21.6 ± 1.5 [132]	23.9 (MP4/6-31G**// MP2/6-31G**)[b] [144]	23.4 (6-31G*) [39]
Cyclobutadiene	—	−59.5[c]	—	−79.2 (MP4/6-31G**// MP2/6-31G**)[d] [144]	−84.6 (6-31G**) [39]
o-Benzyne	5.1 [136]	14.1 (MP2/6-31G* //6-31G*) [136]	−26.0[e] [136]	−43.5 [136] (6-31G)	
Pyridine	65.5 ± 1.6 [132]	60.7 (6-31G*) [132][f]		25.4 (6-31G*) [132]	21.8 (6-31G(5D)) [147]
Pyrimidine	80.4[g]	69.2 (3-21G) [131]			
Pyrazine		63.8 (3-21G) [131]			
1,3-Diazete		−64.5 (3-21G) [135]		−95.0 (4-31G) [135]	
[10] Annulene				26.0 (MP4/6-31G*) [134] 12.0	

			(MP4/STO-3G)[h] [139]
[18] Annulene (D_{3h} structure)			
Cyclopentadienyl cation	− 4	−5 (3-21G) [131]	
Tropylium cation	73[g]	80 (3-21G) [131]	
Silabenzene		46.84 (3-21G*) [138]	16.02 (3-21G*)[i] [138]
Pyrrole		43.80 (3-21G*) [138]	5.62
Furan		35.19 (3-21G*) [138]	5.0 (3-21G*) [138]
Thiophene		32.54 (3-21G*) [138]	9.80 (3-21G*) [138]

[a] 67.2 kcal/mol (MP2/6-31G*) [136].

[b] 24.8 (6-31G*).

[c] Calculated using the 6-31G* total energy of cyclobutadiene, taken from [39]; −66.5 (4-31G) [135], for the correction on the strain energy see the text.

[d] MP2/6-31G*//6-31G* calculation yields HSE = −81.4 kcal/mol [139]; MP2/6-31G*//MP2/6-31G* calculation gives HSE = −107 kcal/mol [146] and after a correction for the ring strain energy HSE = −49 kcal/mol [139].

[e] o-Benzyne + 2 ethylene + acetylene → 2 1-buten-3-yne + trans-1, 3-butadiene.

[f] 71.1 (3-21G) [131].

[g] Taken from [131].

[h] 5.0 kcal/mol (6-31G*) [137].

[i] Corrected for scaled ZPVE, uncorrected HHSE = 17.97 kcal/mol [138]; at HF/6-31G(5D), HHSE = 17.6 kcal/mol [147].

31

either the scheme of the isodesmic reactions — Eq. (2.3) — or that of the homodesmotic reaction — Eq.(2.4). The former belongs to the class of so-called bond separation reactions [27, 131], in which all formal bonds between nonhydrogen atoms in a given molecule are separated due to the formation of the simplest (two-heavy-atom) molecules containing bonds of the same types. So for cyclobutadiene one may write

$$\square \quad + \quad 4CH_4 \longrightarrow 2CH_2{=}CH_2 \; + \; 2CH_3{-}CH_3 \qquad (2.35)$$

In an isodesmic reaction, the number of bonds of each formal type is retained, even though the environment in which these bonds are located has been altered. For any molecule that may be represented with the aid of a classical valence structure, a unique isodesmic bond-separation reaction may be thought of. The energies of some such reactions are given in Table 2.2. The stabilization energy serves as the estimate of the total energy of conjugation rather than of only the cyclic (bond) delocalization energy. In other words, the determined RE corresponds to the Hückel RE. Other components, for example, the strain energy, may also play a significant role. In order to separate these contributions, a combination of two isodesmic reactions may be used. For instance, to determine the antiaromatic destabilization of cyclobutadiene, one has, along with reaction (2.35), to consider reaction (2.36):

$$\text{cyclobutene} + 4CH_4 \rightarrow 3CH_3{-}CH_3 + CH_2{=}CH_2 \qquad (2.36)$$

Whereas the energy of reaction (2.35) calculated with the 3-21G basis set is -70 kcal/mol, for reaction (2.36) $\Delta E = -28$ kcal/mol with the same basis set [132], which gives the strain energy of the four-membered cycle. Hence the destabilization due to the antiaromaticity effects is 52 kcal/mol. This estimate is rather crude, seeing that, for example, in reaction (2.3) three $C(sp^2){-}C(sp^2)$ bonds in benzene "transform" to a $C(sp^3){-}C(sp^3)$ bond in three ethane molecules showing that also the σ-effects of this transformation should be considered [133].

A better agreement between the bond types may be achieved by use of the scheme of homodesmotic reactions [28]. In them, reactants and products contain equal numbers of carbon atoms in the corresponding states of hybridization; moreover, there is the matching of the carbon–hydrogen bonds in terms of the number of hydrogen atoms joined to the individual carbon atoms. For example, in reaction (2.4) (see Eq. (2.3)) all carbon atoms possess sp^2 hybridization and the number of these linked to one or two hydrogen atoms (CH and CH_2, respectively) on the left-hand side of Eq. (2.4) is the same as on the right-hand side. In this reaction, the formally single $sp^2{-}sp^2$ σ-bonds are, on the left, part of a cyclic conjugated molecule, while on the right they are included in a conjugated acyclic system. It is for this reason that such reactions enable the contribution from the cyclic (bond) electron delocalization to be singled out, and the REs determined from them prove to be analogs of the DRE [39]. Comparison of Eqs.

(2.3) and (2.4) show that the homodesmotic reactions are, in fact, a subclass of the isodesmic reactions; in other words, they all are isodesmic.

Note that ΔE of the homodesmotic reaction reflects not exclusively the effect of the cyclic (bond) delocalization. The reference structure is hypothetical and one cannot write the equation of a reaction, where a cyclic and an acyclic structure participate, for which the difference between the energies of products and reactants as determined by a single factor, namely, the aromatic stabilization (antiaromatic destabilization) [28]. What we attempt is a possibly close approximation of this energy, as may become evident when we rewrite Eq. (2.4) as Eq. (2.37) and compare it with the definition of the DRE — Eq. (2.6):

$$\bigcirc \quad \longrightarrow \quad 3(trans\text{-}CH_2{=}CH{-}CH{=}CH_2 - CH_2{=}CH_2) \qquad (2.37)$$

Although the C—H bonds in benzene may be quite closely matched with the C—H bonds on C2 and C3 atoms in butadiene, it does not necessarily mean that the contributions to the total molecular energy, corresponding to these bonds, should be equal [28].

In constructing schemes of homodesmotic reactions for polycyclic benzenoid hydrocarbons, one cannot restrict oneself to butadiene and ethylene. Thus for naphthalene **11** , two homodesmotic reactions may be written [28]:

$$\textbf{11} + 5CH_2{=}CH_2 \rightarrow 2CH_2{=}CH{-}CH{=}CH_2 + 2DVE \qquad (2.38)$$

$$\textbf{11} + 4CH_2{=}CH_2 \rightarrow 2CH_2{=}CH{-}CH{=}CH_2 + TVE \qquad (2.39)$$

where DVE is 2-vinylbutadiene and TVE is 2,3-divinylhexatriene.

When estimating the aromaticity of naphthalene and other benzenoid hydrocarbons, it should be kept in mind that the energy values of homodesmotic reactions do not exclusively reflect the cyclic (bond) delocalization effect both in virtue of the above-mentioned factors and because DVE and TVE have, apparently, nonplanar structures [28]. As is evident from Eqs. (2.38) and (2.39), several schemes of homodesmotic reactions are conceivable for one and the same polycyclic molecule. The homodesmotic reactions scheme has gained wide acceptance for evaluating the aromatic stabilization (antiaromatic destabilization) of carbocyclic and heterocyclic molecules [132–139]. In calculations of their homodesmotic stabilization energies (HSEs) close results are obtained when experimental values of ΔH_f are used [12, 28, 132], or when the ΔE values are calculated by semiempirical [134] and *ab initio* [134–139] methods. This findings opened up the possibility of also determining by *ab initio* calculations the heats of formation of subject molecules [140–142]; hence the accuracy of such calculations of ΔE for homodesmotic reactions has been thoroughly tested and compared with available experimental data [141–143]. For relatively unstrained hydrocarbons, the deviations of ΔH_f (298 K), calculated on the basis of the homodesmotic reactions, from the experimental values are less than

3 kcal/mol (RMP2/6–31G* with zero-point energy corrections), and the root-mean-square error is 1.3 kcal/mol [142]. A somewhat larger error in the HSE value is present in the case of benzene (Eq. (2.4)) – 3.7 kcal/mol and 7.9 kcal/mol at the HF/6–31G* and RMP2/6–31G* levels, respectively [142, 143]. Studies of the basis set dependence of benzene's RE and of the effect the electron correlation has on its value have shown [137, 142] that in going from the 6–31G* to the 6–31G** basis set ΔE is reduced from 24.7 to 23.6 kcal/mol [143], while the experimental value for Eq. (2.4) is 21.6 ±1.5 kcal/mol [12, 132].

When the electron correlation is included and the procedure is confined to the second order of the Møller–Plesset perturbation theory (RMP2), the result gets even worse: ΔE(RMP2/6–311G**) = 28.0 [143] and ΔE(RMP2/6–31G*) = 28.6 kcal/mol, while the HF/6 –31G* calculation gives ΔE = 24.8 kcal/mol [137].[9] An improvement is achieved with higher orders, for instance, ΔE(RMP4SDTQ/6 –31G**/MP2(full)/6–31G**) = 23.9 kcal/mol [144].

While the HSE values estimate the contribution by the cyclic (bond) delocalization, and the ΔE values for the isodesmic reaction (ISE) [134] refer to the stabilization energy associated with the conjugated as a whole, clearly, the latter values turn out appreciably larger—compare HSE (Eq. (2.3)). Indeed, for acyclic polyenes (e.g., hexatriene), the value of the HSE is close to zero: –0.1 (experiment) and 0.4 (HF/6–31G* calculation); whereas ISE = 26.7 (experiment) and 22.7 (HF/6–31G* calculation)—all in kcal/mol [143].

Note also that an appropriate reference structure is selected. For example, if 90° 1,3-butadiene (twisted 90° relative to its molecular plane) is taken for this purpose, HSE for benzene will be 42.7 ± 3.0 kcal/mol [145] (cf. HSE in the case of Eq. (2.4)). The scheme of the hyperhomodesmotic reactions [39], as in Eq. (2.40), is an analog of the HSRE scheme; it takes account of the distinctions between various kinds of bonds on a more subtle way than the HSE scheme does. Indeed, three $H_2C{=}CH_2$ and three $HC{=}CH$ bonds on the left-hand side of Eq. (2.4) are, on the right, replaced with the bond $CH{=}CH_2$. This incongruity can be avoided in Eq. (2.40) [39]:

$$\text{benzene} + 3CH_2{=}CH-CH{=}CH_2$$
$$\rightarrow 3CH_2{=}CH-CH{=}CH-CH{=}CH_2 \qquad (2.40)$$

For this reaction, the stabilization energy (HHSE) equals 23.4 kcal/mol (HF/6–31G*), thus differing insignificantly from HSE = 24.7 kcal/mol (HF/6–31G*)[142]. Note, however, that such close agreement between the HHSE and HSE values is by no means certain to occur in all cases [39]. At MP4SDTQ/6–31G**/MP2(full)/6-31G**, HHSE of benzene is 20.3 kcal/mol and the HSE value is 23.9 kcal/mol [144].

The estimation of the aromaticity (antiaromaticity) of various compounds from the values of the ISE, HSE, and HHSE will be discussed in the respective

---

[9]A corrected value is given; see [143].

sections. For the present, it is only necessary to note that these values (some of which are given in Table 2.2) correlate with those of the HSRE and TRE.

2.2.10 Comparison Between Energetic Criteria

In discussing various schemes for the calculation of resonance energies, it was observed that the estimates of aromaticity (antiaromaticity) obtained by use of these different schemes are close in value. This is hardly surprising since they, for the most part, are based on the Dewar-type definition of the resonance energy. All RE schemes may be divided into two groups [108]. In one of these, numerical values of REs are derived through quantum chemical calculations based on a variety of approaches from the HMO method to *ab initio* calculations [24, 38, 39]. Electron correlation may be included [139], and, in addition to the MO theory, the VB theory may be invoked [118, 148, 149]. Another group comprises graph-theoretical and combinatorial methods for obtaining algebraic expressions for resonance energies [56, 89–94]. The schemes of the former category give more or less exact values, while the latter, though not so accurate, reveal trends in the evolution of REs and, accordingly, of the aromaticity (antiaromaticity) for a broad spectrum of compounds. Clear-cut correlations have been established between the RE values calculated by means of different schemes. In a most detailed fashion these correlations have been studied for polycyclic benzenoid hydrocarbons [91, 113, 150].

Thus the HSRE value for alternant hydrocarbons calculated per one π-electron correlates with the corresponding value of the TRE [151] as follows:

$$TREPE \approx HSREPE + (0.69/N) \ln k \qquad (2.41)$$

where k is the number of Kekule structures and N is the number of π-electrons.

The DRE value (also per one π-electron) strongly correlates (Eq. (2.43)) with the so-called stability index (SI) for benzenoid hydrocarbons defined by [150]

$$SI = k^{2/N} \qquad (2.42)$$

where k is the number of Kekule structures and N is the number of carbon atoms.

$$DREPE \approx 0.5925 \ln SI \qquad (2.43)$$

A correlation has been found between RE values and such characteristics of the reactivity as free energies of activation and rate constants. For example, the difference between the SRTRE values of the aromatic reactants and the products for the Diels–Alder reaction of benzenoid aromatic hydrocarbons with maleic anhydride is linearly related with the free energy of activation (in kcal/mol) as follows [10]:

$$\Delta G^{\#} = 27.8 - 0.55(\Delta SRTRE) \qquad (2.44)$$

All three types of RE, namely, HSRE, DRE, and TRE, exhibit a close correlation with the logarithm of the rate constant k_B for the Diels–Alder addition of maleic anhydride to dehydro[n]annuleno[c]furans [152]; for example,

$$\log k_B = 60.5 \text{ REPE} - 0.475 \tag{2.45}$$

When comparing various RE schemes, calculation results are analyzed for compounds with quite a diverse structure. As for differences between the schemes, preferably the cases are discussed when estimates are conflicting, with, for example, one scheme pointing to aromatic stabilization, while the other suggests antiaromatic destabilization and , consequently, instability of a given compound. Attempts [91, 153], were made to reveal subtler distinctions by comparing RE values calculated for a highly homogeneous set of structures—the benzenoid hydrocarbons. The values of R_1, R_2, R_3 (see Section 2.2.6) obtained through the calculations of the REs of benzene, naphthalene, and anthracene in terms of various models were checked to see whether they met the minimum conditions for the intrinsic consistency of the parameters R_n (Eqs. (2.46) and (2.47)):

$$R_1 > R_2 > R_3 \tag{2.46}$$
$$R_2/R_1 = 0.28 \tag{2.47}$$

Some of the results obtained are given in Table 2.3. The HMO-based RE schemes, namely, the AS (aromatic stabilization) scheme, which is a straightforward HMO version of the DRE model (Eq. (2.48)) [148], the HSRE (Eq.

TABLE 2.3   The R_n ($n < 3$) Parameters of Conjugated Circuits Model (CCM) Calculated from the RE Values Found Using Various Schemes [91, 53]

RE Scheme Parameters	RE			CCM		
	Benzene	Naphtha-lene	Anthra-cene	R_1	R_2	R_3
DRE, eV (Eq. (2.6))	0.869	1.323	1.600	0.869	0.247	0.100
HRE, β (Eq. (2.5))	2.000	3.683	5.314	2.000	1.525	1.579
HSRE, β (Eq. (2.9))	0.390	0.550	0.658	0.390	0.045	0.056
AS, β (Eq. (2.48))	0.440	0.563	0.634	0.440	−0.036	0.019
TRE, β (Eq. (2.25))	0.276	0.390	0.476	0.276	0.033	0.058
SRTRE, eV(Eq. (2.27))	0.84	1.35	1.60	0.84	0.35	−0.01
LM, eV (Eq. (2.30))	0.821	1.302	1.643	0.821	0.311	0.201
RE (VB), J (Eq. (2.49))	0.409	0.648	0.624	0.409	0.154	0.190
ISE^a, eV (Eq. (2.3))	0.918	1.451	1.858	0.918	0.341	0.281
HSE^b, eV (Eq. (2.4))	0.925	1.478	1.853	0.925	0.367	0.197

^{a}ISE taken from [28].
^{b}Using the atomization heats calculated in [13].

(2.9)) , and TRE (Eq. (2.25)) approaches do not satisfy the criteria of Eqs. (2.46) and (2.47):

$$AS = E_\pi\,(HMO) - (n_{C-C}\,E_{C-C} + n_{C=C}\,E_{C=C}) \qquad (2.48)$$

where $E_{C-C} = 0.52\beta$ and $E_{C=C} = 2.00\beta$.

The empirical VB model represented by the SRTRE scheme (Eq.(2.27)) and the logarithmic model (Eq. (2.30)) meet the criterion of Eq. (2.46)) but not that of Eq. (2.47). Neither Eq. (2.46) nor (2.47) is satisfied by the RE value obtained from the VB model involving a large CI matrix computation [149]. In last case, Eq. (2.49) is used to calculate the RE (in J units, J is the exchange parameter in the VB Hamiltonian) of benzenoid hydrocarbons:

$$RE(VB) = E(VB) - E_{reference} = E(VB) - (2n_{C=C} + 0.232n_{C-C}) \qquad (2.49)$$

The schemes ISE and HSE meet the criterion of Eq. (2.46) if they are based on empirical heats of formation or SCF calculations rather than on the HMO method. At the same time, it should be noted that even though an RE scheme may not satisfy the strict criterion of Eq. (2.47), this does not mean that there can never be a good qualitative agreement between the CCM scheme and other schemes. For example, a correlation is observed between the CCM values (in eV) and the numerical predictions (in β units) of the TRE model [90], as, for example, in the case of polycyclic conjugated dianions. Moreover, although the HSRE scheme, based on the HMO model, fails the tests of Eq. (2.46) and (2.47) in regard to benzenoid hydrocarbons [153], the correlation between the second-order rate constants for the Diels–Alder reaction of aromatic hydrocarbons with maleic anhydride and the HSRE(product) – HSRE(reactant) difference is even better [149] than for the SRTRE values [120, 154], which do meet condition (2.46).

Thus there are at present six main schemes for the calculation of resonance energies, namely, DRE, HSRE, TRE, CCMRE, SRTRE, and HSE (HHSE), all of which have been considered above. There is not only a qualitative agreement among the RE values calculated by these schemes (assignment of a compound to aromatic, nonaromatic or antiaromatic systems) but also, in some cases, a quantitative relationship.

The HSE (HHSE) scheme has been used increasingly in recent years: apparently, it has become the most dependable tool for obtaining numerical values of RE. This is partly explained by the growing amount of experimental thermodynamical data on organic compounds, but a still more important reason lies in the rapid development of the *ab initio* methods that enable the HSE to be calculated even when some experimental data lack. Note that such HSE, HHSE calculations may be made for as broad range of compounds, for example, on numerous annulenes [139, 141, 144] and monosubstituted heteroanalogs of benzene [138, 147]. The TRE, CCMRE schemes also have their advantages: they may provide the possibility for deriving analytical expressions that could predict the trends in the variation of the RE.

We have not treated some other approaches to the evaluation of the energy of aromatic stabilization based on empirical REs; for example, the difference between the energies of various isomers. Many of these the reader may find in the literature [1, 3, 29, 130, 155, 156]; some examples will be mentioned below in connection with specific types of aromaticity (antiaromaticity) and relevant molecular structures.

2.3 STRUCTURAL CRITERIA

Viewed from the angle of the energetic criterion of aromaticity, the structural criteria have to reflect those features in the molecular geometry which lead to the stabilization of a cyclic conjugated system. Historically, the formulation of the structural criteria rested essentially on the idea that the π-delocalization is the factor that causes the aromatic stabilization. The following manifestation of the π-delocalization are considered in this connection: the planar geometry of the ring as a factor dictated by the requirement for a better overlap of the p_π-orbitals, equalization of the lengths of the carbon–carbon bonds in the ring, and the correspondence of the completely symmetrical structure to a minimum on the PES. In the most concentrated form these attributes are present in the benzene molecule and structural criteria should be the functions that determine the degree of closeness of these attributes relative to the ones of benzene (with regard to bond lengths, planarity, stability to distortions of high-symmetry structures)..

The structural indices constructed in this fashion are, in essence, phenomenological, and one is entitled to ask whether the specific features in the geometry of the aromatic and antiaromatic molecules used to work out such indices are indeed determined, and if so, to what degree, by the cyclic electron (bond) delocalization. Hence the discussion of the structural criterion will be started with an analysis of these features including the factors that determine them.

2.3.1 Distinguishing Characteristics in the Geometry of Aromatic and Antiaromatic Molecules

The aromatic, antiaromatic molecules and acyclic polyenes have dissimilar geometries; primarily they differ in bond lengths. It is these differences that serve as a basis for the structural indices of aromaticity reflecting the degree of alternation of bond lengths in a ring. Before turning to consider these indices, some specifications of the determination of the bond lengths by different experimental methods will be mentioned.

The parameter determined in gas-phase electron diffraction (ED) experiments, the quantity r_g, is the interatomic distance averaged over all occupied vibrational states at a given temperature $T \neq 0$. The gas-phase microwave spectroscopy (MW) permits the determination of the parameters, averaged over one, usually the main, vibrational level. They are r_s (effective distance derived from measurements on a number of isotopic molecules) and r_o (effective distance obtained through minimization of differences between the experimental and theoretical moments of inertia) [157]. The $r_g - r_s$ difference yields 0.01 ± 0.01 Å with no systematic deviations registered. The quantity r_e represents the internuclear distance in a rigid model, that is, the distance at the minima of potential energy curves or energy hypersurfaces. The distance between nuclei found by the quantum chemical calculation using the Born–Oppenheimer approximation must, generally, approximate the value of r_e [158, 159]. The r_s structure is closer to the r_e structure, and, as a rule, $r_s < r_g$ [157]. The quantity r_z is distance between averaged positions of nuclei for the main vibrational level (T = 0); the values of r_z must be the same in the ED and MW determinations, that is, $r_z > r_e$ [157, 158]. Simultaneous refinement of the ED and MW data gives the value of r_{av}. Whereas in the ED experiment the average distance between atoms is found, the X-ray diffraction (XD) technique determines distances between averaged atomic positions, which is why the ED and XD distances will, in principle, be different [160].

According to experimental data, the benzene molecule has a structure of D_{6h} symmetry with the equal lengths of the CC bond. As may be seen from Table 2.4, the results of the *ab initio* calculations are in good agreement with experi-

TABLE 2.4   Comparison Between Carbon–Carbon Bond Lengths (in Å) in Linear Polyenes, Benzene, and Cyclobutadiene, with Experimental Values (r_g) Found by the Gas Phase Electron Diffraction Method[a] [161, 166, 167] and Calculated Data by *Ab Initio* Calculation [139, 140–143, 164]

Molecules	$H_2C=CH$ Experimental	Calculated	$HC=CH$ Experimental	Calculated	$HC—CH$ Experimental	Calculated
Ethylene	1.336(9) ± 0.001(6)	1.373(3)	—	—	—	—
Butadiene	1.343(9)	1.323(0)	—	—	1.467(2)	1.467(4)
Hexatriene	1.3373	1.324(0)	1.3678	1.329(5)	1.4576	1.463(1)
Benzene	—	—	1.3902[b]	1.3896[c]	1.3902	1.3896[c]
Cyclobutadiene	—	—	1.441(2)[a]	1.3435[d]	1.527(2)[a]	1.5639[d]

[a]In the case of cyclobutadiene, the X-ray data (123 K) are given for tetra-*tert*-butylcyclobutadiene [168].

[b]IR spectroscopy of isotopically substituted benzenes, $^{12}C_6D_6$, $^{13}C_6D_6$, and $^{13}C_6H_6$ [165]. Other values of the CC bond length benzene from data of various experimental methods are: 1.3965 (8) (r_{av}, ED [161]), 1.392(1) (Crystal X-ray study at 270 K, corrected for libration [162]), 1.398 (crystal neutron diffraction at 138 K, averaged value corrected for libration [163]), 1.3967 (9) (r_z, MW [161])

[c]MP2/TZ2P + f data [164]. The TZ2P + f basis set is the TZ ($10s6p15s4p$) basis set on carbons and the TZ($5s/3s$) basis set on hydrogens. For C, there are two d sets (six components) and a 10-component f set. For H, there are two p sets [164].

[d]MP2/6-31G** calculations [144]. At MP2/6-31G**, for comparison, R(CC) in benzene is 1.3947 Å; the calculations with the same basis set without electron correlation give the benzene R(CC) to be 1.386 Å [141].

ment. Investigations of basis set effects on the geometry of benzene have shown that even with split-valence basis sets satisfactory results may be obtained (cf. 1.385 Å with 3-21G [39, 158], 1.385 with 4–21G [169], and 1.388 with 6–31G [137]) [170].

In contrast to benzene, the alternation of the bond lengths is characteristic for acyclic polyenes [171] (Table 2.4). The data of Table 2.4 show that in benzene (an aromatic hydrocarbon) the lengths of the CC bonds turn out intermediate between those of the CH—CH and the HC=CH bonds in acyclic polyenes, namely, 1,3-butadiene and 1,3,5-hexatriene. For antiaromatic molecules, the alternation is even more pronounced that in the case of acyclic polyenes.

Unlike the aromatic molecules, a high-symmetry structure of the lowest singlet state of the antiaromatic molecules does not correspond to a minimum on the PES (Fig. 2.1). For example, the $^1B_{1g}$ state structure (**18**) of cyclobutadiene with equal CC bond lengths corresponds to a transition state for the topomerization of rectangular D_{2h} structures (**19**) [172] (Eq. (2.50)) and the D_{3h} structure of the singlet $^1E'$ state of the cyclopropenide anion $C_3H_3^-$ to a hill top on the PES [173]:

$$\square \;=\; \bigcirc \;=\; \square \qquad (2.50)$$

19   **18**   **19a**

The instability of geometry configurations of the antiaromatic molecules having equal CC bond lengths may give rise to structures with alternating bond lengths, in contrast to high-symmetry structures of aromatic molecules (Fig.

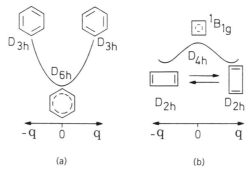

(a)   (b)

Figure 2.1   Potential energy curves for the molecules of (a) benzene and (b) cyclobutadiene along the coordinate corresponding to the π-bond shift. For the D_{6h} benzene structure, the experimental value of the CC stretching frequency (b_{2u}) is 1309 cm^{-1} [174] (1310 cm^{-1} [175], CISD/6-31G calculations give 1334 cm^{-1} [169]). In contrast, the D_{4h} $^1B_{1g}$ structure of cyclobutadiene is not a minimum but rather a transition state for the automerization of the D_{2h} structure. The calculated barrier for this automerization is 6.2 kcal/mol (MCSCF) [176]; an experimental estimate yields a lower bound of this barrier equal to 1.6 kcal/mol [177] (for details see Chapter 4).

2.1). The former geometric feature should naturally be associated with antiaromatic destabilization of the high-symmetry structure and the intention to avoid it, while the latter has to do with aromatic stabilization. It is this dividing line between the ground-state geometries of the aromatic and antiaromatic molecules on which structural criteria were based.

D_{6h} $b_{2u}(D_{6h} \rightarrow D_{3h})$ D_{3h}

10   **20**   **21**

However, in recent years this basis was undermined twice. The first attempt to question the structural criteria whose reference point is represented by the structure of benzene was prompted by the unexpectedly detected temperature dependence of the X-ray data for tetra-*tert*-butylcyclobutadiene (**21**). It turned out that at room temperature these data indicated a very insignificant difference between the intercyclic CC bond lengths (1.482 and 1.464 Å) [178]. At first, such a nearly square structure was thought to be due to the influence of the four *tert*-butyl groups [179], but experiments at 123 K revealed substantial alternation of the bond lengths (0.086 Å) [168] and this unexpected temperature dependence of the structure was ascribed to the disordering phenomenon effect with residual disorder possibly present even at 123 K. The analysis of the anisotropic displacement parameters of ring carbon atoms in **21** has shown that the data measured at room temperature, indicating the nearly square structure, in actual fact correspond to an averaged superposition of two mutually perpendicular rings with the bond alternation [180].

These findings gave rise to a critical reappraisal of experimental data on the benzene structure, which, surprisingly, showed that a rigorous experimental proof of the generally accepted D_{6h} structure of benzene in which the structural criteria of aromaticity are based is actually nonexistent! It turned out that the X-ray structural data for benzene are compatible not only with the crystallographically ordered D_{6h} structure but also with disordered D_{3h} model associated with superposition of Kekule-type benzene molecules rotated by 60° with respect to each other about the threefold axis [181]. The possible causes of the disorder are conceivable: statistical distribution of D_{3h} structures in a crystal cell or dynamical disordering through either the interconversion of D_{3h} forms via the D_{6h} form (double-well potential similar to that for cyclobutadiene) or a hindered rotation of the benzene molecules about the threefold axis. It has been shown by very simple calculations that if the difference between the C—C and C=C bond lengths in the D_{3h} form is 0.10 Å (which means the superimposed carbon atoms are only 0.058 Å apart) the disorder contribution is as small as 0.0008 Å², which does not allow one to discard a possible D_{3h} disordered model on experimental grounds even at very low temperatures.

Other studies, such as infrared and Raman spectra of gaseous benzene, neutron diffraction studies of crystalline benzene, gas–electron diffraction studies, and gas–rotational spectroscopy are equally incapable, according to critical analysis [181], of resolving unambiguously the D_{3h}–D_{6h} structural dilemma of the benzene molecule. Furthermore, decisive conclusions could not be drawn from photoelectron spectra or ^{1}H NMR measurements of benzene molecules in a liquid crystal environment. The latter experiments merely indicate that the average life time of a D_{3h} structure (if it appears on the PES) is less than 10^{-4}s, corresponding to the energy barrier of the $D_{3h} \rightarrow D_{6h} \rightarrow D_{3h}$ interconversion of approximately 12 kcal/mol.

Therefore, quite paradoxically, one is confronted with the fact that contrary to the commonly accepted opinion, the alternative D_{6h} and D_{3h} models of the benzene molecule cannot be distinguished on experimental grounds, and the former structure model had rather be assumed to interpret experimental data.

Bearing all this in mind, a special role in resolving the D_{3h}/D_{6h} problem ought to be assigned to its reliable quantum mechanical treatment. According to results of *ab initio* calculations presented in Table 2.5, the D_{3h} structure (**20**) of benzene possesses a higher energy than the D_{6h} structure (**10**). It is noteworthy that the D_{3h} structure does not correspond to a stationary point on the potential energy surface and in calculating this structure a different model geometry was used (see Table 2.5).

$$E_{\text{distort}} \qquad (2.51)$$

$$D_{6h} \qquad\qquad D_{3h}$$

$$\textbf{10} \qquad\qquad\qquad \textbf{20}$$

Thus the *ab initio* calculations unequivocally point to the D_{6h} structure of benzene. Important in this connection is the question of whether this structure

TABLE 2.5   Difference Between the Energies of the D_{3h} and D_{6h} Benzene Structures Calculated by *ab initio* Methods for Various Lengths of the C—C and C=C Bonds in the D_{3h} Benzene Structure

$R(\text{C—C})$, Å	$R(\text{C=C})$, Å	$\Delta E\,(D_{3h} - D_{6h})$, kcal/mol	Basis Set	References
1.409	1.325	4.5	3-21G*	182
1.483^a	1.339^a	8	3-21G*	182
1.462	1.334	6.8 (8.6)^b	6-31G*	141
1.4627	1.34	7.2	6-31G	25, 184

^{a}Lengths of the central C—C bond in butadiene and of the C=C bond in ethylene.
$^b\Delta E$ (MP2/6-31G*) is given in parentheses.

originates from benzene's aromatic π-electron system. Attempts to answer it deepened the insight into the true nature of the structural criterion of aromaticity based on the requirement for equality of the carbon–carbon bond lengths, whereby the validity of this criterion was once again subjected to a severe test.

Views were developed at the *ab initio* level [25, 26, 184–187], which had first been formulated in the late 1950s without attracting, at the time, the attention they deserved. It has already [188] been shown on the basis of the simplest variable-β Hückel calculations that in going from the benzene structure with equal CC bond lengths to the one with alternating lengths the total π-electron energy is lowered and that the D_{6h} structure of benzene is dictated by the σ-skeleton, which, as distinct from the π-system, is stable against the b_{2u} distortion. That conclusion was supported by Berry's results [189], published at about the same time. In this work an analysis was performed of the experimental b_{2u} vibration mode for benzene, which is characterized by a large value of the classical zero-point amplitude of motion for the carbon atom (0.031 Å). This analysis showed that the D_{6h} structure of benzene is determined not by the π-electrons at all; rather, the function of a "corset" for the regular hexagon ring structure is fulfilled by the σ-electrons. The significance of the conclusions reported in [188, 189] was underlined by the fact that they were included in the well-known book

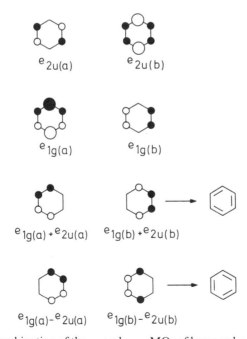

Figure 2.2   One combination of the e_{1g} and e_{2u} π-MOs of benzene leads to a Kekule type D_{3h} structure, whereas the opposite combination leads to another Kekule type structure. (Adapted from [191].)

[130] in which it was explicitly stated that "the tendency to distort already exists in benzene" [130].

A Kekule distorted structure may be regarded as one induced by the mixing of the e_{1g} and e_{2u} π-orbitals of benzene (Fig. 2.2). Owing to the considerable difference between the energies of the ground $^1A_{1g}$ state and the excited $^1B_{2u}$ state of benzene, their mixing is small and the lowering of the π-electron energy (ΔE^{π}_{dis}) proves insufficient to offset the rise of the σ-electron energy (ΔE^{σ}_{dis}) due to the b_{2u} distortion [130]. Thus, in the end, benzene owes its high-symmetry structure to the σ-skeleton.

As the difference between the energies of the ground and the lowest excited singlet state is reduced, the distortion developing in consequence of the second-order Jahn-Teller effect [192] may produce a value of ΔE^{π}_{dis} that would exceed that of ΔE^{σ}_{dis}. As a result, a structure with alternating CC bond lengths will correspond to the ground state, as indeed is observed in linear polyenes and large [n]annulenes. These results have been worked out in recent studies [25, 184–187] based on the curve-crossing diagram model (Fig. 2.3). The coordinate Q (abscissa in Fig. 2.3) corresponds to alternations in the geometry when passing from a Kekule configuration into its mirror image ($Q_1 \rightarrow Q_2$). The point Q_0 refers to the high-symmetry structure of the delocalized species D with equal bond lengths. The symbols K_1^0 and K_2^0 denote the two localized Kekule structures at their optimum geometries, while K_1^* and K_2^* designate the two excited Kekule forms having the geometry of either the form K_1^0 or K_2^0 but of the same bond-pairing type as in the ground state with which the given form correlates along the interchange coordinate (Fig. 2.3). The total deformation energy needed to reach the crossing point (see Fig. 2.3) makes only a fraction of the magnitude of the energy gap G, which represents the initial difference between the energies of the K^0 and K^* states: $\Delta E_{def} = fG$ with $f < 1$.

TABLE 2.6   Distortion Energies ΔE_{dis} and QMREs (in kcal/mol) for Benzene, Cyclobutadiene, and Their Heteroanalogs According to Calculations [25, 185, 187][a]

Molecule	$\Delta E^{\pi\sigma}_{dis}$	QMRE	ΔE^{σ}_{dis}	ΔE^{π}_{dis}	ΔE^{π}_{dis}/PB[b]
Benzene, $(CH)_6$	7.2	85	16.3	−9.1	−3.03
Cyclobutadiene (singlet)[a] $(CH)_4$	−3.4	30	7.6	−11.0	−5.50
Hexazine, [d] N_6	0.4	103	13.7	−13.4	−3.35
Tetrazet, N_4	−5.5	108	9.2	−14.7	−7.35
Hexasilabenzene, $(SiH)_6$	3.2	42	5.3	−2.1	−0.07
Tetrasilacyclobutadiene, $(SiH)_4$	0.6	18	2.7	−2.1	−1.05
Hexaphosphabenzene[d]	1.0	44.1	3.8	−2.8	−0.93

[a]Distortion ΔE_{dis} were calculated with 6-31G basis set and inclusion of π-space CI; the same basis set was used for QMRE calculations.
[b]Per π-bond.
[c]D_{4h} structure with $R(CC) = 1.463$ Å, D_{2h} structure with $R(C=C) = 1.393$ Å; $R(C-C) = 1.5162$ Å according to *ab initio* calculations [172].
[d]The D_{6h} structures were considered for hexazine and hexaphosphabenzene [185, 187]. In fact, these molecules have twist-boat D_2 structures [144, 190].

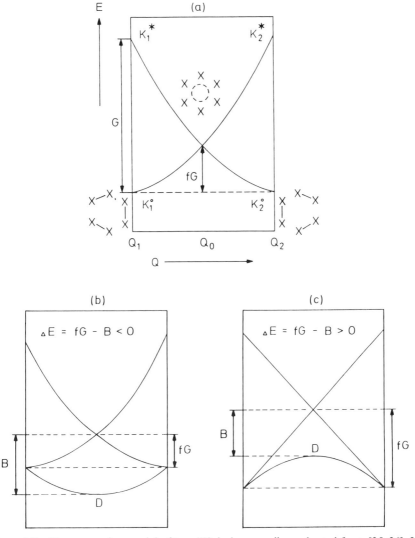

Figure 2.3 Curve-crossing model of two "Kekule curves" as adapted from [25, 26]. In diagrams (b) and (c) the lower states after avoided crossing are shown by dashed lines. The G, K_i, B, and D symbols are explained in the text.

The quantity B in Eq. (2.52) is the so-called avoided-crossing resonance interaction; it corresponds to the QMRE (see Scheme 2.1). The magnitude of the energy of the delocalized state D relative to the localized states K_1° and K_2° is ΔE ($K^\circ \to D$); hence the stability of the species D to distortion is determined by the balance of two counteracting factors, ΔE_{def} and B:

$$\Delta E(K^\circ \to D) = fG - B \quad (B = QMRE) \qquad (2.52)$$

Thus, for $\Delta E < 0$ (Fig. 2.3b), the delocalized state will be stable and at $\Delta E > 0$ (Fig. 2.3c) it will be unstable. As may be concluded from Fig. 2.3 and Eq. (2.52), given the appropriate value of G (and, consequently, of fG), a π-delocalized system will be unstable with respect to the distortion into a Kekule-type K° structure despite the large value of the QMRE [25, 26].

The analysis of how the values of B and f change depending on the value of G in the isoelectronic series X_n, whereby the atomic identity of X is changed, has shown that with only a small value of G the electronic delocalization may prove to be the decisive factor dictating the characteristic bonding and structure, and the inequality $\Delta E < 0$ will be valid.

For the systems with, on average, one electron for every center, such as benzene ($6e/6c$) or cyclobutadiene ($4e/4c$), the value of G is given [26] by

$$G = n(3/4)\,\Delta E_{\text{S}-\text{T}} \tag{2.53}$$

where n is the number of X–X dimeric units in a Kekule ground state (see Fig. 2.3) and $\Delta E_{\text{S}-\text{T}}$ is the singlet–triplet excitation energy of X_2. If the bond in X_2 is strong, the corresponding value of $\Delta E_{\text{S}-\text{T}}$ will, as a rule, be quite high.

As we are now concerned with the π-electron systems, trends will be examined in some detail of the changes in the quality G for the π-components in σ-π systems. Since for π-bonds formed by atoms of the first-row elements $\Delta E_{\text{S}-\text{T}}(\pi\pi^{*})$ takes on values in the 4.5–6.2 eV range [193–195], with ethylene and acetylene used as a model for the π-bond (CC), the values of G must be large and the corresponding delocalized π-systems either altogether unstable or inadequately stable with respect to a localizing distortion (only π-components are considered). By contrast, the second-row atoms form less strong π-bonds [196, 197] and the corresponding delocalized π-systems will have weaker distortive propensity or even prove fully stable [187].

Results of *ab initio* calculations listed in Table 2.6 support these qualitative conclusions. The D_{6h} hexagonal structure of benzene is indeed related to the value of $\Delta E_{\text{dis}}^{\sigma}$ exceeding that of $|\Delta E_{\text{dis}}^{\pi}|$. The lower σ-resistance of hexazine N_6 and, at the same time, higher π-distortivity involve lesser stability of the D_{6h} structure, which is in agreement with *ab initio* calculations reported by Saxe and Schaefer [198]. (The D_{6h} structure corresponds to a minimum on the PES only when the DZ + P basis set is used.)

In calculations [25, 185] ΔE_{dis} for the D_{3h} benzene structure, a geometry was assumed that would arise as a result of distortion of the D_{6h} symmetry structure, with nuclear repulsions between carbons remaining constant ($R(\text{C}=\text{C}) = 1.34$ Å and $R(\text{C}-\text{C}) = 1.4627$ Å). This approach was criticized, particularly on the grounds that a more rational choice would be that of the average length for internal $\text{C}=\text{C}$ bonds in conjugated linear polyenes, for example, in 1,3,5-hexatriene ($r_g(\text{HC}=\text{CH}) = 1.3678$ Å [166]), which would lead to substantial changes in the values of $\Delta E_{\text{dis}}^{\pi}$ and $\Delta E_{\text{dis}}^{\sigma}$ [199]. However, the results of Hiberty and co-workers [25, 185, 186, 200] were later borne out by the *ab initio* calculations [183] with geometry optimization, where the present values were only those of the HCC angles imitating the effect of the Mills–Nixon type deformation of the benzene ring (e.g., see [3, 200]). In going from the D_{6h} to the D_{3h} structure (for model with six HCC angles equal to 90° using the 3–21G basis set $R(C=C) = 1.319\,\text{Å}$ and $R(C-C) = 1.562\,\text{Å}$, while with MP2/3-21G one has $R(C-C) = 1.346\,\text{Å}$ and $R(C-C) = 1.579\,\text{Å}$), the total energy rises by 145.8 kcal/mol, whereas the π-electron energy (defined as $\Sigma_i n_i \varepsilon_i$) drops by 4.5 kcal/mol (3–21G) [183].

Authors [25, 184, 186] have arrived at the following important conclusion: "the connection between aromaticity–antiaromaticity and geometry is not meaningful in a broad sense" [26], having in mind that the aromaticity or antiaromaticity is reflected in the values of the QMRE[10] rather than in π-distortivities.

In our view, this conclusion ought not to be that categorical. Compare aromatic and antiaromatic molecules, for example, benzene and cyclobutadiene. For the latter, the value of $|\Delta E^{\pi}_{\text{dis}}|$ is larger; when the values are compared normalized with respect to the number of π-bonds, which is more correct procedure (see Section 2.2), this value is 1.8 times that for benzene (Table 2.6). This behavior becomes understandable in the light of the variation of the cyclic electron delocalization energy, estimated from the RE, that takes place in going from the D_{6h} to the D_{3h} structure of benzene and from the D_{4h} to the D_{2h} structure of cyclobutadiene. Glukhovtsev et al. [69], using for the D_{3h} structure of benzene and D_{2h} structure of cyclobutadiene the same ring geometry given as by Shaik et al. [25], found that the TRE calculation shows the following. Even though in both cases the distortion to the Kekule structure is attended by the lowering of the total π-electron energy (increase in absolute value), for benzene the change in the π-energy component attributable to the cyclic electron delocalization is reflected in a decrease of the TRE values: for the D_{6h} structure TRE = 0.276, while for D_{3h} it is 0.220 (in β_0 units with β_0 corresponding to $R(CC) = 1.40$ Å). This character of the variation in the cyclic π-electron delocalization energy is, for example, seen in the appreciable lowering of the QMRE value for the Kekule-type benzene structure (85 kcal/mol for **10**, 53.5 kcal/mol for **20**, 6–31G). As opposed to it, for cyclobutadiene the change of this component parallels that of the total π-electron energy (Scheme 2.3).

Clearly, since with the growing ring the energy of the cyclic electron delocalization per one π-electron or one π-bond is lowered (see REPE in Table 2.1) while the magnitude of (G/n) remains unchanged (Eq. (2.53)), for conjugated molecules with a large ring, such as large-size annulenes, the structure with bond alternation may prove energetically more advantageous [1, 3, 130, 201].

Thus, even though the value of the change in the cyclic π-electron delocalization energy may not be the biggest (in absolute value) component of $|\Delta E^{\pi}_{\text{dis}}|$, its role may prove crucial, being one factor determining the fulfillment of the

---

[10] It will be recalled that the energy of the cyclic electron delocalization is only one component of the QMRE (Section 2.2); hence the QMRE, being an analog of the HRE, will be a positive value also for antiaromatic molecules [26] (cf. the HRE of cyclooctatetraene [1.3]).

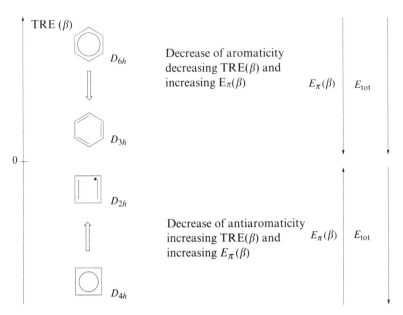

Scheme 2.3 $E_\pi(\beta)$ is π electron in β units, E_{tot} is total energy (negative) of a molecule.

inequality $|\Delta E_{dis}^\pi| < |\Delta E_{dis}^\sigma|$ for benzene and $|\Delta E_{dis}^\pi| > |\Delta E_{dis}^\sigma|$ for cyclobuta-diene. Thus the distortion into a Kekule-type structure, which leads to rising of the total $(\pi + \sigma)$ energy ($E_{tot.}$ is negative), results in the opposite changes of the aromatic cyclic π-electron delocalization energy (e.g., for benzene) estimated by TRE (β) and of the total π-electron $E_\pi(\beta)$ (this energy is given in β units β is negative) and, vice versa, it gives rise to the qualitatively similar changes of the antiaromatic cyclic π-electron energy (an increasing of TRE (β)) and of the total π-electron energy $E_\pi(\beta)$; see Scheme 2.3. This behavior pattern bears witness to the validity of the structural criteria verified for a good deal of quite diverse structures of aromatic and antiaromatic molecules. Next, we move on to examine the main types of these criteria.

2.3.2   Indices of Aromaticity Based on Estimates of Bond Length Alternation in a Ring

The ideas of equality of carbon–carbon lengths as a characteristic attribute of stable aromatic molecules put forward, in particular, in [202, 203] were then embodied in the form of the aromatic stability index of Julg and Francois [204]:

$$A = 1 - \frac{225}{n}\sum_{r}^{n}\left(\frac{d_r - \bar{d}}{\bar{d}}\right)^2 \tag{2.54}$$

where d_r is the length of the rth bond and d is the average length of n peripheral bonds (in Å). Later this expression was modified to take account of the requirement for uniform distribution of the electron density along the molecular ring bonds [205]:

$$A = \left[1 - \frac{225}{n}\sum_r^n\left(\frac{d_r - \bar{d}}{\bar{d}}\right)^2\right]\prod_r^n\left[1 - \left(\frac{\Delta q_r}{d_r}\right)^2\right] \qquad (2.55)$$

where $\Delta q_r/d_r = (q_i - q_j)/d_r$, is the charge gradient over the rth CC bond between the atoms i and j. Equation (2.55) proved useful for estimating the aromaticity of heterocyclic molecules (see [206]).[11]

At the same time, some shortcomings of this index have been pointed out [156]. Since it takes into account only peripheral CC bonds, it is incapable of a subtler differentiation of the aromaticity of polycyclic molecules and cannot be applied in the case of such conjugated cyclic compounds as, for example, 1,3,5-triazine. In Eqs. (2.54) and (2.55), the mean value of the CC bond was taken as the reference bond length, but it should be kept in mind that, for example, for polyacenes the value of $\bar{d}$ grows with the increase in the molecular size [207]. These drawbacks were circumvented in the aromaticity index HOMAS (harmonic oscillator model of aromatic stability) [208] as well as in the ΔN [156] and V indices [209–211] based on estimates of the variation of bond orders of the heterocycles. In the HOMAS, in place of d the optimal value of the bond length d_{opt} is used, determined from experimental data on CC bond lengths in ethane (s—"pure" single bond) and in ethylene (d— "pure" double bond) and on k, the force constant ratio for stretching modes of "pure" single and double bonds (2.56) [212]. The value of d_{opt} (CC) equals 1.397 Å; for each other bonds it is 1.338 (CN), 1.308 (NN), and 1.300 (CO)Å [212].

$$d_{opt} = \frac{s + kd}{1 + k} \qquad (2.56)$$

$$\text{HOMAS} = 1 - \frac{98.89}{n}\left\{\sum_{r=1}^{n_{CC}}(d_{opt}^{CC} - d_r)^2 + \sum_{r=1}^{n_{CX}}(d_{opt}^{CX} - d_r)^2\right.$$

$$\left. + \sum_{r=1}^{n_{CY}}(d_{opt}^{CY} - d_r)^2 + \sum_{r=1}^{n_{XY}}(d_{opt}^{XY} - d_r)^2\right\} \qquad (2.57)$$

where n_{CC}, n_{CX}, n_{CY}, and n_{XY} are the numbers of π-bonds of the corresponding type; the total number of all π-bonds is $n = n_{CC} + n_{CX} + n_{CY} + n_{XY}$. In the case of benzene, HOMAS equals unity for the D_{6h} structure and zero for the D_{3h} structure. The values of HOMAS (2.57) correlate well with the HSRE for polyacenes,

---

[11] The values of A for furan and pyrrole given in [205] were subsequently refined [156] to 0.93 and 0.98, respectively.

[n]annulenes, and nonalternant conjugated hydrocarbon [212–214]. The HOMAS index proved convenient in the description of aromaticity of individual rings in polycyclic molecules [212]. When in place of the bond lengths the bond orders P_r are used, the HOMAS$_p$ index is obtained:

$$\text{HOMAS}_p = 1 - \frac{3.66}{n} \sum_r^n \left(\frac{2}{3} - P_r \right)^2 \tag{2.58}$$

Subsequently, the HOMAS scheme was modified by introducing an expression that correlated the bond lengths with the quantum chemical index B of the bond [215]

$$\text{HOMAS}_{\text{W(M)}} = 1 - \frac{1}{N} \left[a \sum_{kl} (B^o - B_{kl})^2 + b \sum_{mn} (B^o - B_{mn})^2 \right] \tag{2.59}$$

where B_{kl} and B_{mn} are the bond indices of the single and double bonds between the atoms k, l, and m, n, respectively. When B is represented by the Wiberg index, Eq. (2.59) corresponds to the HOMAS$_W$ index, and in case B is the Mulliken bond overlap population (θ), this equation corresponds to the HOMAS$_M$ index [215]. The constants a and b refer to the single and double bonds, respectively [216]. The HOMAS$_{\text{W(M)}}$, as opposed to the HOMAS (2.57), indices are more sensitive to specific features of the structure, including nonplanarity and strain [215].

The values of bond lengths of aromatic conjugated molecules may be used to evaluate energetically the aromaticity. This approach is represented by the HOSE (harmonic oscillator stabilization energy) index [217]. This index is determined as the negative value of the energy E_{distort} required to distort the real molecule into its Kekule (or resonance) structure—see, for example, Eq. (2.51). It is found from experimental data on the length of the π-bonds R_r^s and R_r^d in the real molecule and using the Hooke law in which the corresponding values of the k_r^s and k_r^d force constants are calculated by means of the empirical equation based on the linear proportionality of k_r with respect to R_r; namely, $k_r = a + bR_r$ [217, 218]:

$$\text{HOSE} = -E_{\text{distort}} = 71.98 \left[\sum_{r=1}^{n_1} (R_r^s - R_0^s)^2 (a + bR_r^s) + \sum_{r=1}^{n_2} (R_r^d - R_0^d)^2 (a + bR_r^d) \right] \tag{2.60}$$

In Eq. (2.60) n_1 and n_2 are the number of single and double bonds in the Kekule structure; the constants a and b in Eq. (2.60) are determined from the values of k_r and R_r for the reference single (R_0^s) and double (R_0^d) bonds in conjugated acyclic molecules [215, 216]. For example, using the values of the $H_2C=CH$ and $HC-CH$ bond lengths in 1,3-butadiene (ED data) we have $a = 44.39 \cdot 10^5$ dyn/cm and $b = -26.02 \cdot 10^3$ dyn/cm^2 [218].

When a π-conjugated molecule can be described by means of several Kekule structures, each ith structure will correspond to a separate HOSE$_i$ value and the total value of the HOSE will be

$$\text{HOSE} = \sum_{i=1}^{N} c_i \text{HOSE}_i \qquad (2.61)$$

where the summation is made over all Kekule structures (more generally, resonance structures) and c_i is the contribution from the ith resonance structure

$$c_i = (\text{HOSE}_i)^{-1} / \sum_j (\text{HOSE}_j)^{-1} \qquad (2.62)$$

where the summation is performed over all resonance structures. From Eqs. (2.61) and (2.62) we get

$$\text{HOSE}^{-1} = N^{-1} \sum_{j=1}^{N} (\text{HOSE}_j)^{-1} \qquad (2.63)$$

The HOSE_i value is the energy by which the real molecule is more stable than its ith resonance structure. In other words, the more destabilized is the resonance structure relative to the real molecule, the greater is the magnitude of HOSE_i and the smaller the contribution by the ith resonance structure to the description of a given molecule [218].

The c_i contributions determined in this purely empirical way from experimental geometry ($c_i \cdot 100\%$) correlate quite well with the contributions from the Kekule structures calculated by means of the CCM scheme (see Section 2.2.6) [218–220]. Also the HOSE values exhibit good correlation with the HSRE [218]. Making use of the HOSE index, one may successfully describe the stability of benzenoid hydrocarbons, EDA-complexes of tetracyanoquinodimethane (TCNQ) [217, 218, 220] (also determining the percentage of contributions from the quinoid and benzenoid structures [218, 220]), substituted benzoquinones [217], and dimers of carbocyclic acids with hydrogen bonds [217], as well as the effect of various substituents giving rise to either the quinoid or antiquinoid geometry of the benzene ring [221].

An important question connected with the use of HOSE concerns the effect of the precision of experimental determination of the HOSE value. Thus the difference between the HOSE values, calculated from X-ray data on benzene (270.15K), corrected and uncorrected for libration amounts to 25% [218]. This difference is substantially smaller when neutron diffraction data are employed, obtained, incidentally, at a lower temperature (at 135.15K for benzene HOSE (corr.) = 12.37, HOSE (uncorr.) = 13.02 kcal/mol [218]).

For deriving the aromaticity index proposed by Pozharskii [156, 222, 223] the sum is calculated of the absolute values of all differences between the bond orders of n skeletal bonds including those with equal values of the orders and it is normalized with respect to the number of those differences equaling that of the twofold combinations of n elements (C_n^2).

$$\Delta \bar{N} = \frac{\Sigma |\Delta N|}{C_n^2} \qquad (2.64)$$

When the percentage of aromaticity is to be calculated, benzene is the 100% reference structure ($\Delta\bar{N} = 0$). With the aid of the $\Delta\bar{N}$ index, the aromaticity of a separate ring in a polycyclic molecule may be estimated [156].

A similar index was suggested by Bird [209–211]. In this case, the variation of bond orders is described as follows:

$$V = \frac{100}{N}\left(\sum \frac{(N - \bar{N})^2}{n}\right)^{1/2} \tag{2.65}$$

where $\bar{N}$ is the arithmetic mean of various bond orders and n is the number of bonds. For the D_{6h} benzene structure $V = 0$, while in the case of the Kekule structures of the five- and six-membered rings with bond alternation $V_K = 35$ and 33.3, respectively. With this notation the aromaticity index I may be represented by [209]

$$I = (1 - V/V_K)100\% \tag{2.66}$$

A unified aromaticity index I_A is defined as

$$I_A = I_6 = 1.235I_5 = 2.085I_{5,6} \tag{2.66a}$$

where I_6, I_5, and $I_{5,6}$ are the corresponding indices for benzene, cyclopentadienide anion, and indenyl anion, respectively [211].

When deriving the indices $\Delta\bar{N}$ and I, for calculating the values of N from experimental data on bond lengths R, the following empirical relationship is employed:

$$N = a/R^2 - b \tag{2.67}$$

The constants a and b are given for various bonds in [156, 209]. A drawback inherent to both these indices is their inapplicability to the estimation of aromaticity of high-symmetry structures, such as 1,3,5-triazine.

The values of various structural indices listed in Table 2.7 are , as a rule, in good agreement among themselves. Note, however, that the $\Delta\bar{N}$ index shows furan to be eight times less aromatic than benzene, while with the I index the reduction is by 2.3 only. Furthermore, according to the I scale, pyridine and pyrimidine are nearly equally aromatic [210], but the ratio between the respective indices $\Delta\bar{N}$ shows the aromaticity of pyrimidine to be 82% that of pyridine [156]. Since both indices are of the same type, such discrepancies are not admissible [155] and appropriate modifications are required.

Jug [224] has proposed an aromaticity index based on the magnitude of the minimal bond order in cyclic molecules. This magnitude corresponds to the weakest link in the ring, which, in turn, sets the upper limit to the magnitude of the ring current. The bond order is defined as the weighted sum of eigenvalues of the two-center parts of the density matrix for a pair of given atoms. With the

TABLE 2.7   Structural Indices of Aromaticity

Compound	HOMAS [212, 213]	HOSE kJ/mol^{-1} [218]	$\Delta\bar{N}$, % [156, 222, 233]	I_A [211]	Ring Current Index (RCI) [224, 226]	λ,β units [230–233]
Benzene	1.000	51.41	100	100	1.751	0.7910
Naphthalene	0.930	75.41	63	142	1.514	0.7742
Anthracene	0.910	109.70	—	—		0.7280
Cyclobutadiene	0.174		—	—	0.980^a	
Cyclooctatetraene	0.609		—	—	1.287^a	
Pyrrole			37	85	1.463	
Furan			12	53	1.430	
Thiophene			45	81.5		
Pyridine			82	86	1.731	0.7839
Pyridazine			65	79	1.716	
Pyrimidine			67	84	1.727	0.7571
Pyrazine			75	89	1.739	0.7860

^{a}For $(CH)_4^{2+}$ 1.509, $(CH)_4^{2-}$ 1.510; for D_{2d} structure of $(CH)_4^{2+}$ 1.626 [224], for $(CH)_8^{2-}$ 1.694 [226].

bond orders so defined, this index (ring-current index, RCI) may be applied to nonplanar molecules as well. When the magnitudes of the bond orders in ethane (1.254) and ethylene (2.155) are taken as the reference points, the conjugated cyclic components are classified into the aromatic (1.775–1.694), moderately aromatic (1.548–1.332), nonaromatic (1.297–1.212), moderately antiaromatic (1.176–1.140), and antiaromatic (1.042–0.98) ones. The values of this index calculated by the SINDO1 method are presented in Table 2.7. The virtue of this index lies in the fact it may be applied to a wide spectrum of compounds, including excited states and unstable compounds for which no reliable experimental data on molecular geometry are available, as well as some hypothetical structures [225, 226].

The results obtained are, by and large, in agreement with experiment as well as with other aromaticity indices (e.g., TRE). At the same time, for certain compounds the results prove clearly unsatisfactory. Thus pyrazole (1.463) turns out nonaromatic, equally nonaromatic is the highly unstable butalene (**22**) with the index (1.406) close to the above value [224]. Furan (1.430) is more aromatic than imidazole (1.423). The cyclobutadiene dianion of which the planar D_{4h} structure does not correspond to a minimum on the PES [227, 228] should, however, according to this index (1.510), be assigned to aromatic compounds. Such as assignment is not compatible with the HSRE value either (Table 2.1).

22

2.3.3 Aromaticity Gauged by Stability of a High-Symmetry Structure Against Distortion into Kekule-Type Structures with Bond Alternation

The criteria considered in Section 2.3.2 are based on estimates of the bond alternation from experimental or calculational data on molecular geometry; that is, they are in essence static. In other words, the aromaticity of a molecule is evaluated not for its high-symmetry configuration but rather in terms of a geometry corresponding to a minimum on the PES, though with an aromaticity lowered on account of bond length alternation. However, in some cases, such as the D_{4h} structure of the $^1B_{1g}$ state of cyclobutadiene, the high-symmetry structure does not have any alternation, and the same is true of certain opposite types of aromatic structures, for example, D_{6h} of benzene. Thus the pertinent information provided by the structural indices of aromaticity treated in the preceding section is derived from the extent of distortion of the geometry that a high-symmetry structure has to undergo in order to reduce its antiaromaticity.

An approach different in principle may be based on the estimation of the stability of high-symmetry structures against distortion into structures with bond-alternation. In structural terms, the aromaticity is associated with the stability of a high-symmetry structure with equal bond lengths in regard to distortions into Kekule-type structures with bond alternation, while the antiaromaticity has to do with the instability of high-symmetry structures and the energetic preferability of precisely the structures of the Kekule-type (Fig. 2.1). Clearly, the criteria based on estimates of structural stability[12] of high-symmetry geometry configurations should be quite useful. Such criteria will indicate the character of the PES and show whether dynamic π-bond shift isomerization processes can occur for a given compound.

As has already been noted, the energy difference between the ground and the lowest excited singlet state is an important factor governing relative stability of the high-symmetry D_{nh} structure of $(4k + 2)\pi$-electron [n]annulene with equal bond lengths. With increasing n this difference gets smaller and, owing to the second-order Jahn-Teller effect, a structure with bond-length alternation may prove energetically advantageous even in the case of [n]annulenes that satisfy the Hückel rule [130] (see Chapter 4). Whether this effect will indeed lead to the distortion of a given high-symmetry structure of a conjugated molecule into one with bond-length alternation may be checked using the schemes suggested by Binsch et al. [230–233] and Nakajima et al. [234–237]. Their approach, known as the "theory of double-bond fixation in conjugated molecules," is based on the estimation of the value of the force constant for the normal vibration Q_i corresponding to a distortion that deforms the high-symmetry structure.

If Q_i is a small displacement along the ith normal coordinate corresponding to the distortion of a high-symmetry nuclear configuration, then, making use of

---

[12] Structural stability implies the presence of a minimum on the PES corresponding to a given structure [229]

the second-order perturbation theory, we get for the ground-state energy [234]:

$$E(Q_i) = E_0 + \left\langle \Psi_0 \left| \frac{\partial H}{\partial Q_i} \right| \Psi_0 \right\rangle Q_i + \frac{1}{2} \left\{ \left\langle \Psi_0 \left| \frac{\partial^2 H}{\partial Q_i^2} \right| \Psi_0 \right\rangle \right.$$

$$\left. - 2 \sum_{n(\neq 0)} \frac{\left[\left\langle \Psi_n \left| \frac{\partial H}{\partial Q_i} \right| \Psi_0 \right\rangle \right]^2}{E_n - E_0} \right\} Q_i^2 \qquad (2.68)$$

where E_0 is the energy corresponding to the initial high-symmetry configuration, $\Psi_0, \Psi_1,...,\Psi_n$ are the unperturbed electronic wave functions, and $E_0, E_1,...,E_n$ are the corresponding eigenvalues. If the ground-state wave function Ψ_0 is nondegenerate, the first-order term (second term in Eq. (2.68)) will be nonzero only for completely symmetrical nuclear displacements. The displacements not altering the symmetry of the nuclear configuration correspond to a first-order bond fixation [230].

To estimate the changes in energy, no longer associated with the symmetrical first-order distortion (E_0^1), but with the second-order bond fixation, one has to analyze the terms in braces in Eq. (2.68) that correspond to the expression for the force constant k_0^i of the normal vibration Q_i. If the condition of the π-σ separability is fulfilled, Eq. (2.68) assumes the form (2.69):

$$E(Q_i) = E_0^{(1)} + \left\{ k_\sigma + \left\langle \Psi_0 \left| \frac{\partial^2 H_\pi}{\partial Q_i^2} \right| \Psi_0 \right\rangle - 2 \sum_{n \neq 0} \frac{\left| \left\langle \Psi_n \left| \frac{\partial H_\pi}{\partial Q_i} \right| \Psi_0 \right\rangle \right|^2}{E_n - E_0} \right\} Q_i^2 \qquad (2.69)$$

Neglecting the second derivative of the resonance integral $\beta(r)$ with respect to r (whereby the second term in the braces in Eq. (2.69) turns into zero [234], we obtain for k_0^i

$$k_0^i = k_\sigma - 2 \sum_{n \neq 0} \frac{\left[\left\langle \Psi_n \left| \frac{\partial H_\pi}{\partial Q_i} \right| \Psi_0 \right\rangle \right]^2}{E_n - E_0} \qquad (2.70)$$

where k_σ is the force constant for the σ-bond between the approximately sp^2-hybridized carbon atoms CC. Thus the force constant for Q_i can be negative if the second term, which is positive, exceeds the first term in magnitude. To estimate the value of k_0^i, Binsch and co-workers made use of the bond–bond polarizabilities [230–233]. With this approach, Eq. (2.70) is reduced to (Eq. (2.71)) [234]:

$$k_0^i = k_\sigma + 2\lambda_i(\beta')^2 \qquad (2.71)$$

where λ_i are the eigenvalues of the bond–bond polarizability matrix and β' is the first derivative of $\beta(r)$. If the greatest positive (in units of β_0^{-1}) eigenvalue of λ_{max} exceeds a certain critical value λ_{crit}, then the quantity k_0^i will be negative, and the high-symmetry structure will be subject to distortion into one of lower symmetry [230, 233]:

$$\lambda_{max} > \lambda_{crit} = -\frac{k\beta_0}{2(\beta')^2} \qquad (2.72)$$

So the quantity λ_{max} may serve as an index of aromaticity. The conjugated cyclic molecules are nonaromatic if their high-symmetry geometry configurations are characterized by $\lambda_{max} > \lambda_{crit}$ ($1.8\beta^{-1}$ or $1.22\beta^{-1}$ in the HMO and PPP calculations, respectively) and, consequently, they are unstable with respect to distortion into a structure of lower symmetry (Fig. 2.1) [232, 233]. In other words, a strong first-order or a second-order bond fixation does not exist for aromatic molecules [232, 233] (Table 2.7). By contrast, for pentalene $\lambda_{max} = 3.1539\beta^{-1} > \lambda_{crit}$ [233] and the D_{2h} structure (**23**) of the ^1A_g state does not correspond to a minimum on the PES but rather to a transition state of the bond-shift isomerization [238].

$$(2.73)$$

$$C_{2h} \qquad\qquad \mathbf{23}, D_{2h} \qquad\qquad C_{2h}$$

Studies conducted by Nakajima et al. [234–237] have shown that if for a high-symmetry configuration the lowest singlet excitation energy ΔE_1 calculated by the PPP method is less than the empirically found critical value of ~1.2 eV, then the force constant k_0^i for a given nonsymmetrical in-plane nuclear vibration should be negative — see Eq. (2.70) — and this structure will not correspond to a minimum on the PES and will be subject to a Q_i distortion. For example, for pentalene $\Delta E_1 = 0.35$ eV [234]; that is, the D_{2h} structure does not correspond to a minimum on the PES — see Eq. (2.73) — which is in agreement with calculated λ_{max}. For [18] and [30]annulenes, the calculation of ΔE for the lowest $^1A_{1g} \rightarrow {}^1B_{2u}$ transition (assuming D_{6h} geometry) gives 1.6 and 0.95 eV [234]; that is, $C_{18}H_{18}$ is stable against distortion into a structure with bond alternation, while $C_{30}H_{30}$ is not. This conclusion is in agreement with experimental data on [18]annulene and dehydro[30]annulene (e.g., see [1]).

These approaches to the estimation of the stability of high-symmetry structures have also been employed in the case of the excited state [237] and radical [235] structures. The instability in question is regarded as a manifestation of the lattice instability [239, 240]; one points in this connection to the similarity with the well-known case of the one-dimensional lattice instability studied by Peierls [241].

As is apparent from Eq. (2.70), the chief factor on which the appearance of instability of a high-symmetry configuration depends is the energy difference ΔE

between the ground and the lowest singlet electronic state; as a rule, the main contribution to the latter comes from the HOMO-to-LUMO electron transfer; that is, the HOMO–LUMO energy gap is an essential factor in its own right (e.g., see [242]). At the same time, these two values determine the manifestation of instability of the restricted Hartree–Fock (RHF) solutions [243, 244], as may be seen from expressions for diagonal elements of the $(A + B)$ matrix of singlet and triplet instability of RHF solutions[13] [244]:

$$(A^s_{ij, ij} + B^s_{ij, ij}) = e_j - e_i - J_{ij} + 3K_{ij} = {}^1\Delta E_{ij} + K_{ij} \qquad (2.74)$$

$$(A^t_{ij, ij} + B^t_{ij, ij}) = e_j - e_i - J_{ij} - K_{ij} = {}^3\Delta E_{ij} - K_{ij} \qquad (2.75)$$

In Eqs. (2.74) and (2.75), i and j denote the occupied and the virtual orbitals, e_i and e_j are the energies of these orbitals, J_{ij} and K_{ij} are the Coulomb and the exchange integrals, and ${}^1\Delta E_{ij}$ and ${}^3\Delta E_{ij}$ are the energies of excitation from the ground state to the singlet and triplet excited states, respectively.

Another solution in the case of the singlet instability, mentioned in the footnote, is the so-called off-diagonal charge-density wave (CDW), or bond-order alternation wave (BOAW), solution. As has been pointed out in [249], the BOAW solution involves occurrence of the nonzero Hellman–Feynman forces giving rise to a distortion of the nuclear framework. In other words, there exists an interrelation between the presence of a BOAW solution and the lattice instability leading to bond alternation:

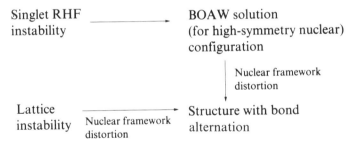

The relationship between the last term in braces in Eq. (2.68), k^{rel} (whose magnitude determines the possibility of a negative value of the force constant for the normal vibration Q; for the case of the π-system see Eq. (2.69), and the eigenvalues of the instability matrix has been established in [245]; see also [248,

---

[13] Various types of instability of the HF solutions were analyzed in [245–248]. The HF instability is distinguished into internal and external [245] depending on whether the possible variations keep the wavefunction within the limitations of a given HF method or may extend it outside these. The singlet RHF instability is characterized by wavefunction variations under which the double occupancy of the orbitals is retained; that is, it is internal. The triplet instability involves rejection of the double occupancy condition and thus transition to the unrestricted HF method. The singlet RHF instability indicates the existence within the same method of another solution with lower energy and may be associated with some real feature of the molecular system.

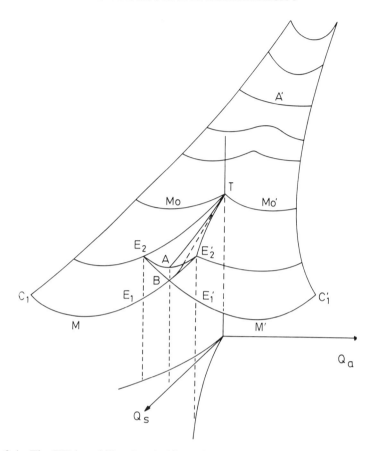

Figure 2.4   The HF instability threshold on the two-dimensional energy surface. The upper ATA' curve runs to the threshold (point T) along the ridge and continues in the valley $E_2AE_2'T$ of the unstable HF solution. The lower double degenerate curve TB corresponds to stable solutions. The M_0 and M_0' minima of the curve in the perpendicular section across T lie on the bottom of the potential surface valleys originating from point A'. After passing the instability threshold T, two stable HF solutions emerge ($E_2BE_1'C_1'$ and $E_2'BE_1C_1$), with the corresponding nuclear equilibrium configurations possessing lower symmetry than configuration A. For more details see [245].

250]. A detailed analysis carried out there has led to the following conclusion. If a singlet RHF solution instability has been ascertained for the symmetrical nuclear equilibrium configuration A (see Fig. 2.4), it must be preceded by a real lattice instability of a certain other symmetrical nuclear configuration A' (preceding configuration A on the symmetrical coordinate Q_s) with respect to deformations along the nonsymmetrical coordinate Q_a. These deformations have the same symmetry as the variations of the electron density matrix that give rise to the instability of the solution for the symmetrical nuclear configuration. Thus approaching the threshold of the closed-shell HF instability results in the negativity of at least one of the force constants [241]. This means that if the HF solution for a symmetrical nuclear equilibrium configuration is unstable, there exists for it at least one more solution, albeit of lower energy. The PES corresponding to this solution has, generally, a minimum for a different nuclear configuration (in Fig. 2.4 two minima M_1 and M_1' are shown that correspond to configurations of lower symmetry).

The relation between the lattice instability and the HF solution instability is given by the following expression derived in [251]:

$$k^{\text{rel(HF)}} = -2(C\bar{C})\left(\frac{AB}{B\,A}\right)\left(\frac{C}{C}\right) \qquad (2.76)$$

where the matrix element $C_{ij} = \langle \varphi_i | \dfrac{\partial \varphi_j}{\partial Q} \rangle$ and A, B are the matrices constituting the corresponding $(A + B)$ and $(A - B)$ HF instability matrices; see Eqs. (2.74) and (2.75).

Thus in some cases, the lattice instability of the high-symmetry nuclear configuration can be accompanied by the singlet instability of the corresponding RHF solution. As may be apparent from the comparison between Eqs. (2.68) and (2.74), for the singlet instability to arise the conditions have to be more severe than in the case of the lattice instability recall that in Eq. (2.74) the form of only diagonal terms is given; for details see [252, 253]). PPP calculations indicate that the singlet instability of the RHF solution for high-symmetry configurations of nonalternant hydrocarbons arises for $\Delta E_1 < 0.2$ eV, while the lattice instability appears when $\Delta E_1 < 1.2$ eV [239].

As the lattice instability and the instability of the closed-shell RHF solution are interrelated, one may expect [37] that the magnitude of the least eigenvalue $^s\lambda_+$ of the singlet RHF instability matrix $(^sA + {^sB})$ could serve as an aromaticity index similarly to the λ_i eigenvalues of the bond–bond polarizability matrix considered earlier in connection with the lattice instability; see Eqs. (2.71) and (2.76). Indeed, according to PPP calculations on [n]annulene, the singlet instability of the RHF solution arises starting with [26]annulene only [254, 255]. Even though [26]annulene has not been obtained so far, experimental data on dehydro[26]annulene point to a structure with bond alternation [3]. By contrast, for the first members in the series of $(4k + 2)$ π-electron [n]annulenes, the singlet RHF instability appears at very small (in absolute value) unrealistic magnitudes of the resonance integral β [252, 256] (for benzene $\beta_{\text{crit}}^s = -0.298$ eV while for [26]annulene $\beta_{\text{crit}}^s = -2.719$ eV [254]). The lattice instability of the D_{6h} benzene structure leading to b_{2u} distortion into a D_{3h} Kekule-type structure is manifested only for $|\beta| < 1.6$ eV [257]. Thus the singlet instability of the closed-shell RHF solution may serve to determine the critical size of the ring of $(4k + 2)$ π-electron [n]annulene at which it loses its aromatic character [257]. At the same time, the detection of such instability for a high-symmetry configuration leads one to expect that a structure with bond alternation will have lower energy. On the other hand, for high-symmetry structures of some antiaromatic molecules, such as **24–26** [258, 259] and **27** [261], the singlet instability of the RHF solution is manifested even at standard values of β:

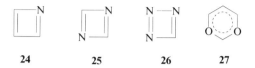

This instability was found also in *ab initio* calculations on the D_{4h} structure of cyclobutadiene ($^1B_{2g}$ state) [261].

It should be emphasized that when an RHF instability (triplet, nonsinglet, singlet) is detected, it is advisable to take into account correlation effects [37, 262], which would enable correct results to be achieved, indicating preferability of high-symmetry structures, in calculations on [10], [14], [18]annulenes [262, 263] and azulene [242, 263].

2.3.4 Effects of Nonplanarity

The planarity of the molecular structure is usually regarded as a distinguishing feature of the π-aromatic species [1, 3], while the preferability of nonplanar structure is often taken to be a manifestation of the antiaromaticity of the planar form, unstable with respect to nonplanar distortions [264, 265]. Based on the trends in geometry, a structural criterion of aromaticity and antiaromaticity could be set up on the condition that the planarity (nonplanarity) be determined above all by the aromaticity (antiaromaticity) of a given molecule.

Convenient targets for studying the relationship between the aromaticity of a molecule and the stability of its geometry with respect to out-of-the-plane distortions are cyclophanes. Resonance energy (RE) calculations from the energy of homodesmotic reactions (i.e., from the HSE, see Section 2.2.9) have shown that the destabilization of these molecules is directly proportional to the ring-bending angle φ [4, 266, 267]. The geometries and REs were calculated using the DZ basis set (C($9s5p/4s2p$), H($4s/2s$)); for $n = 8$, the minimal basis set STO-3G was employed. In the case of benzene (D_{6h} structure), the DZ value of ΔH of the corresponding homodesmotic reaction is 28.1 kcal/mol [267].

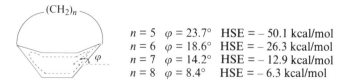

$n = 5$ $\varphi = 23.7°$ HSE $= -50.1$ kcal/mol
$n = 6$ $\varphi = 18.6°$ HSE $= -26.3$ kcal/mol
$n = 7$ $\varphi = 14.2°$ HSE $= -12.9$ kcal/mol
$n = 8$ $\varphi = 8.4°$ HSE $= -6.3$ kcal/mol

As is apparent from these HSE values, *para*-cyclophanes are strongly destabilized relative to benzene, by, respectively, 78.2 and 54.4 kcal/mol for [5]- and [6]-*para*-cyclophanes. To evaluate the effect of the nonplanar distortion of the benzene ring on those values, calculations have been made on the benzene molecule with a model structure in which the ring has been frozen in conformations characteristic of the [5] and [6]-*para*-cyclophanes and two hydrogen atoms lie in the same direction as the first carbon atoms of the chains in *para*-cyclophanes. The comparison of the determined destabilization energies (60.1 and 33.5 kcal/mol, respectively) with the difference between the HSEs of *para*-cyclophanes and benzene has shown that the negative values of the HSEs for [5], [6]-*para*-cyclophanes are due mainly to the nonplanarity of the benzene ring rather than to any other effect, such as the strain within the $(CH_2)_n$ chain [266, 267].

These values of the HSE indicate energetic disadvantage of nonplanar distortions for aromatic molecules, which may cause the loss of aromaticity determined by the energy criterion. This conclusion is supported by calculations of changes in the energy for the benzene molecule that occur upon folding the benzene ring about the 1–4 axis; for the folding angle of 160° the energy rises by 15 kcal/mol relative to the planar structure (6-21G basis set) [265].

Now another question arises. Is such a destabilization chiefly caused by the lessening aromaticity or it is mainly due to the total π-energy being minimal precisely when the ring geometry is planar? This question does not have one simple answer. One may not categorically declare that nonplanar distortions of the benzene ring necessarily involve substantial diminution of aromaticity, not to mention its total loss.

The fact of the matter is that the loss of aromaticity in a nonplanar distortion identified, for example, by an energy criterion, does not always abolish all manifestations of aromaticity indicated by structural (bond alternation) or magnetic (NMR, see Section 2.4) criteria. For example, based on the latter criteria, [*n*]*para*-cyclophanes ($n = 5 – 7$) should be classified as aromatic notwithstanding the considerable nonplanar distortions of the benzene ring [4, 266, 267]; see also Section 2.6 and Chapter 1. Another example is given by 8,11-dichloro[5]-*meta*-cyclophane. According to X-ray data [268], it has a strongly bent benzene ring (unsymmetrical boat conformation, bending angles of 26.8° and 12°). Nevertheless, the bond fixation that would point to the loss of the aromaticity is not observed and the CC bond lengths are equal within experimental accuracy (1.393 ± 0.007 Å).

The use of a correlation of the *ortho*-benzylic coupling constants with the square of the SCF bond order for *para*-cyclophanes showed that in the species the benzene ring can undergo severe distortions of the σ-framework without disruption of the π-system [269]. Only the [6]-*para*-cyclophane may have a detectable amount of π-electron distortion.

Just as the planar structure of a conjugated cyclic molecule (such as tropone [270]) cannot be regarded as the sufficient condition for a conclusion in favor of its aromaticity, so also the nonplanar structure of the ground state of such a molecule will not be a reliable sign of antiaromaticity, and in each particular case an analysis of the factors determining its geometry is required.

The prototype of the antiaromatic molecules, cyclobutadiene, has a planar

D_{2h} structure, while the D_{2d} structure (**28**) possesses a higher energy and does not correspond to the minimum on the PES [271]. Cyclooctatetraene has a nonplanar structure of D_{2d} symmetry (**29**); however, the π-electron energy of this molecule is minimal in the case of the planar geometry [272]. The nonplanar geometry of the carbon skeleton of cyclooctatetraene and of its derivatives results from the angular strain arising in the σ-system for the planar ring configuration (their contribution to the enthalpy of transition into the D_{2h} structure is $\sim$ 85%) [273].

28, D_{2d} 29, D_{2d} 30   31, D_{4h}

32, D_2 X = N, P   32a, D_2

A molecular structure classified as antiaromatic can have only slight deviations from planarity, as in 1,5-bisdehydrol[12]annulene (**30**) [274], on the other hand, specific features of a σ-system may give rise to nonplanar geometry of a molecule regarded as aromatic. Thus the dication and dianion of cyclobutadiene, though satisfying the $(4n + 2)$ Hückel rule, possess nonplanar structures, as has been shown by *ab initio* calculations [227, 275–277]. An important part in the destabilization of a planar D_{4h} structure of the cyclobutadiene dianion (structure **31**) may belong to the electron redundancy of the ring [227, 228]. Hexazine, N_6, and hexaphosphabenzene, P_6, have nonplanar twist-boat D_2 structures (**32**) [190]. While antiaromatic cyclobutadiene has the planar D_{2h} structure, the tetraphosphacyclobutadiene, P_4, which is less antiaromatic than cyclobutadiene, possesses the nonplanar D_2 structure (**32a**) [278].

There is no straightforward and simple quantitative relationship between the trend toward distortion of a completely symmetrical structure to a Kekule-type with bond-length alternation and the effects of the cyclic electron delocalization that determine the manifestations of aromaticity (see Section 2.3.1). For this reason, none of the above-considered structural indices, phenomenological in essence, may claim to represent a general quantitative scale of aromaticity (antiaromaticity). This would, apparently, be possible only on condition that out of numerous factors the effect that the aromaticity (cyclic electron delocalization) exerts on the alternation of bond lengths in a given ring could be singled out and evaluated. This task is solved in a more satisfactory manner in terms of the energy criteria. The structural criteria are adequate for estimating quantitative variation of aromaticity in the series of molecules of the same type.

2.4   MAGNETIC CRITERIA

2.4.1   Concept of the Ring Current

The magnetic criteria of aromaticity are based on the model of interatomic ring currents induced in conjugated cyclic molecules by external magnetic fields. The idea of the ring currents was put forward by L. Pauling [279]. It is assumed that in such molecules the π-electrons move freely along the closed contour in the ring plane. When an external magnetic field is applied perpendicular to this plane, the π-electrons form diamagnetic ring current about the direction of the field. This current gives rise to a secondary field, which may be approximated by the field of a dipole μ located in the ring center and directed antiparallel with respect to $\mathbf{H}_0$ (Fig. 2.5). In accordance with this model, the ring current I

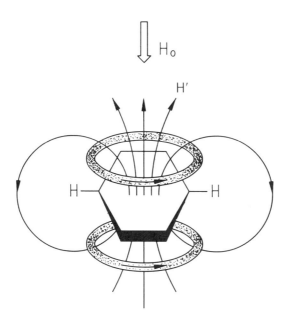

Figure 2.5   Schematic representation of the dipole model of the secondary magnetic field of intensity $\mathbf{H}'$ originated from π-electron ring current in benzene upon application of an external magnetic field $\mathbf{H}_0$.

induced in the benzene ring is defined as follows [280]:

$$I = -\frac{2e^2}{9c\hbar^2}\beta SH_0 \tag{2.77}$$

where S is the area of the benzene ring and β is the resonance integral for the CC π-bond in benzene. Analogous, though more complex, expressions defining the ring current induced in a given π-electron ring have been derived for the general case of the polycyclic molecules [280].

The ring currents cannot be directly determined by experimental methods; however, comparison of experimental values of magnetic susceptibilities, their exaltations, and anisotropies, as well as of ^{1}H NMR chemical shifts with the respective data calculated·from the ring-current model points to the adequacy of this model for the interpretation of experiment.

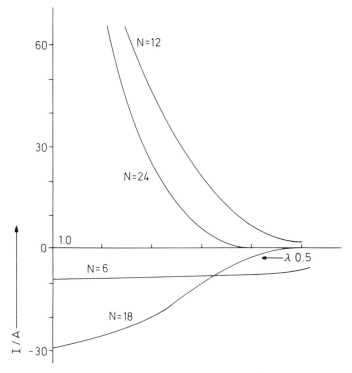

Figure 2.6 Dependence of the ring current in C_NH_N annulenes per unit magnetic field on the value λ describing bond alternation: $\lambda = \beta_1/\beta_2$, where β_1 and β_2 are the resonance integrals corresponding to the long and short CC bond in the annulene structure with bond length alternation [281]. $A = \pi^2 e^2 \beta_0 R_{CC}^2/4\hbar^2 c$, where R_{CC} is the CC bond length in annulene.

Next, we consider certain specific features of the model, as applied to conjugated cyclic systems, which are important for working out various magnetic criteria of aromaticity.

Figure 2.6 represents the dependence of the induced ring current per unit magnetic field on the bond alternation λ for the monocyclic conjugated hydrocarbon $C_N H_N$ with a regular planar polygon geometry derived by the HMO method [281]. The magnetic susceptibility associated with the ring current, known as the London susceptibility, is given by

$$\chi = \frac{IS}{cH_0} \tag{2.78}$$

For aromatic annulenes with the number of the π-electrons $N = 4n + 2$, where n is the natural number, for any value of the bond alternation λ a negative contribution to the susceptibility takes place (diamagnetic ring current). By contrast, in the case of antiaromatic $C_N H_N$ molecules with $N = 4n$ the contribution is positive (paramagnetic ring current). Thus, in aromatic and antiaromatic annulenes, ring currents of opposite types are induced. Note that owing to an increase in the bond alternation the paramagnetic ring current diminishes more rapidly than the diamagnetic one (Fig. 2.6). One may recognize here a direct analogy with the lessening of the antiaromaticity, determinable from the RE values, as the bond alternation is increased and with the leveling of differences in the properties of $C_N H_N$ annulenes as n grows (see Section 2.2).

For [4n]annulenes, with diminishing bond alternation the values of paramagnetic ring currents tend to infinity [251], which can be explained by the incorrect description of the system with a degenerate incompletely filled to highest level.

In the case of $4n$ π-electronic monocyclic systems whose symmetry group is distinct from D_{4h}, D_{8h} (D_{4kh}), for the configuration e^2 there are four electronic states, namely, the ^3A triplet, the degenerate ^1E singlet (the $^1B_1$ and $^1B_2$ degeneracy is lifted by external magnetic field), and the ^1A singlet state. As has been shown in [282], in all these states of $4n$ polyenes of this type (i.e., charged $4n$ π-electron systems with the number of atoms in the ring $N \neq 4n$, such as $(CH)_3^-$, $(CH)_5^+$, and $(CH)_6^{2+}$) the induced ring currents are equal, finite, and diamagnetic. The assignment of these structures to the antiaromatic class is dictated only by the presence and orientation of a nonzero induced orbital magnetic moment [282]. As opposed to it, for $4n$ π-electron monocyclic neutral polyenes (D_{4h} structure of cyclobutadiene, D_{8h} structure of cyclooctatetraene, etc.), the orbital magnetic moment equals zero and in the lowest singlet state finite induced paramagnetic currents flow in the ring, which has been confirmed by PPP calculations with the electron correlation taken fully into account [283, 284].

The molecules that sustain the diamagnetic ring current induced by an external magnetic field are termed diatropic and classified on the basis of the ring current model as aromatic. By contrast, the molecules are antiaromatic if the ring currents are paramagnetic (so-called paratropic systems). The assignment of monocyclic $4n$ π-electron conjugated ions is made in accordance with the above-mentioned conclusions of [282].

The question to be answered is whether the ring-current model criterion is compatible with the chief energy criterion of aromaticity and antiaromaticity. The answer will be positive if a relationship is revealed between ring currents and resonance energies, which are quite dissimilar characteristics of aromaticity. Desirable would be the existence of a ring current versus resonance energy dependence so that the latter, upon application of a magnetic field, would change differently for the diatropic and the paratropic systems.

As has been shown by Aihara [9], the influence of the magnetic field on the total π-electron energy of a molecule manifests itself in the change of only those contributions to the magnitude of the coefficients of the characteristic polynomial $P(X)$ which are due to the presence of rings . The contributions of this category represent a monotonically decreasing function of H, which means that their values are smaller than the values of such contributions to the $P(X)$ coefficients for a field-free conjugated system. Since these contributions contain, as it were, "encoded" effects of the electron cyclic delocalization, the external magnetic field reduces the magnitude of these effects and, accordingly, the aromatic stabilization (antiaromatic destabilization). Since the energy of the reference structure remains unchanged, the corresponding changes occur in the REs. Thus, by reducing the aromaticity or antiaromaticity, the magnetic field destabilizes aromatic molecules and ions while stabilizing antiaromatic ones [9, 282, 285]. For example, let us turn to [N]annulenes, making use of the topological index Z^* of their total π-electron energy $E_{\pi} \approx 6.0846 \log Z^*$ [286]. This index represents a modification of Hosoya's index Z, to be discussed in Chapter 3; it is a function of all coefficients of the characteristic polynomial $P(X)$: $Z^* = |P(i)|$, $i = \sqrt{(-1)}$ [287]. Taking into account the external magnetic field, the quantity Z^* is defined for [N]annulenes as

$$Z^* = \frac{1}{\sqrt{5}}\left(\frac{1+\sqrt{5}}{2}\right)^{N+1} - \frac{1}{\sqrt{5}}\left(\frac{1-\sqrt{5}}{2}\right)^{N+1} - 2(-1)^{N/2}\cos\theta_N H \quad (2.79)$$

where

$$\theta_N = eS_N/\hbar c \quad (2.80)$$

Here S_N is the area of the N-membered ring. The action of the magnetic field reduces Z^* and, consequently, E_{π} at $N = 4n + 2$ and increases them (in β units) at $N = 4n$. An expression analogous to Eq. (2.79) was also derived for general-type polycyclic systems [289]. Since in a monocyclic system the decrease in RE (in absolute value for antiaromatic molecules) due to the magnetic field is proportional to the susceptibility attributable to the ring current [9], one may conclude that the sign of the susceptibility of that ring and by virtue of Eq. (2.78), the sign of the ring current are opposite to that of the resonance energy [9, 288]. In other words, the aromatic rings possessing positive resonance energies will have, because of ring currents, negative London susceptibilities, while the antiaromatic rings for which the REs are negative will have positive susceptibilities. For a monocyclic conjugated molecule the relationship between the RE and the London susceptibility χ is given by Eq. (2.81) [9, 288]:

$$\chi \approx - \text{ RE} \cdot \theta_N^2 \qquad (2.81)$$

Thus the ring currents are indeed directly related with the REs indicating compatibility of the model in question with the energy criterion of aromaticity. Moreover, the conclusions as to aromaticity drawn from calculated values of the ring currents are in accord with those derived from a set of experimental parameters. For example, the enhanced ring current of charged aromatic annulenes [283, 284] is in line with the conclusion that $(4n+2)$ π-electron charged annulene possesses greater aromaticity relative to isoelectronic neutral annulene (e.g., the planar diatropic structures of the cyclooctatetraene dianion and cyclononatetraenide anion are more aromatic than the nonplanar structure of [10]annulene) [3].

However, the quantitative determination of aromaticity (antiaromaticity) from the ring-current model may be complicated by at least two problems. First, experimentally observable values of magnetic susceptibilities, their exaltations and anisotropies, as well as the ^{1}H NMR chemical shifts, are not necessarily determined exclusively by the ring currents; hence all other effects have to be identified and removed from consideration. Naturally, for this model to work, the contribution by the ring current must be predominant. Another problem is that the calculational results on ring-current intensities for molecules from the diatropic-paratropic border area may vary qualitatively depending on the method of calculation [289]. We examine these problems in the following sections when dealing with the experimentally observable quantities that reflect the presence of the ring currents and may serve as a basis for corresponding scales of aromaticity.

2.4.2 Magnetic Susceptibility Exaltation and Anisotropy

Since the magnetic susceptibility anisotropy $\Delta\chi$ is a characteristic attribute of the aromatic molecules [130, 290], its value could play the role of an aromaticity index. The bulk susceptibility can be represented as the sum

$$\chi = 1/3 \ (\chi_{aa} + \chi_{bb} + \chi_{cc}) \qquad (2.82)$$

where χ_{aa}, χ_{bb}, and χ_{cc} are the diagonal elements of the magnetic susceptibility tensor, and $\Delta\chi$ is given by

$$\Delta\chi = \chi_{cc} - 1/2 \ (\chi_{aa} + \chi_{bb}) \qquad (2.83)$$

with c being the out-of-plane axis for the planar molecule [130, 291].

Direct application of $\Delta\chi$ for the quantitative evaluation of aromaticity is, however, not practicable since its magnitude is determined not only by ring currents. Quite substantial may prove the local contributions by the π-bond anisotropy, the anisotropy of CC and CH σ-bond magnetic susceptibility, and the anisotropy due to local paramagnetic currents (for more details see [130]). For example, about half the magnetic susceptibility anisotropy of benzene is attributable precisely to the local contributions [292, 293].

Despite fairly large values of $\Delta\chi$ for tropone, 2-pyrone, and 4-pyrone, they are assigned to the nonaromatic molecules since the magnitudes of $\Delta\chi$ are in these cases determined mainly by local effects [294].

Thus the conclusion is obvious: the local and nonlocal contributions to $\Delta\chi$ ($\Delta\chi^{local}$ and $\Delta\chi^{nonlocal}$) must be sorted out and only the latter can be aromaticity index [295, 296]. The values of the $\Delta\chi^{nonlocal}$ can be derived from the comparison of experimental values of $\Delta\chi$ with the local ones calculated for a hypothetical localized structure [296].

To obtain the experimental values of $\Delta\chi$, various means and procedures may be used: the high-resolution Zeeman microwave spectroscopy [292]; growing of single crystals and determination of the susceptibility provided that the structure of the crystal is known and the orientation of the molecules does not compensate their anisotropies [297]; and the Cotton–Mouton effect [298] or the high-field NMR spectra of fully deuterated compounds [299]. Note that direct experimental techniques are applicable to a limited number of compounds, while the indirect methods yield somewhat uncertain results on account of corrections needed. Clearly, all this obstructs wide-scale use of the $\Delta\chi$ index.

Another quantitative characteristic of the magnetic manifestation of aromaticity is represented by the exaltation of the total magnetic susceptibility[14] Λ [290, 300]. For conjugated compounds, this parameter is given by the difference between χ_M and χ'_M. These are the experimentally measured molar susceptibility and the molar susceptibility, respectively, calculated by an additive scheme:

$$\Lambda = \chi_M - \chi'_M \qquad (2.84)$$

More precisely, Eq. (2.84) represents the difference between the magnetic susceptibility of a cyclic conjugated system and that of a hypothetical cyclic system with localized double bonds in which the ring current vanishes. A molecule is aromatic when $\Lambda > 0$, antiaromatic when $\Lambda < 0$, and at $\Lambda \simeq 0$ it is nonaromatic [290, 300–302]; see Table 2.8.

Thus the determination of aromaticity by means of Eq. (2.84) is analogous to

---

[14] In the literature, the term "diamagnetic susceptibility" is frequently used instead of "total magnetic susceptibility." The point is that (in the case of aromatic compounds) the negative diamagnetic contribution to the total value of the magnetic susceptibility often turns out to be greater in absolute value than the positive paramagnetic contribution. It should, however, be kept in mind that the total susceptibility accounts as a rule for only about 10% of the value of either the diamagnetic or paramagnetic component [295].

TABLE 2.8   Magnetic Susceptibility Anisotropies ($\Delta\chi$), Exaltations of the Magnetic Susceptibility (Λ, $\Delta\chi$, and λ are in 10^{-6} cm^3/mol units) and related ρ Index of Aromaticity

Compound	$\Delta\chi$ [292]	Λ [156, 290, 300]	ρ [305]
Benzene	59.7[a]	13.7(15.1)[b]	1.00
Naphthalene	119.9[a]	30.5 (31.1)	0.901
Anthracene	182.4[a]	48.6 (49.1)	0.665
Cyclopentadienide anion	—	—	1.165
Azulene	—	29.6 (36.0)	0.899
Heptalene	—	– 6.0 (–3.8)	—
Pyridine	57.2	13.4 (18.3)	—
Pyrrole	42.4	10.2 (14.5)	—
Furan	38.7	8.9 (13.9)	—
Thiophene	50.1	13.0 (17.8)	—
[8]Annulene	—	– 0.9	—
[16]Annulene	—	– 5.0	—

[a]Found by single-crystal measurements; the rest are obtained by microwave spectroscopy.
[b]Calculated using the parameters of Haberditzl [304]; in parentheses using Pascal's constants (see [156, 303]).

the scheme for calculating the RE. Several systems of additive parameters are known for calculating χ'_M for a reference structure [290, 303, 304].

It is on the value of Λ that the index of aromaticity ρ is based [305]:

$$\rho = k\,\frac{n\Lambda}{S^2} \tag{2.85}$$

where n is the number of π-electron, S is the area of the ring, and k is the scaling factor enabling benzene to be taken as a reference compound ($\rho = 1$ for benzene).

Calculations of the values of the London susceptibility for $(4n + 2)$- and $4n$-membered circuits in polycyclic hydrocarbons have shown [285, 306] that for the magnetic susceptibility a rule can be formulated analogous to the generalized Hückel electron-count rule (see Chapter 3). Note that this rule for the Hückel annulenes is opposite to that for the Möbius annulenes [307]. Namely, a polycyclic conjugated molecule containing only the $(4n + 2)$-membered circuits exhibits the diamagnetic London susceptibility, while a polycyclic hydrocarbon with only the $4n$-membered circuits will have the paramagnetic London susceptibility. As opposed to the Hückel annulenes, in the corresponding Möbius annulenes the $4n$ π-electron molecules have diamagnetic susceptibilities and the $4n + 2$ π-electron molecules are characterized by the paramagnetic London susceptibilities [307].

At the same time, the London susceptibilities and Λ have a number of shortcomings in the role of aromaticity indices. For example, some nonaromatic molecules have a value of Λ that would warrant their assignment to the aromatic class. In particular, for cyclopentadiene $\Lambda = 6.5$ and for cycloheptatriene $\Lambda = 8.1$ [291] (units as in Table 2.8). 7-*Tert*-butylcycloheptatriene, whose energy and structural criteria do not justify its classification under either the aromatic or homoaromatic type, has, all the same, an exaltation ($\Lambda = 14.8$) exceeding in absolute value Λ of benzene [308]. Another problem is that, for polycyclic compounds, the contribution to the value of χ, due to the ring currents, is proportional to the corresponding ring area squared [9, 288]:

$$\chi \approx - \sum_i RE^{(i)} \theta_i^2 \qquad (2.86)$$

In Eq. (2.86), the summation is performed over all rings, $RE^{(i)}$ is the resonance energy of the ith ring, and θ_i proportional to the area of the ith ring and is defined by expression (2.80). As a result, larger rings will make larger contributions even if they have smaller resonance energies. Ultimately, the magnitude of the London susceptibility for the molecule as a whole will be determined by the contributions from larger rings, while the total magnitude of the RE will depend on those from the smaller ones. This leads to the situation where some polycyclic compounds that are diatropic have, according to the energy criteria, to be considered as antiaromatic. For example, butalene (**33**) and bicyclo[6.2.0] decapentaene (**34**) belong to such molecules [9, 288, 309] (however, the RE estimates for **34** are contradictory (see Chapter 4).

33	**34**

Thus we see that diatropicity and antiaromaticity may coexist in polycyclic systems of this type [288, 309].

There is one more problem, which we mentioned earlier. Antiaromatic molecules are, as a rule, characterized by a small energy gap between HOMO and LUMO (see Section 2.5). This gives rise, in the case of such molecules and the borderline aromaticity structures, to an essential dependence of calculated London susceptibility values on the method of calculation. Thus for pyracyclene (**35**), the results of the calculation of London susceptibilities differ qualitatively depending on the method used [310].

Another example is given by calculations of the π-electron magnetic susceptibility for the icosahedral structure of the earlier-mentioned carbon cluster C_{60} (buckminsterfullerene (**36**)) [33, 34] in which a fragment can be singled out structurally close to pyracyclene. By using a modification of London's method applicable to nonplanar structures, it has been found that for equal magnitudes of the resonance integrals β_1 and β_2, that is, in the case of a D_{6h} structure of benzene rings in **36**, C_{60} possesses a weakly paramagnetic overall π-electron ring-

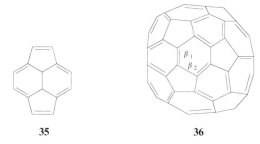

35   36

current susceptibility [311]. However, when the β_1 / β_2 ratio is changed by a mere 2%, the sign before the ring-current magnetic susceptibility is reversed and the contribution from this ring-current component to the total molecular susceptibility of **36** becomes probably weakly diamagnetic (see also [312]).

2.4.3 Nuclear Magnetic Resonance Chemical Shifts

As is apparent from Fig. 2.5, the secondary field H' induced by the ring current deshields the protons lying in the molecular plane outside the ring. By contrast, the protons above and under the ring plane where the total field $(H_0 + H')$ is smaller are strongly shielded. The deshielding of the outer protons and stronger shielding of the inner ones in aromatic annulenes as well as the opposite effects in antiaromatic annulenes (Table 2.9) [2, 313] attributable to diamagnetic or paramagnetic ring currents[15] offer an attractive opportunity for devising a scale of aromaticity and antiaromaticity [1–3, 314]. It should, however, be kept in mind that the value of the shielding constants σ of the nucleus A is determined by several different contributions [130]. They include the contributions σ_d^{AA} and σ_p^{AA} stemming from the diamagnetic Langevin-type currents and the paramagnetic currents induced in the atom A itself, respectively; the contribution $\sum_{B(\neq A)}\sigma^{AB}$ due to the diamagnetic and paramagnetic currents induced in the atoms B bordering on atom A; and the contribution $\sigma_{\text{ring curr}}^A$ coming from the interatomic ring currents:

$$\sigma = \sigma_d^{AA} + \sigma_p^{AA} + \sum_{B(\neq A)} \sigma^{AB} + \sigma_{\text{ring curr}}^A \qquad (2.87)$$

The contributions of the first three types are practically local in character; they are close in value for two protons with the similar structural environment, such as ethylenic and aromatic types protons. It is only the last term in Eq. (2.87) that defines the values of the chemical shifts characteristic of aromatic or antiaromatic compounds.

The data of Table 2.9 show that the difference between chemical shifts for the outer and inner protons are quite sizeable; for the aromatic (diatropic) and

---

[15] This question has been considered in detail in [1–3, 130, 314] (see also [289, 313, 315]). We will not delve further into it here.

TABLE 2.9 Chemical Shifts in ^{1}H NMR Spectra of [n] Annulenes and Their Dianions [2] (δ, ppm)

Aromatic Systems			Antiaromatic Systems		
Annulene or Its Ion	Outer Protons	Inner Protons	Annulene or Its Ion	Outer Protons	Inner Protons
[6]	7.37	—	[8]	5.68	—
[14]	7.6	0	[12]	5.91	7.86
[18]	9.28	− 2.99	[16]	5.40	10.43
[22]	8.50 to 9.65	− 0.4 to − 1.2	[20]	4.1 to 6.6	10.9 to13.9
[12]$^{-2}$	6.23	− 4.6	[18]$^{-2}$	− 1.13	28.1
	6.98				29.5
[16]$^{-2}$	7.45	− 8.17			
	8.83				

the antiaromatic (paratropic) systems they are of opposite sign. The large value of this difference for charged annulenes is in accord with the enhanced ring current in these species [283, 284]. However, it should be pointed out that in the case of multiple charged species, other factors, such as non-uniform π-electron charge densities, are likely to make as significant a contribution to the experimentally observed ^{1}H NMR chemical shifts in these systems as the ring-current effects themselves [315].

The effective use of the difference between chemical shifts of the outer and inner protons of [n]annulenes as a quantitative index of aromaticity is impeded by the following circumstances. The contribution by the $\sigma^A_{\text{ring curr}}$ is obscured by three other aromaticity-independent contributions—see Eq. (2.87)—and by the fact that the ^{1}H NMR spectra depend on the temperature [1–3, 313]. Furthermore, the monocyclic systems of small and medium size have no inner protons at all. Nor should one neglect the substantial contribution to deshielding of the outer protons in aromatic systems coming from the local anisotropic effects at the carbon atoms [316] due to the deviation of the electronic cloud from the spherical symmetry caused by chemical bonding [316, 317].

In the [4n]annulenes structures, the bond alternation leads to a substantial weakening of the ring-current effects so that the deshielding of the inner protons, relative to the outer ones (see Table 2.9), is largely determined not by these effects but precisely by the local anisotropic contributions since the inner protons lie closer to a greater number of the carbon atoms than the outer ones [316].

The role of the local effects is even more important in the case of the heterocyclic molecules in which the values of chemical shifts depend on the nonuniform distribution of the electron density and on the anisotropy of the heteroatom. This may bring about a situation when estimates of the relative aromaticity based on values of the chemical shifts will prove wrong. For example [155], for pyrrole the proton signals ($\delta(2H) = 6.68$, $\delta(3H) = 6.22$) lie in a stronger field than those of furan ($\delta(2H) = 7.42$, $\delta(3H) = 6.37$), even though, according to energetic and some other criteria, pyrrole should have a greater aromaticity as compared to furan (e.g., see [318]). In the pyridine–bismabenzene series, the proton H_α signal is shifted downfield [319, 320], while the energies of isodesmic reactions unambiguously point to a lessening of the aromatic character in this series [138]:

X = N($\delta(H_\alpha)$ = 8.29), P($\delta(H_\alpha)$ = 8.61),
As($\delta(H_\alpha)$ = 9.68), Sb($\delta(H_\alpha)$ = 10.94),
Bi($\delta(H_\alpha)$ = 13.25),
at X = CH ($\delta(H)$ = 7.37)

37

In the carbon-13 NMR spectroscopy, the shielding constant σ for the nucleus in the magnetic field should be represented as a sum of three terms, Eq. (2.88) [314]; compare with the Eq. (2.87). The first of these, σ_d, corresponds to the local diamagnetic contribution, the second, σ_p, to the local paramagnetic contribution, while σ' represents the set of long-range shielding effects originating from neighboring atoms and groups and caused largely by the magnetic anisotropy and electric field effects.

$$\sigma = \sigma_d + \sigma_p + \sigma' \qquad (2.88)$$

The magnitude of σ' accounts only for approximately 10% of the total shielding constant [321], and the ring-current effects having in the ^{13}C NMR spectroscopy the same order of magnitude as in the ^1H NMR technique are, when two compounds are compared, altogether obscured by variations in the local contributions σ_p and σ_d [314].

As for the effects of ring currents, they can be isolated and estimated only through a comparison of a given compound with appropriately selected reference compounds [321]. Consider, for example, the transformation of the 14 π-electron system (38) into the 16 π-electron one (39) in which upfield shifts of 15.7 and 9.9 ppm are observed for central carbons [322]. In order to estimate which part of the total value of these upfield shifts should be attributed to the ring-current effects, one may take naphthalene, pleiadiene (40), and acenaphthalene (41) as reference molecules [322]:

	38	**39**		**40**	**41**
a	126.8	133.9		138.9	128.9
b	127.4	133.9		139.2	128.6

The calculated values of δ turn out to be greater by 6.5 and 7.1 ppm than the observed ones. These upfield shifts for the central atoms in (**39**) can be regarded as being due to the peripheral ring current. As has been shown by Vysotsky and Mestechkin [323], the inclusion of the ring currents when calculating ^{13}C chemical shifts for alternant hydrocarbons and azines enables a much better agreement with experiment to be achieved compared to calculation without such inclusion. For dications of benzenoid hydrocarbons, a correlation has been shown to exist between the values of average carbon-13 chemical shifts and those of the CCMRE [324].

2.5 OTHER CRITERIA OF AROMATICITY AND ANTIAROMATICITY

In the preceding sections, the principal criteria of aromaticity and antiaromaticity, namely, the energetic, structural, and magnetic ones, have been examined. Their connection with the main factor of the aromatic stabilization, that is, the electron cyclic (bond) delocalization, can clearly be traced in most cases. Of course, the distinguishing features of the aromatic and antiaromatic compounds are reflected in a larger set of physicochemical characteristics, many of which can serve as a basis for specific criteria of aromaticity. Admittedly, these characteristics are often not directly related to the electron cyclic (bond) delocalization, or this relation may be obscured by some other effects.

Such criteria are qualitative; they cannot claim to evaluate quantitatively the degree of aromaticity. By and large, conclusions based on them are consistent with those obtained by means of the principal criteria; even quantitative agreement is not rare. However, their applicability is confined to rather limited spectrum of compounds. The merit of these criteria consists primarily in the fact that they, as it were, enlarge the mirror reflecting the properties of aromatic and antiaromatic compounds so as to make the reflection possibly more diversified.

Two groups of these criteria can be singled out. The first of these is based on the historical interpretation of aromaticity of a compound as its propensity for undergoing substitution reactions with retention of the structural type. The criteria making up this group rest on specific features that characterize the reactivity of aromatic compounds, and some of them are interesting today only from the historical viewpoint. The criteria of the second group are rooted in the characteristic features of the electron distribution in aromatic and antiaromatic compounds. These criteria were formulated not long ago and in recent years they have been the subject of considerable attention.

2.5.1   Retention of the Structural Type

Since the aromaticity is defined as the stabilization due to the cyclic electron (bond) delocalization, the data on thermodynamics and kinetics of various reactions leading to damage in the cyclic delocalization system (or, conversely, to its formation) may be used for assessing aromatic stabilization or antiaromatic destabilization.

   This assessment may be based on the so-called empirical resonance energies determined from the thermodynamic parameters of reactions characterized by retention of the structural type or on various indices of the structural stability and reactivity, such as the HOMO–LUMO energy gap.

2.5.1.1   Empirical Resonance Energies   The assessment of the aromatic stabilization from heats of hydrogenation reactions, the values of pK_a, and the constants of tautomeric equilibria is practiced very frequently. In addition to aromaticity (antiaromaticity), numerous other factors are operative in this case, so these estimates cannot measure up to the status of a quantitative approach. Still, they may be quite useful in just diagnosing the effects of the aromatic stabilization or antiaromatic destabilization.

Heats of reactions   Starting with the classic experiments of Kistiakowsky et al. [325], the endothermicity of the first step, Eq. (2.89), of successive hydrogenation of benzene, namely, the transformation into cyclohexa-1,3-diene, unlike the exothermicity of the reactions of addition of the following hydrogen molecules, was thought to be largely associated with the loss of the aromatic stabilization:

$$+ \quad H_2 \longrightarrow \qquad \Delta H = 5.6 \text{ kcal/mol} \qquad (2.89)$$

For cyclobutadiene, the value of ΔH of the analogous reaction calculated by the MINDO /3 method is, by contrast, negative [326]:

$$+ \quad H_2 \longrightarrow \qquad \Delta H = -61.9 \text{ kcal/mol} \qquad (2.90)$$

Defining the energy of aromatic stabilization of benzene as the difference between the heats of the reactions (2.91) and (2.92) (were benzene not aromatic, these two heats should be nearly equal), we obtain for it the value of 25.6 kcal/mol [326]:

$$2H_2C = CH_2 \longrightarrow \text{/\/} + H_2 \quad \Delta H = 1.1 \text{ kcal/mol} \quad (2.91)$$

$$\text{(cyclohexadiene)} \longrightarrow \text{(benzene)} + H_2 \quad \Delta H = -24.5 \text{ kcal/mol} \quad (2.92)$$

Then ΔH(Eq. (2.93)) = ΔH(Eq. (2.89)) – ΔH(Eq. (2.90)) = 67.5 kcal/mol can be regarded as the difference between the energies of the aromatic stabilization of benzene and the antiaromatic destabilization of cyclobutadiene, which gives for the latter the value of – 41.9 kcal/mol [326].

$$\text{(benzene)} + \text{(cyclobutadiene)} \longrightarrow \text{(benzene)} + \text{(cyclobutadiene)} \quad (2.93)$$

The reaction represented by Eq. (2.93) may formally be considered as the dehydrogenation of cyclobutadiene with benzene. Here this reaction has served for comparing the aromatic and antiaromatic compounds, but a reaction of this type may also be used for ascertaining the relative degree of aromaticity. An example is given by the dehydrogenation of 4H-thiapyran with a pyrylium cation:

$$\text{(thiapyran, S)} + \text{(pyrylium, O}^+\text{)} \longrightarrow \text{(thiapyrylium, S}^+\text{)} + \text{(pyran, O)} \quad (2.94)$$

The equilibrium of this reaction is considerably shifted to the right, which may be regarded as evidence for greater aromaticity of the thiapyrylium cation over the pyrylium cation [156].

When estimating the aromatic stabilization from the data on the heats of hydrogenation, the contribution comes from the difference between strain energies of interconverting cyclic structures. Partial compensation of the ring strain effects can be achieved through combination of, for example, Eqs. (2.90) and (2.95), which leads to ΔH(Eq. (2.96)) = ΔH(Eq. (2.90)) – ΔH(Eq. (2.95)). For Eq. (2.96) $\Delta E = -38.2$ kcal/mol (6-31G* //6-31G*) (see [327]).

$$\text{(cyclobutene)} + H_2 \longrightarrow \text{(cyclobutane)} \quad (2.95)$$

$$\text{(cyclobutadiene)} + \text{(cyclobutane)} \longrightarrow 2 \text{(cyclobutene)} \quad (2.96)$$

To what extent the inclusion of one such factor as the ring strain may influence the result is seen from the difference between the estimates of the antiaromatic destabilization based on the heats of reactions represented by Eqs.

(2.97) [326] and (2.98) [327]. In the former case the energy amounts to – 68.0 kcal/mol[16], while in the latter it is a mere – 3.5 kcal/mol, which cannot be attributed solely to the use of different methods of quantum chemical calculations.

$$\text{(2.97)}$$

$$\text{(2.98)}$$

For the cyclopentadienyl cation $(CH)_5^+$, the value of the antiaromatic destabilization energy (AADE) calculated by the MINDO/3 method is –14.5 kcal/mol [328].

The degree of destabilization of the $(CH)_5^+$ relative to the cyclopentyl cation can also be estimated from a comparison between the heats of reactions (2.99) and (2.100). The values of the heats of formation of the $C_5H_9^+$, $C_5H_7^+$, and $C_5H_5^+$ cations needed for the calculation of the above heats have been found from the ionization potentials of the respective radicals determined with the aid of an electron monochromator–mass spectrometer combination [329]:

$$+ \ H \ + \ \bar{e} \qquad \Delta H = 11.60 \ \text{eV} \qquad \text{(2.99)}$$

$$+ \ H \ + \ \bar{e} \qquad \Delta H = 10.56 \ \text{eV} \qquad \text{(2.100)}$$

$$+ \ H \ + \ \bar{e} \qquad \Delta H = 11.93 \ \text{eV} \qquad \text{(2.101)}$$

As is apparent from the ΔH values of these reactions, the inclusion of one double bond into the C_5 ring stabilizes the cation relative to the neutral form by roughly 1.04 eV; as opposed to it, the introduction of the second double bond leads to destabilization by 1.37 eV. Thus, compared to the cyclopentyl cation, the cyclopentadienyl cation is destabilized by 0.33 eV (7.6 kcal/mol) [329].

Activation Parameters of Valence Isomerizations   When in the structure of a transition state of isomerization a ring is formed of aromatic or antiaromatic character, a comparison between the activation parameters of this isomerization and those of an analogous isomerization not attended by the formation of such a ring can give the experimental estimate of the RE. For example, the dif-

---

[16] The value given in [326] was refined in [328] to – 63.0 kcal/mol.

ference between the values of activation barriers of the thermal valence isomer-izations (2.102) and (2.103) is thought to be associated with the formation in the former case of structure **43** destabilized because of the presence of an antiaro-matic cyclobutadiene ring. From the values of $\Delta G^{\neq}$ obtained by means of ^{1}H NMR spectroscopy, it may be inferred that norcaradienes **42** and **42a** are more stable by about 14 kcal/mol than **43** [330]:

42   **43**   **42a**

(2.102)

(2.103)

When the REs of 3,4-dimethylenecyclobutadiene and 1,2-divinylcyclobutadi-ene calculated by an *ab initio* method are taken into account, the ultimate esti-mate of the RE for cyclobutadiene comes out at – 21 kcal/mol. This is two times less (in absolute value) than the above-mentioned antiaromatic destabilization energy of cyclobutadiene determined from the calculated heats of hydrogena-tion.

Equilibrium Constants   The estimation of aromaticity may be made on the basis of equilibrium constants of reactions in which one of the interconverting forms contains a cyclic system of conjugated bonds. For example, triphenylcy-clopropene (**44**) has $pK_a \geq 51$ (indirect estimation via a thermodynamic cycle); it is less acidic by 20 pK units than triphenylmethane (**45**) for which $pK_a = 31$. This fact may be accounted for by the antiaromatic destabilization of the triphenyl-cyclopropenide anion (**46**) formed as a result of the deprotonation of **44** [331, 332].

44   **46**

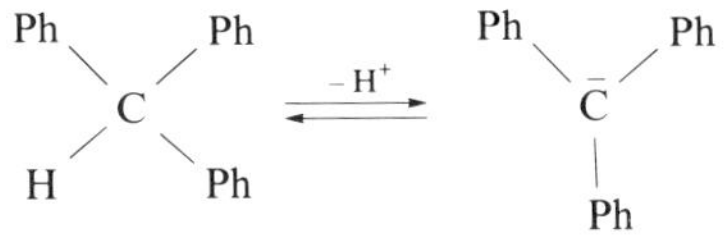

45

The estimation of the antiaromatic destabilization of the cyclopropenide anion from the pK_a data gives the value of 28 kcal/mol [332], which differs appreciably from the above-given value of AADE (– 63.0 kcal/mol).

As the constants of tautomeric equilibria are determined by the pK_a values of individual tautomers, the former may also be used for estimating the empirical resonance energies, as has been shown in the review article by Cook et al. [333]. In particular, when two tautomers HA and AH form the cation HAH$^+$, the tautomeric equilibrium constant K_T can be expressed through the basicity constants K_B and K_A of these tautomers $K_T = K_A/K_B$ [333]:

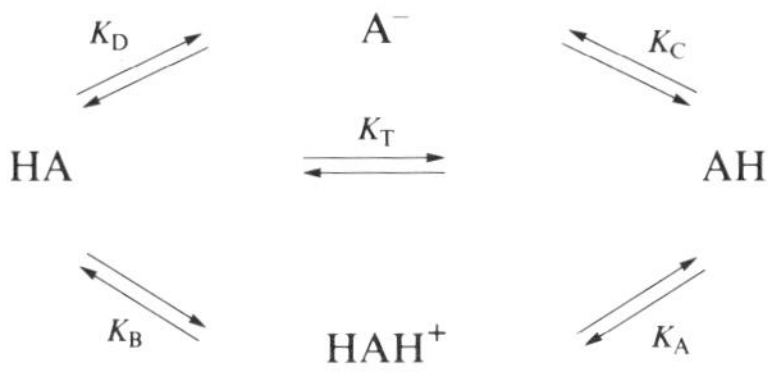

$$K_T = K_A/K_B = K_C / K_D$$

Since in the equilibrium experiment the difference between the energies of two forms is determined, one can directly ascertain only the difference between aromatic stabilization energies. In this case, the influence of such effects has to be taken into account as the solvation and the differences in stability, which are characteristic of the tautomeric functional groups. The latter effect can be included if in determining the difference between the aromatic stabilization energies (ASE) of, for example, the tautomers **47** and **48**, the system **49** $\rightleftharpoons$ **50** is also examined:

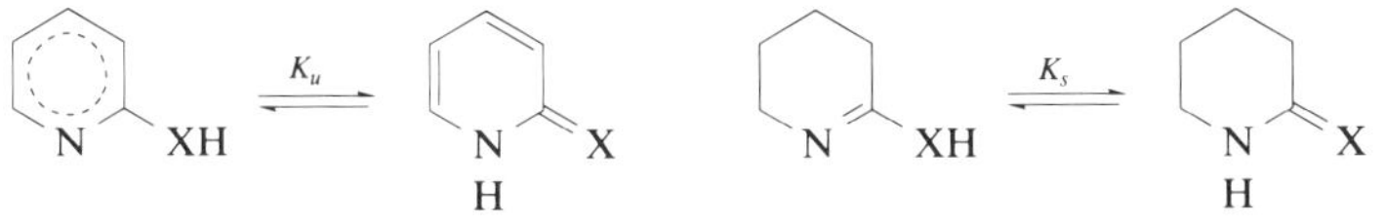

| 47 | 48 | 49 | 50 |

Then ASE(**47**) – ASE(**48**) = $\Delta H_s - \Delta H_u$. In order to pass from the ΔG values derived from the tautomeric equilibrium constants to the ΔH values, some approximations are introduced; in particular, one assumes $\Delta G \approx \Delta H$ or $\Delta H = 1.3$ ΔG [333]. Using the above scheme, the change in the difference between the ASEs with the variation in X may be traced. Thus 2-pyridone has a 7.5 kcal/mol smaller ASE compared to that of 2-oxypyridine (X = O). For X = CH_2, the difference between the ASEs is much greater (18 ± 3 kcal/mol) [333].

When one of the tautomers explicitly lacks cyclic conjugation, the difference between the stabilization energies of tautomers HA and AH arising from conjugation effects including the effect of the cyclic electron delocalization may serve to evaluate the ASE of the tautomer with cyclic conjugation.

For example, let us assume that the ASE of benzene can be approximated by the ASE value of phenol **51**. To evaluate the latter, two equilibria have to be considered:

Having estimated for the phenol **51** $\rightleftharpoons$ 2,4-cyclohexadienone (**52**) equilibrium the value of $pK_T \approx 9.5$, we can find that ΔH for **53** $\rightleftharpoons$ **54** equals 29 kcal/mol. In order to determine the ASE of **51** and, accordingly, the ASE of benzene, one has to allow for the stabilization of 2,4-cyclohexadienone in consequence of the conjugation (5 kcal/mol); then the latter is estimated to be 34 kcal/mol. Thus this value obtained for benzene within the framework of an approximate approach differs essentially from the estimates found with the aid of various RE calculation schemes (see Table 2.1).

The examples adduced show that the estimation of the aromatic stabilization (antiaromatic destabilization) energy based on thermodynamic characteristics of different reactions may yield for the same compound quite dissimilar values. As has already been pointed out, these discrepancies stem from the fact that the cyclic electron delocalization is only one component of the overall effect, with other constituents subject to considerable variations depending on the reaction type.

2.5.1.2 Reactivity Indices as a Measure of Aromaticity

The HOMO–LUMO Energy Gap   The value of the HUMO–LUMO energy separation, Δ_{HL}, may serve as an index of structural stability [334] and reactivity depending on the measure in which the HOMO and LUMO take part in driving chemical reactions [335].

For benzenoid hydrocarbons whose data on reactivity are often used for correlation with other aromaticity indices [116], an approximate formula has been derived for calculating Δ_{HL} (in β units) [150];

$$\Delta_{HL} = 2(-2.90611(2M/N)^{1/2} + 3.91744\ K^{2/N}) \qquad (2.104)$$

where M and N are the numbers of the CC bonds and carbon atoms, respectively, and K is the number of the Kekule structures. By comparing Eq. (2.104) with the stability index (SI) represented by Eq. (2.42), which correlates with the DRE, Eq. (2.6), we can see that Δ_{HL} is related not only to SI but also to the mean vertex degree ($2M/N$) of the molecular graph. Thus the hydrocarbons may possibly have the same values of SI but different Δ_{HL}. This means that the value of Δ_{HL} should be used not as an index of thermodynamic stability but as a measure of reactivity [150].

Such association of the aromaticity with the low reactivity estimated from the Δ_{HL} value is reflected in the aromaticity criterion based on the concept of relative hardness η_r [336], which is defined as the difference between the values of the absolute hardness for a given molecule (η) and for the corresponding acyclic reference structure (η_a):

$$\eta_r = \eta - \eta_a \qquad (2.105)$$

In the molecular orbitals theory, the absolute hardness is given by [336, 337]

$$\eta = (e_{LUMO} - e_{HOMO})/2 \qquad (2.106)$$

Hence, making use of the corresponding roots x_i of the characteristic polynomial $P(G,X)$ and the acyclic reference polynomial $R(G,X)$—see Eqs. (2.20) and (2.23)—the values of η and η_a can be calculated with the aid of Eqs. (2.107) and (2.108):

$$\eta = \beta\, (x_{LUMO} - x_{HOMO})/2 \qquad (2.107)$$

$$\eta_a = \beta\, (^{ac}x_{LUMO} - {}^{ac}x_{HOMO})/2 \qquad (2.108)$$

It is not surprising that the η_r values correlate well with the TREPE. A high value of the absolute hardness (i.e., the large HOMO–LUMO energy gap) is a measure of high stability and low reactivity [334]; consequently, the corresponding η-scale reflects the lowering of the reactivity of aromatic compounds.

For [$4n+2$]annulenes, a direct relationship between the RE and the HOMO–LUMO energy gap has been found [338]. However, in some cases, for example, in annelated [14]- and [18]annulenes, the aromatic stability determined from the TRE values does not always correlate with the Δ_{HL} values [302].

The HOMO–LUMO energy difference is equally related with the magnetic criteria of aromaticity. For example, a linear correlation has been found between the values of Δ_{HL} and ^1H NMR paratropic shifts of the $4n$ π-electronic polycyclic benzenoid dications (2.109) and dianions (2.110) [339]:

$$\Delta\delta(\text{ppm}) = -1.23\,\Delta_{HL}(\text{eV}) + 4.31 \qquad (2.109)$$

$$\Delta\delta(\text{ppm}) = -2.50\,\Delta_{HL}(\text{eV}) + 6.40 \qquad (2.110)$$

The values of paramagnetic ring currents are inversely proportional to the difference between the energies of the ground and the excited states [130]. Assuming that the energy of transition to the lowest excited state is largely determined by the HOMO–LUMO transition, one may expect that for structurally similar molecules, such as **55** and **56**, the values of the paramagnetic ring currents will primarily depend on the HOMO–LUMO energy gap Δ_{HL} [340].

As for antiaromatic compounds, they are characterized by the closeness in energy of the ground state and the lowest excited one [37, 339]. As is apparent from Eqs. (2.70) and (2.76), the energy gap between the ground and the lowest excited states determines structural manifestations of the aromaticity and antiaromaticity and gives rise to the instability of HF solutions.

As has been shown by DRE [341] and TRE [79] calculations, the aromatic (antiaromatic) character is often inverted in the lowest excited state. Therefore, for molecule with the aromatic ground state, one may expect antiaromatic destabilization of the lowest excited state and a sizeable energy gap between them. Conversely, for molecules with the antiaromatic ground state, this gap will be much smaller.

This means that the antiaromatic compounds should exhibit deep coloring (compound **56** is needed deep green [340]) and low energy of the long-wave transitions. Evidently, this energy can qualitatively diagnose antiaromaticity (in the case of aromatic molecules, the lowest singlet transition has an energy not much different from that of the lowest transition in an olefinic molecule of the same size [342]). For example, in the case of molecules **57–60** containing an eight π-electron ring of 1,4-dihydropyrazine, which is an antiaromatic species [343], the PPP calculations show very low long-wave transition energies [344]:

Here the wavelengths (λ_{max}) are given in angstrom units (Å).

Reactivity Indices for Electrophilic and Nucleophilic Substitution Reactions In an electrophilic or nucleophilic attack upon a conjugated cyclic molecule, the formation of a corresponding σ-complex **61** or **62** leads to the disappearance of the effect of the cyclic electron (bond) delocalization. In order to estimate, under these conditions, the energy of such delocalization, one may have recourse, depending on the reaction type, to the energy of the cation L_μ^+ or anion L_μ^- localization.

Hafelinger [345] has suggested the following criteria of aromaticity: $L_\mu \geq 2.00$ β, $F_\mu < 0.46$, and $\pi_{\mu\mu} < 0.44$ (Hückel theory; for benzene $L_\mu^+ = 2.536$, $F_\mu = 0.399$), where μ is the index of the carbon atom of the aromatic ring to which the electrophilic or nucleophilic agent is added in the limiting step of the reaction. The last two quantities are taken because the sum of the orders of the π-bonds broken in the formation of the σ-complex is linearly related with the free-valency index F_μ, which in aromatic hydrocarbons, is proportional to the atom–atom polarizability $\pi_{\mu\mu}$.

Ab initio calculations (MP2/6-31G**, with geometries of transition state structures optimized by the MNDO method) of activation energies for the reactions of the H^+ abstraction from and H_2O addition to the cyclobutenylium ion (**63**) and cyclohexadienylium ion (**62a**) illustrate basic differences between the reactivities of these species [346].

Whereas for **63** the activation barrier of the H^+ abstraction is higher by 35.9 kcal/mol compared to that of the H_2O addition, in the case of **62a** it is, on the contrary, lower by 7.6 kcal/mol. Calculation of the rate constant ratio for

H$^+$ abstraction versus H$_2$O addition yields the values of $2.123 \cdot 10^{-24}$ and $1.212 \cdot 10^9$ (25°C) for **63** and **62a**, respectively [346]. This preferability of the aromatic substitution over the addition reaction and the opposite trend in the case of antiaromatic compounds calculated by *ab initio* methods [346] are reflected in the aromaticity index A', Eq. (2.111), proposed earlier [347]. This index was constructed on the assumption that the H$^+$ abstraction from the σ-complex is easier, the greater the positive charge is in the CH$_2$ group orbital *h*; and that the rate of the competitive addition process is determined by the maximal charge density in the carbon p_π atomic orbital:

$$A' = \frac{\lambda c_h^2 - c_m^2(\text{max})}{\sum c_r^2} \qquad (2.111)$$

where c_r is the coefficient of the *r*th carbon p_π-orbital, c_m is the maximal coefficient, and c_h is the coefficient of the pseudo-π-orbital of the CH$_2$ group. When the parameter λ is taken to be unity, we have for benzene $A' = 1$. For antiaromatic annulenes C$_{4n}$H$_{4n}$, $A' = -1/2n < 0$; for linear polyenes C$_{2n}$H$_{2n+2}$, $A' = 0$; and for aromatic annulenes C$_{4n+2}$H$_{4n+2}$, $A' = 3/(2n + 1) > 0$ [347].

However, the kinetic data on the reactivity of aromatic compounds is not, generally speaking, a fully reliable source for estimating relative aromaticity. The reactivity depends on a variety of factors, among which the aromatic stabilization of the ground-state structure is not necessarily predominant (for details see [1, 154, 156]). In other words, the thermodynamic stability of aromatic cyclic conjugated molecules attributable to the cyclic electron (bond) delocalization will not always correlate with their kinetic stability. Such correlation is usually observed in benzenoid hydrocarbons. On the other hand, nonbenzenoid hydrocarbons, which are assigned to aromatic compounds, can vary greatly in kinetic stability and possess high reactivity [348]. Thus, although calicene (**64**) has a fairly high per electron HSRE value (0.043β [16]), its reactivity is also high, which is apparently the main reason for the failure of all attempts to isolate the parent molecule (only few calicene derivatives are known to be substituted with strong electron-donating and/or electron-withdrawing groups at the three-membered and/or five-membered ring [2]).

64

At the same time, chemical stability characteristic of the aromatic compounds can also be possessed by species, which, according to the main criteria, should be classified as antiaromatic. An example is given by the earlier mentioned naphtho[1,8-*cd*:4,5-*c'd'*]bis[1,2,6] thiadiazine (**56**), whose stability is attributed to inertness of the –NSN– linkage [349]. The propensity for retaining the structural type can also be manifested by nonaromatic compounds: for example, in chemical reactions of λ^5-phosphorines (**65**) regeneration of the ylide structure is observed [350].

65

All the same, the kinetic data may be quite useful in estimating the aromaticity when reactivity of congeneric structures is compared. As a case in point, one may recall (see Section 2.2.10) the correlation between the values of the rate constants for the Diels–Alder reaction of maleic anhydride with benzenoid aromatic hydrocarbons and the RE values: see Eqs. (2.44) and (2.45) [10, 116, 152, 153].

Rather paradoxically, the characteristics specific to the reactivity of the aromatic compounds, which have been responsible for their identification as a special class of compounds, cannot serve as a basis for constructing quantitative indices of aromaticity (antiaromaticity). But then, the propensity towards retention of the structural type can help to quantitatively discriminate between the aromatic and antiaromatic compounds by their reactivity.

2.5.2   Specific Features of the Electron Distribution

The theory of molecular structure based on the topology of molecular charge distribution, developed by Bader and co-workers (see [351, 352] and the literature cited therein), enables certain features to be revealed that are characteristic of the systems with aromatic (antiaromatic) cyclic electron delocalization. To describe the structure of a molecule, it is necessary to determine the number and kind of critical points in its electronic charge distribution, that is, the points where for the gradient vector of the charge density ρ, denoted by $\Delta\rho$, the condition $\Delta\rho = 0$ is fulfilled. The critical point r of the charge distribution is characterized by the rank and signature of the Hessian matrix of $\rho(r)$. The rank equals the number of nonzero eigenvalues of the Hessian matrix and the signature is the algebraic sum of the signs of the eigenvalues ($\lambda_1, \lambda_2, \lambda_3$). The elements of the molecular structure are characterized by the following types of nondegenerate critical points of $\rho(r)$: the positions of nuclei, the $(3, -3)$ critical points (maximum) r_a; the bond , the $(3, -1)$ bond critical point r_b (saddle point in three dimensions, a minimum on the path of the maximum electron density (MED) path, $\lambda_1, \lambda_2 < 0, \lambda_3 > 0$); the ring, the $(3, +1)$ ring critical point r_r (saddle point, $\lambda_1 < 0, \lambda_2, \lambda_3 > 0$); and the cage, the $(3, +3)$ cage critical point r_c (minimum, $\lambda_1, \lambda_2, \lambda_3 > 0$, the electron density is a local minimum at the position of r_c) [351, 352]. Note that the bond path length R_b may differ from the equilibrium value R_e of the internuclear distance, for example, in the case of small cyclic molecules, and the interpath angles β_i may differ from the geometrical angles α_i [353] (see Chapter 7).

The amount of "π-character" of a bond is determined by ellipticity of ρ_b (ρ_b is the electron density (e/$\mathring{A}^{-3}$) at r_b) and the bond order n [353–356]:

$$\varepsilon = \lambda_1/\lambda_2 - 1 \tag{2.112}$$

$$n = \exp[A(\rho_b - B)] \tag{2.113}$$

In Eq. (2.112), λ_1 and λ_2 are the principal curvatures perpendicular to the bond path (for r_b, λ_1, $\lambda_2 < 0$, $\lambda_3 > 0$). The parameters A and B in Eq. (2.113), determined using various basis sets, are given by Bader et al. [354]. For the CC bond in ethane, λ_1 and λ_2 are degenerated and $\varepsilon = 0$. By contrast, for the CC bond in ethylene, $\lambda_1 \neq \lambda_2$ and the eigenvectors corresponding to the eigenvalues λ_2 and λ_1 define a pair of axes perpendicular to the bond path, namely, the major and the minor axes. Along the former, the value of the negative curvature of ρ_b is minimal (λ_2) (soft direction), while along the latter it is maximal (λ_1) (steep direction) [356]. The direction of the soft curvature (λ_2) is taken as the "direction" of the bond ellipticity and is indicated by a double-headed arrow (Fig. 2.7).

The conjugation between the single and double bonds will be manifested by the fact that for a formally single CC bond (e.g., C2C3 in butadiene) $n > 1$ and $\varepsilon > 0$.

A convenient parameter for the quantitative analysis of conjugation effects is the relative π-character η (in per cent) of the CC formal double or single bonds determined with reference to the bond of ethylene [356].

$$\eta = 100\varepsilon(\text{CC}) / \varepsilon(\text{ethylene}) \tag{2.114}$$

For ethylene $\eta = 100\%$, for ethane 0%. In the case of acyclic and cyclic polyenes, the π-conjugation of double bonds giving rise to the π-electron delocalization will be manifested in the lowering of ε, n, and η for formal double bonds relative to the CC bond of ethylene and in their increase for formal single bonds com-

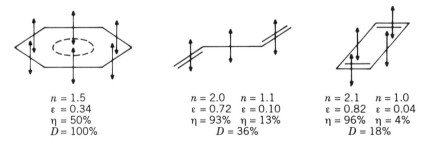

$n = 1.5$	$n = 2.0$ $n = 1.1$	$n = 2.1$ $n = 1.0$
$\varepsilon = 0.34$	$\varepsilon = 0.72$ $\varepsilon = 0.10$	$\varepsilon = 0.82$ $\varepsilon = 0.04$
$\eta = 50\%$	$\eta = 93\%$ $\eta = 13\%$	$\eta = 96\%$ $\eta = 4\%$
$D = 100\%$	$D = 36\%$	$D = 18\%$

Figure 2.7   STO-3G calculated bond order n, bond ellipticities ε (major axes are shown by the double headed arrows), π-characters η, and delocalization parameter D for benzene, butadiene, and cyclobutadiene [354–356].

pared with the CC bond of ethane. The degree of delocalization in these polyenes is given (in per cent) by the index D, Eq. (2.115). It is derived using the notions of the averaged π-character η_{av} (Eq. (2.116)) and standard deviation σ_η (Eq. (2.117)) with k being total number of bonds between carbon atoms in the conjugated polyene [356]:

$$D = 100(1 - \sigma_\eta) \qquad (2.115)$$

$$\eta_{av} = (1/k) \sum_i^k \eta(CC)_i \qquad (2.116)$$

$$\sigma_\eta = (1/k) \sum_i^k [\eta(CC)_i - \eta_{av}]^2 / \eta_{av}^2 \qquad (2.117)$$

The aromatic electron delocalization must bring the n, ε, and η values closer together for all CC bonds in a ring [353–356]. In benzene these values are indeed all equal for all these bonds (Fig. 2.7), indicating complete delocalization ($D = 100\%$). In the prototype antiaromatic molecule of cyclobutadiene, the π-characters of the CC bonds differ substantially (96% and 4%); D equals only 17.7% thus being twice as low as in butadiene (Fig. 2.7). Therefore comparison of the ε, n, η, and D values for a given cyclic molecule with the same values for the reference acyclic structure leads to estimates of the aromaticity and antiaromaticity. In the case of polycyclic molecules, this approach can estimate the aromaticity of individual rings as well as that of the peripheral ring. The ellipticities of CC bonds are useful characteristics for the description of the homoaromaticity [357, 358] (Chapter 6) and σ-aromaticity [353, 359] (Chapter 7).

A less accurate scheme is based on the examination of the degree of delocalization in π-electron-type localized MOs (LMOs). The Mulliken population analysis of such orbitals shows that the planar D_{4h} structure of cyclooctatetraene and D_{2h} structure of cyclobutadiene should be assigned to the antiaromatic class if the LMOs of hexatriene are taken as the nonaromatic reference point [360]. This reference model is analogous to that for the DRE scheme (see Section 2.2.3). However, when a system of noninteracting ethylene bonds is employed as the reference model (analogously to the HRE scheme, Section 2.2.2), this absolute aromaticity scale shows cyclooctatetraene to be aromatic, while cyclobutadiene persists in its antiaromaticity.

The LMOs scheme may also be used to determine the value of the resonance energy [361]. To this end, the LMOs are truncated (TLMOs) and even subjected to symmetric orthogonalization (OTLMOs), after which the variational energy E(OTLMO) is calculated, permitting straightforward calculation of the total delocalization energy $DE = E$(OTLMO) $- E$(SCF). Next, hyperconjugation – delocalization corrections (HDC) and geometry-change corrections (GCC) are introduced and the RE calculated. For example, in the case of benzene $DE(\pi) \approx 90$ kcal/mol, for the CC π-bond HDC ≈ 8 kcal/mol and making use of the GCS = 23 kcal/mol, one arrives at RE $\approx 90 - 3(8) - 23 = 43$ kcal/mol.

The cyclic electron delocalization is manifested in aromatic annulenes in the closeness of the values of the π-bond orders P_{ij}, while for antiaromatic annulenes alternation of these values is observed. This being the case, the information content IC (information index), which is an inverse proportional measure of the electron delocalization, may be used as an index of aromaticity [362, 363]:

$$IC = \text{lb}\left(\frac{k}{A}\right) + \frac{1}{A}\sum_{i,j=1}^{k} P_{ij}\text{lb}(P_{ij}) \tag{2.118}$$

Here, P_{ij} is the π-bond order, $A = \Sigma\,\Sigma_{i \leq j}P_{ij}$, k is the total number of P_{ij} elements of the density matrix P ("tight-binding" approximation is used in which P_{ij} for nonneighboring atoms are omitted), and lb is the logarithm to the base 2 for measuring IC in bits (binary digits). Calculations on [n]annulenes within the HMO approximation have shown that the more even the charge–bond order distribution in a molecule, the higher is its aromatic character and the lower its information content. For example, for benzene (D_{6h}) IC = 0.02905 and in the case of cyclobutadiene (D_{2h}) IC = 0.41503 (in bits) [363]. For aromatic, antiaromatic and nonaromatic compounds we have, respectively, IC < 0.06, IC > 0.095, and $0.06 \leq$ IC ≤ 0.095 [362]. The values of IC for [n]annulenes correlate with HSRE and TRE [363].

2.5.3 Anisotropic Optical Polarizability as an Index of Aromaticity

The degree of the π-electron delocalization may be ascertained from the optical polarizability tensor b_{ij} of a molecule, which is an important anisotropic characteristic of the electron distribution. This tensor depends on the degree of the cyclic electron delocalization (for aromatic compounds the highest polarizability is observed in the plane of the molecule [364]). The index of aromaticity I_1 proposed by Bulgarevich et al. [365] is based on the anisotropic polarizability:

$$I_1 = (b_{aa} + b_{bb})^{\pi}(1/n) \,/\, ((b_{aa} + b_{bb})^{\pi}_{\text{benzene}} (1/6)) \tag{2.119}$$

where n is the number of endocyclic bonds. The value of $(b_{aa} + b_{bb})$ representing the difference between the total polarizability in the molecular plane and the contributions coming from the σ-system reflects the aptness of the electrons to shift under an applied field, which, in turn, depends on the degree of the electron delocalization. This index is constructed in such a fashion that for benzene $I_1 = 1.0$. It should be noted that for acyclic conjugated molecules I_1 is nonzero; that is, it does not describe exclusively the cyclic π-electron delocalization as is required of the aromaticity indices. Another index proposed in the same work [365] is given by the relation of the longitudinal polarizability of the formally single C—C bond to that of the formally double C=C bond:

$$I_2 = b_L(\text{C—C})/b_L(\text{C=C}) \tag{2.120}$$

TABLE 2.10   Aromaticity Indices I_1 and I_2 Based on the Anisotropic Polarizability

Compound	I_1	I_2
Benzene	1.00	1.00
Pyridine	0.978	0.991
Pyrrole	0.866	0.865
Furan	0.821	0.751
Thiophene	0.836	0.805

The values of the I_1 and I_2 indices (Table 2.10) are in accord with the estimates of aromaticity from the values of $\Delta\chi$, Λ, and resonance energies (Tables 2.1 and 2.8).

Summing up, we should like to emphasize that it was not our task to present an all-embracing review of manifold aromaticity indices. Rather, we wish to formulate the main principles underlying the quantitative description of aromaticity represented by the energetic, structural, and magnetic criteria and to trace interrelations among these. The ultimate goal is to present a possibly complete picture where a general idea could unify many fragmentary pieces. Next , we go into those interrelations and try to establish whether an integrated picture of aromaticity can be put together.

2.6   INTERRELATION BETWEEN VARIOUS TYPES OF AROMATICITY INDICES

In discussing various indices of aromaticity developed from different criteria (energetic, structural, magnetic), we noted a correlation between indices based not only on one type of criteria but also on different ones. Is the interrelation among different criteria always clear-cut and convincing? Will the aromaticity inferred from, say, a magnetic criterion be confirmed by an energetic one? This point may be illustrated by the problem of the aromaticity of [n]-*para*-cyclophanes (see Section 2.3.4). The ^{1}H chemical shifts indicate for [5]-*para*-cyclophanes the retention of aromaticity of the benzene ring despite its bending [366]. A similar conclusion in favor of aromaticity of [5]-*meta*-cyclophane, in which the benzene ring also deviates considerably from planarity, has been drawn from values of the magnetic susceptibility anisotropy $\Delta\chi$ [367]. These findings are, however, in stark contradiction with estimates for [5]-*para*-cyclophane [4] and other members of this series [267, 266] based on the energetic criterion HRE (see Sections 2.2.9 and 2.3.4). This criterion shows that [n]-*para*-cyclophanes ($n = 5$–7) are not aromatic molecules and the major reason for the loss of the aromaticity is seen, within this criterion, precisely in the nonplanarity of the benzene ring.

Some interesting conclusions on the interrelation among aromaticity indices drawn from energetic, structural, and magnetic criteria may be found in Katritzky et al. [5], where the so-called principal component analysis is applied. The scheme of principal components is given by

$$X_{ik} = X_{ik} + \sum_{a=1}^{A} t_{ia} p_{ak} + e_{ik} \tag{2.121}$$

where X_{ik} is the mean scaled value of the experimental quantities (variables), t_{ia} are the scores, p_{ak} are the loadings, e_{ik} are the residuals, i is the chemical compound, k is the experimental measurement, and a is the principal component.

The first principal component is defined as the best summary of the linear relationships exhibited in the data. The second component is defined analogously after removing from the data the effect of the first. The principal components have definite values (t_{1i}, t_{2i}, etc., the "scores") for every compound under consideration and are taken in certain proportions (p_{1k}, p_{2k}, etc., the "loadings") for each type of characteristic.

This analysis conducted for nine compounds (benzene, pyridine, pyrimidine, pyrazine, thiophene, furan, pyrrole, pyrazole, and imidazole) has shown that 83% of the variation of 12 characteristics of the compounds (energetic, geometrical, and magnetic data) is described by the first three principal components. Relationships between various characteristics may be revealed by examining the numerical values of their principal component loadings [5]. According to these values, three main groups of characteristics may be identified. The first group, comprising $I_{5,6}$, ΔN, DRE, and HSRE (Table 2.11), has large values of the p_1 loadings and small-to-moderate ones of the p_3 loadings. The p_1 value may be regarded as the measure of the "classical aromaticity" [5]. The second group, orthogonal to the first, includes the magnetic indices χ_M (molar magnetic susceptibility) and Λ (exaltation) for which p_1 loadings are very small, while p_2 (positive value) and p_3 (negative value) are quite large (Table 2.11). These indices describe the "magnetic" aromaticity, which is almost completely orthogonal to the "classical". This is why correlation between them is not generally to

TABLE 2.11   Principal Component Loadings Obtained by Principal Component Analysis [5] for Some Aromaticity Indices

Aromaticity Indices	p_{1k}	p_{2k}	p_{3k}
$I_{5,(6)}$	0.3574	−0.0088	−0.0133
ΔN	−0.3431	−0.0175	0.0274
DRE	0.3066	−0.0318	0.1675
HSRE	0.3362	−0.1203	−0.1501
χ_M	0.0508	0.4072	−0.6116
Λ	0.1075	0.4106	−0.2841
RCI	0.2645	0.1917	0.5202
I_1	0.2394	0.3613	0.0346

be expected. The first group includes the indices RCI and I_1, which refer to both "classical" and "magnetic" aromaticity.

This division of aromaticity into two types [5] may account for uncertainties concerning [n]-*para*-cyclophane; however, it cannot be accepted without qualification. Indeed, one ought to examine the effects resulting from changes in the set of various indices (energetic, structural, magnetic) as well as from the situation when, in addition to the aromatic heterocyclic molecules, a broader spectrum of compounds is inspected representing various types of aromaticity and antiaromaticity.

One should not forget that there are certain cases of interrelation between indices of one type and those of another, such as REs and magnetic susceptibilities due to ring currents [9, 285, 291, 302, 307, 309, 368]. Admittedly, the proportionality of the two quantities may not always be observed in polycyclic systems [9, 307]; the parallelism is also lost in large [4n]annulenes [284]. A number of analytical relationships have been derived, for example, for annulenes, between the REs and magnetic indices. In the case of [4n + 2]annulenes, a simple dependence has been obtained of the RE (the infinite ring size was the reference structure) on the ring current [7, 76]. Linear correlations have also been found of the HSREs per electron (HSREPE) with the differences between the chemical shifts of the outer and inner protons $(\tau_o - \tau_i)$ in annulenes [369, 370], such as

$$\frac{\tau_a - \tau_i}{S} = k \cdot \text{HSREPE} \tag{2.122}$$

The material presented in this chapter warrants the conclusion that the main test of aromaticity and antiaromaticity is represented by the energetic criterion realizable within the framework of various schemes for calculating resonance energies. In most cases, it correlates with the structural and magnetic criteria; moreover, it often accords well with manifestations of numerous properties of compounds, which, being regarded as attributes of the aromaticity, make its very concept substantially broader. Indeed, the concept of aromaticity claims an increasing number of types of compounds and requires a more and more sophisticated classification.

However, before entering upon examination of particular types or aromaticity, we wish to devote the next chapter to modes of electron delocalization, to the corresponding orbital models and to the electron-count rules based on these, which permit simple, pencil-and-paper calculations to ascertain the presence of the aromatic stability or antiaromatic instability of a molecule.

Note added in proofs
The basis set superposition error correction makes the MP4SDTQ/6-31G*//HF/6-31G* calculated HSE of benzene (Eq. (2.4), also see Section 2.2.9), to be 21.35 kcal/mol [371], that is very close to the experimental value of 21.6 kcal/mol (see, e.g. [12]).

REFERENCES

1. D. Levis and D. Peters, *Fact and Theories of Aromaticity*, Macmillan, London, 1975.

2. D. Lloyd, *Non-Benzenoid Conjugated Carbocyclic Compounds*, Elsevier, Amsterdam, 1984.

3. P. J. Garrat, *Aromaticity*, Wiley, New York, 1986.

4. J. E. Rice, T. J. Lee, R. B. Remington, W. D. Allen, D. A. Clabo, and H. F. Schaefer, *J. Am. Chem. Soc.*, **109**, 2902 (1987).

5. A. R. Katritzky, P. Barczynski, G. Musumarra, D. Pisano, and M. Szafran, *J. Am. Chem. Soc.*, **111**, 7 (1989).

6. D. Lloyd and D. R. Marshall, *Agnew. Chem. Int. Ed. Engl.*, **11**, 404 (1972).

7. R. C. Haddon, *J. Am. Chem. Soc.*, **101**, 1722 (1979).

8. J. I. Aihara, *J. Am. Chem. Soc.*, **103**, 5704 (1981).

9. J. I. Aihara, *Pure Appl. Chem.*, **54**, 1115 (1982).

10. W. C. Herndon, *J. Chem. Soc. Chem. Commun.*, 817 (1977).

11. J. I. Aihara and H. Ichikawa, *Bull. Chem. Soc. Jpn.*, **61**, 223 (1988).

12. P. George, C. W. Bock, and M. Trachtman, *J. Chem. Educ.*, **61**, 225 (1984).

13. M. J. S. Dewar and C. deLlano, *J. Am. Chem. Soc.*, **91**, 789 (1969).

14. N. C. Baird, *J. Chem. Educ.*, **48**, 509 (1971).

15. M. J. S. Dewar and N. Trinajstić, *Theoret. Chim. Acta*, **17**, 235 (1970).

16. B. A. Hess and L. J. Schaad, *J. Am. Chem. Soc.*, **93**, 305 (1971).

17. L. J. Schaad and B. A. Hess, *J. Chem. Educ.*, **51**, 649 (1974).

18. B. A. Hess and L. J. Schaad, *J. Am. Chem. Soc.*, **93**, 2413 (1971).

19. B. A. Hess, L. J. Schaad, and C. W. Holyoke, *Tetrahedron*, **28**, 3657, 5299 (1972).

20. J. I. Aihara, *J. Am. Chem. Soc.*, **98**, 2750 (1976).

21. J. I. Aihara, *J. Am. Chem. Soc.*, **98**, 6840 (1976).

22. I. Gutman, M. Milun, and N. Trinajstić, *J. Am. Chem. Soc.*, **99**, 1692 (1977).

23. N. Trinajstić, *Int. J. Chem. Symp.*, **11**, 469 (1977).

24. H. Kollmar, *J. Am. Chem. Soc.*, **101**, 4832 (1979).

25. S. S. Shaik, P. C. Hiberty, J. M. Lefour, and G. Ohanessian, *J. Am. Chem. Soc.*, **109**, 363 (1987).

26. S. S. Shaik, P. C. Hiberty, G. Ohanessian, and J. M. Lefour, *J. Phys. Chem.*, **92**, 5086 (1988).

27. W. J. Hehre, R. Ditchfield, L. Radom, and J. A. Pople, *J. Am. Chem. Soc.*, **92**, 4869 (1970).

28. P. George, M. Trachtman, C. W. Bock, and A. M. Brett, *Theor. Chim. Acta,* **38**, 121 (1975).

29. P. George, *Chem. Rev.*, **75**, 85 (1975).

30. A. Streitwieser, *Molecular Orbital Theory for Organic Chemists*, Wiley, New York, 1961.

31. A. D. J. Haymet, *Chem. Phys. Lett.*, **122**, 421 (1985).

32. B. A. Hess and L. J. Schaad, *J. Org. Chem.*, **51**, 3902 (1986).

33. T. G. Schmalz, W. A. Seitz, D. J. Klein, and G. E. Hite, *J. Am. Chem. Soc.*, **110**, 1113 (1988).

34. W. H. Kroto, *Angew. Chem.*, **104**, 113 (1992).

35. M. J. S. Dewar, A. J. Harget, and N. Trinajstić, *J. Am. Chem. Soc.*, **91**, 6321 (1969).

36. M. J. S. Dewar and G. J. Gleicher, *J. Am. Chem. Soc.*, **87**, 692 (1965).

37. M. N. Glukhovtsev, B. Ya. Simkin, and V. I. Minkin, *Russ. Chem. Rev.*, **54**, 54 (1985).

38. H. Ichikawa and Y. Ebisawa, *J. Am. Chem. Soc.*, **107**, 1161 (1985).

39. B. A. Hess and L. J. Schaad, *J. Am. Chem. Soc.*, **103**, 7500 (1983).

40. R. C. Haddon and J. J. Starness, *Adv. Chem. Ser.*, **169**, 333 (1978).

41. I. Gutman, B. Ruscic, and N. Trinajstić , *J. Chem. Phys.*, **62**, 3399 (1975).

42. B. A. Hess, L. J. Schaad, and C. W. Holyoke, *Tetrahedron*, **31**, 295 (1975).

43. B. A. Hess and L. J. Schaad, *J. Org. Chem.*, **41**, 3508 (1976).

44. B. A. Hess, L. J. Schaad, and I. Agranat, *J. Am. Chem. Soc.*, **100**, 5268 (1978).

45. L. J. Schaad and B. A. Hess, *J. Am. Chem. Soc.*, **94**, 3068 (1972).

46. I. Agranat, B. A. Hess, and L. J. Schaad, *Pure Appl. Chem.*, **52**, 1399 (1980).

47. B. A. Hess and L. J. Schaad, *Pure Appl. Chem.*, **52**, 1471 (1980).

48. R. B. Bates, B. A. Hess, C. A. Ogle, and L. J. Schaad, *J. Am. Chem. Soc.*, **103**, 5052 (1981).

49. R. S. Mulliken and R. G. Parr, *J. Chem. Phys.*, **19**, 1271 (1951).

50. S. Aono, T. Ohmae, and K. Nishikawa, *Bull. Chem. Soc. Jpn.*, **54**, 1645 (1981).

51. A. Moyano and J. C. Paniagua, *J. Org. Chem.*, **51**, 2250 (1986).

52. N. Trinajstić, Hückel Theory and Topology in G. A. Segal (Ed.), *Semiempirical Methods of Electronic Structure Calculation*, Part A, Plenum Press, New York, 1977, p. 1.

53. I. Gutman and N. Trinajstić, *Topics Curr. Chem.*, **42**, 49 (1973).

54. I. Gutman and O. E. Polansky, *Mathematical Concepts in Organic Chemistry*, Springer–Verlag, Berlin, 1986.

55. A. Graovac, I. Gutman, and N. Trinajstić, *Lecture Notes in Chemistry*, Springer–Verlag, Berlin, 1977.

56. N. Trinajstić, *Chemical Graph Theory*, CRC Press, Boca Raton, FL, 1983.

57. N. Trinajstić, *Croat. Chem. Acta*, **49**, 593 (1977).

58. I. Gutman, M. Milun, and N. Trinajstić, *Croat. Chem. Acta*, **48**, 87 (1976).

59. C. D. Gotsil and I. Gutman, *Acta Chim. Acad. Sci. Hung.* **110**, 415 (1982).

60. I. Gutman and S. Bosanas, *Tetrahedron*, **33**, 1809 (1977).

61. W. C. Herndon, *J. Am. Chem. Soc.*, **104**, 3541 (1982).

62. I. Gutman and W. C. Herndon, *Chem. Phys. Lett.*, **105**, 281 (1984).

63. E. Heilbronner, *Chem. Phys. Lett.*, **85**, 377 (1982).

64. I. Gutman, *Chem. Phys. Lett.*, **66**, 595 (1979).

65. I. Gutman, *Theor. Chim. Acta*, **56**, 89 (1980).

66. J. I. Aihara, *Chem. Phys. Lett.*, **73**, 404 (1980).

67. J. Cioslowski, *Int. J. Quant. Chem.*, **34**, 417 (1988).

68. J. I. Aihara, *Bull. Chem. Soc. Jpn.*, **53**, 2689 (1980).
69. M. N. Glukhovtsev, B. Ya. Simkin, and V. I. Minkin, *Zh. Org. Khim.*, **25**, 673 (1989).
70. I. Gutman and B. Mohar, *Chem. Phys. Lett.*, **77**, 567 (1981).
71. P. Ilic and N. Trinajstić, *J. Org. Chem.*, **45**, 1738 (1980).
72. J. I. Aihara, *Bull. Chem. Soc. Jpn.*, **50**, 3057 (1977).
73. J. I. Aihara, *Bull. Chem. Soc. Jpn.*, **52**, 2202 (1979).
74. J. I. Aihara, *J. Am. Chem. Soc.*, **100**, 3339 (1978).
75. I. Gutman, *Bull. Soc. Chim. Beograd*, **43**, 191 (1978).
76. J. I. Aihara, *Bull. Chem. Soc. Jpn.*, **53**, 1163 (1980).
77. N. Mizoguchi, *J. Mol. Struct.* (THEOCHEM), **181**, 245 (1988).
78. N. Mizoguchi, *J. Am. Chem. Soc.*, **107**, 4419 (1985).
79. J. I. Aihara, *Bull. Chem. Soc. Jpn.*, **51**, 1788 (1978).
80. J. I. Aihara, *Bull. Chem. Soc. Jpn.*, **58**, 3617 (1985).
81. V. I. Minkin, M. N. Glukhovtsev, and B. Ya. Simkin, *J. Mol. Struct.* (THEOCHEM), **181**, 93 (1988).
82. M. N. Glukhovtsev, B. Ya. Simkin, and I. A. Yudilevich, *Teor. Eksp. Khim.*, 229 (1990).
83. J. I. Aihara and H. Hosoya, *Bull. Chem. Soc. Jpn.*, **61**, 2657 (1988).
84. J. C. Panigua and E. C. A. Moyano, *Afinidad*, **44**, 26 (1987).
85. A. Juric, N. Trinajstić, and G. Jashari, *Croat. Chem. Acta*, **59**, 617 (1986).
86. I. Gutman, *Z. Naturforsch.*, **33a**, 214 (1978).
87. A. C. Puiu and O. Sinanoglu, *Chim. Acta Turcica*, **6**, 1 (1978).
88. E. Clar, *Polycyclic Hydrocarbons*, Academic, London, 1964.
89. M. Randic, *J. Am. Chem. Soc.*, **99**, 444 (1977).
90. M. Randic, D. Plavšić, and N. Trinajstić, *J. Mol. Struct.* (THEOCHEM), **185**, 249 (1989).
91. S. Nicolic, M. Randic, D. J. Klein, D. Plavšić, and N. Trinajstić, *J. Mol. Struct.* (THEOCHEM), **188**, 223 (1989).
92. M. Randic, *Tetrahedron*, **33**, 1905 (1977).
93. M. Randic, N. Trinajstić, J. V. Knop, and Z. Jeričević, *J. Am. Chem. Soc.*, **107**, 849 (1985).
94. M. Randic, S. Nicolic, and N. Trinajstić, *Coll. Czech. Chem. Commun.*, **53**, 2023
95. M. Randic, D. Plavšić, and N. Trinajstić, *J. Mol. Struct.* (THEOCHEM), **183**, 29 (1989).
96. M. Randic and H. E. Zimmerman, *Int. J. Quant. Chem. Symp.*, **20**, 185 (1986).
97. M. Randic, *Pure Appl. Chem.*, **55**, 347 (1983).
98. M. Randic, *Int. J. Quant. Chem.*, **17**, 549 (1980).
99. D. J. Klein and N. Trinajstić, *Pure Appl. Chem.*, **61**, 2107 (1980).
100. D. Plavšić, S. Nikolic, and N. Trinajstić, *J. Mol. Struct.* (THEOCHEM), **277**, 213 (1992).
101. M. Randic, *Chem. Phys. Lett.*, **38**, 68 (1976).
102. D. J. Klein, *Int. J. Quant. Chem.*, **20**, 193 (1986).

103. M. Randic, *Chem. Phys. Lett.*, **128**, 193 (1986).

104. M. Randic, V. Solomon, S. C. Grossman, D. J. Klein, and N. Trinajstić, *Int. J. Quant. Chem.*, **32**, 35 (1987).

105. H. Vogler and N. Trinajstić, *J. Mol. Struct.* (THEOCHEM), **164**, 325 (1988).

106. H. Vogler and N.Trinajstić, *Theor. Chim. Acta*, **73**, 437 (1988).

107. D. Plavšić, N. Trinajstić, M. Randic, and C. Venier, *Croat. Chem. Acta*, **62**, 719 (1989).

108. M. Randic, B. M. Gimark, S. Nikolic, and N. Trinajstić, *J. Mol. Stuct.* (THEOCHEM), **181**, 111 (1988).

109. D. J. Klein, T. G. Schmalz, G. E. Hite, and W. A. Seitz, *J. Am. Chem. Soc.*, **108**, 1301 (1986).

110. M. Randic, S. Nikolic, and N. Trinajstić, *Croat. Chem. Acta*, **60**, 595 (1987).

111. D. J. Klein, W. A. Seitz, and T. G. Schmalz, *Nature*, **323**, 703 (1986).

112. W. C. Herndon, *J. Org. Chem.*, **46**, 2119 (1981).

113. W. C. Herndon and L. M. Ellzey, *J. Am. Chem. Soc.*, **95**, 6631 (1974).

114. W. C. Herndon, *J. Am. Chem. Soc.*, **96**, 2404 (1973).

115. W. C. Herndon, *Isr. J. Chem.*, **20**, 270 (1980).

116. D. Biermann and W. Schmidt, *J. Am. Chem. Soc.*, **102**, 3163, 3173 (1980).

117. E. Clar, *The Aromatic Sextet*, Wiley, London, 1972.

118. W. C. Herndon and H. Hosoya, *Tetrahedron*, **40**, 3987 (1984).

119. W. C. Herndon, *Tetrahedron*, **29**, 3 (1973).

120. W. C. Herndon, *J. Chem. Educ.*, **51**, 10 (1974).

121. R. Swinborhe-Sheldrake and W. C. Herndon, *Tetrahedron Lett.*, 755 (1975).

122. H. C. Longuet-Higgins, *J. Chem. Phys.*, **18**, 265 (1950).

123. W. C. Herndon, *J. Am. Chem. Soc.*, **102**, 1538 (1980).

124. M. J. S. Dewar, *The Molecular Orbital Theory of Organic Chemistry*, McGraw-Hill, New York, 1969.

125. M. J. S. Dewar and R. C. Dougherty, *The PMO Theory of Organic Chemistry*, Plenum Press, New York, 1975.

126. R. E. Lehr and A. P. Marchand, *Orbital Symmetry: A Problemsolving Approach*, Academic, New York, 1972.

127. L. A. Yanovskaya, *Contemporary Theoretical Principles of Organic Chemistry*. Khimiya, Moscow, 1978.

128. B. Ya. Simkin, M. E. Kletsky, and M. N. Glukhovtsev, *Problems in Molecular Quantum Theory*, Vysshaja Scola, Moscow, 1994.

129. K. A. Durkin and R. F. Langler, *J. Phys. Chem.*, **91**, 2422 (1987).

130. L. Salem, *The Molecular Orbital Theory of Conjugated Systems*, Benjamin, New York, 1966.

131. W. J. Hehre, L. Radom, P. v. R. Schleyer, and J. A. Pople, *Ab Initio Molecular Orbital Theory*, Wiley, New York, 1986.

132. P. George, C. W. Bock, and M. Trachtman, *Tetrahedron Lett.*, **26**, 5667 (1985).

133. R. Janoschek, *J. Mol. Struct.* (THEOCHEM), **229**, 197 (1991).

134. P. George, M. Trachtman, C. W. Bock, and A. M. Brett, *J. Chem. Soc. Perkin Trans.*, **11**, 1222 (1976).

135. R. C. Haddon, *Pure Appl. Chem.*, **54**, 1129 (1982).

136. C. W. Bock, P. George, and M. Trachtman, *J. Phys. Chem.*, **88**, 289 (1984).

137. R. C. Haddon and K. Raghavachari, *J. Am. Chem. Soc.*, **107**, 289 (1985).

138. K. K. Baldridge and M. S. Gordon, *J. Am. Chem. Soc.*, **110**, 4204 (1988).

139. R. C. Haddon, *Pure Appl. Chem.*, **58**, 129 (1986).

140. R. L. Disch, J. M. Schulman, and M. L. Sabin, *J. Am. Chem. Soc.*, **107**, 1904 (1985).

141. J. M. Schulman and R. L. Disch, *J. Am. Chem. Soc.*, **107**, 5059 (1985).

142. J. M. Schulman, R. C. Peck, and R. L. Disch, *J. Am. Chem. Soc.*, **111**, 5675 (1989).

143. R. L. Disch and J. M. Schulman, *Chem. Phys. Lett.*, **152**, 402 (1988).

144. M. N. Glukhovtsev and P. v. R. Schleyer, not yet published.

145. P. George, M. Trachtman, C. W. Bock, and A. M. Brett, *Tetrahedron*, **32**, 1357 (1976).

146. S. W. Staley and T. D. Norden, *J. Am. Chem. Soc.*, **111**, 445 (1989).

147. C. W. Bock, M. Trachtman, and P. George, *Struct. Chem.*, **1**, 345 (1990).

148. S. Kuwajima, *J. Am. Chem. Soc.*, **106**, 6469 (1984).

149. S. A. Alexander and T. G. Schmalz, *J. Am. Chem. Soc.*, **109**, 6933 (1987).

150. J. Cioslovski, *Int. J. Quant. Chem.*, **31**, 581 (1987).

151. L. J. Schaad and B. A. Hess, *Pure Appl. Chem.*, **54**, 1097 (1982).

152. B. A. Hess and L. J. Schaad, *J. Chem. Soc. Chem. Commun.*, 243 (1977).

153. M. Randic and N. Trinajstić, *J. Am. Chem. Soc.*, **109**, 6923 (1987).

154. B. A. Hess, L. J. Schaad, W. C. Herndon, D. Biermann, and W. Schmidt, *Tetrahedron*, **37**, 2983 (1981).

155. M. V. Gorelik, *Usp. Khim.*, **59**, 197 (1990).

156. A. F. Pozharskii, *Khim. Geterotsikl. Soed.*, 867, (1985).

157. L. V. Vilkov, V. S. Mastryukova, and N. I. Sadova, *Determination of the Geometrical Structure of Free Molecules*, Mir Publishers, Moscow, 1983.

158. K. Kuchitsu, in Hückel Theory and Topology, A. Domenicano and I. Hargittai (Eds.), *Accurate Molecular Structure*, Oxford University Press, New York, 1992, pp. 14–46.

159. P. N. Skancke, *Int. J. Quant. Chem.*, **26**, 729 (1984).

160. U. Burkert and N. L. Allinger, *Molecular Mechanics*, American Chemical Society, New York, 1982.

161. K. Tamagawa, T. Iijima, and M. Kimura, *J. Mol. Struct.*, **30**, 243 (1976).

162. E. G. Cox, D. W. J. Cruickshank, and J. A. S. Smith, *Proc. R. Soc. (London)*, **A247**, 1 (1958).

163. G. E. Bacon, N. A. Curry, and S. A. Wilson, *Proc. R. Soc. (London)*, **A279**, 98 (1964).

164. N. C. Handy, P. E. Maslen, R. D. Atoms, C. W. Murray, and G. J. Laming, *Chem. Phys. Lett.*, **197**, 506 (1992).

165. J. Pliva, J. W. C, Johns, and L. Goodman, *J. Mol. Spectrosc.*, **148**, 427 (1991).

166. M. Traettenberg, *Acta Chem. Scand.*, **22**, 628 (1968).

167. L. S. Bartell, E. A. Roth, C. D. Hollowell, K. Kuchitsu, and J. E. Young, *J. Chem. Phys.*, **42**, 2683 (1965).

168. H. Irngartinger and M. Nixdorf, *Angew. Chem. Int. Ed. Engl.*, **22**, 403 (1983).

169. A. G. Ozkabak, L. Goodman, and K. B. Wiberg, *J. Chem. Phys.*, **92**, 4115 (1990).

170. J. Almlöf and K. Faegri, *J. Chem. Phys.*, **79**, 2284 (1983).

171. F. H. Allen, O. Kennard, D. G. Watson, L. Brammer, A. G. Orpen, and R. Taylor, *J. Chem. Soc. Perkin Trans.*, **II**, S1 (1987).

172. W. T. Borden, E. R. Davidson, and P. Hart, *J. Am. Chem. Soc.*, **100**, 388 (1978).

173. E. R. Davidson and W. T. Borden, *J. Chem. Phys.*, **67**, 2191 (1977).

174. S. Brodersen, J. Christoffersen, B. Bac, and J. T. Nielsen, *Spectrochim. Acta*, **21**, 2077 (1965).

175. S. N. Thakur, L. Goodman, and A. G. Qzkabak, *J. Chem. Phys.*, **84**, 6642 (1986).

176. H. Agren, N. Correia, A. Flores-Riveros, and H. J. A. A. Jensen, *Int. J. Quant. Chem. Symp.*, **19**, 237 (1986).

177. D. W. Whitman and B. K. Carpenter, *J. Am. Chem. Soc.*, **104**, 6473 (1982).

178. H. Irngartinger, N. Riegler, K. D. Malsch, K. A. Schneider, and G. Maier, *Angew. Chem. Int. Ed. Engl.*, **19**, 211 (1980).

179. W. T. Borden and E. R. Davidson, *J. Am. Chem. Soc.*, **102**, 7958 (1980).

180. J. D. Dunitz, C. Krüger, H. Irngartinger, M. F. Maverick, Y. Wang, and M. Nixdorf, *Agnew. Chem. Int. Ed. Engl.*, **27**, 387 (1988).

181. O. Ermer, *Angew. Chem. Int. Ed. Engl.*, **26**, 782 (1987).

182. R. Janoschek, *Angew. Chem. Int. Ed. Engl.*, **26**, 1298 (1987).

183. A. Stanger and K. P. C. Vollhardt, *J. Org. Chem.*, **53**, 4889 (1988).

184. S. S. Shaik and P. C. Hiberty, *J. Am. Chem. Soc.*, **107**, 3089 (1985).

185. P. C. Hiberty, *Topics Curr. Chem.*, **153**, 27 (1990).

186. P. C. Hiberty, S. S. Shaik, G. Ohanessian, and J. M. Lefour, *J. Org. Chem.*, **51**, 3908 (1986).

187. G. Ohanessian, P. C. Hiberty, J. M. Lefour, J. P. Flament, and S. S. Shaik, *Inorg. Chem.*, **27**, 2219 (1988).

188. H. C. Longuet-Higgins and L. Salem, *Proc. R. Soc.* (*London*), **251**, 172 (1959).

189. R. S. Berry, *J. Chem. Phys.*, **35**, 2253 (1961).

190. M. N. Glukhovtsev and P. v. R. Schleyer, *Chem. Phys. Lett.*, **198**, 547 (1992).

191. C. Minot, *J. Phys. Chem.*, **91**, 6380 (1987).

192. I. B. Bersuker, *The Jahn-Teller Effect and Vibronic Interactions in Modern Chemistry,* Plenum Press, New York, 1984.

193. K. K. Sinil, K. D. Jordan, and R. Shepard, *Chem. Phys.*, **88**, 55 (1984).

194. D. E. Dance and I. C. Walker, *Chem. Phys.*, **18**, 601 (1973).

195. A. Lofthus and P. H. Krupenie, *J. Phys. Chem. Ref. Data*, **6**, 113 (1977).

196. K. D. Dobbs and W. J. Hehre, *Organomettallics*, **5**, 2057 (1986).

197. P. v. R. Schleyer and D. Kost, *J. Am. Chem. Soc.*, **110**, 2105 (1988).

198. P. Saxe and H. F. Schaefer, *J. Am. Chem. Soc.*, **105**, 1760 (1983).

199. N. C. Baird, *J. Org. Chem.*, **51**, 3907 (1986).

200. P. C. Hiberty, G. Ohanessian, and F. Delbecq, *J. Am. Chem. Soc.*, **107**, 3095 (1985).

201. K. A.Klingenmith, H. J. Dewey, E. Vogel, and J. Michl, *J. Am. Chem. Soc.*, **111**, 1539 (1989).

202. J. A. Elvidge and L. M. Jackman, *J. Chem. Soc.*, **8**, 859 (1961).

203. F. Sondheimer, *Pure Appl. Chem.*, **7**, 363 (1964).

204. A. Julg and P. Francois, *Theor. Chim. Acta*, **7**, 249 (1967).

205. A. Julg, A New Definition of the Degree of Aromaticity in E. D. Bergmann and B. Pullman (Eds.), *Aromaticity, Pseudo-Aromaticity*, The Israel Academy of Sciences and Humanities, Jerusalem, 1971, pp. 383–385.

206. S. M. van der Kerk, *J. Organomet. Chem.*, **215**, 315 (1981).

207. D. H. Lo and M. A. Whitehead, *J. Can. Chem.*, **46**, 2027 (1968).

208. J. Kruszewski and T. M. Krygowski, *Tetrahedron Lett.*, 3839 (1972).

209. C. W. Bird, *Tetrahedron*, **41**, 1409 (1985).

210. C. W. Bird, *Tetrahedron*, **42**, 89 (1986).

211. C. W. Bird, *Tetrahedron*, **48**, 335 (1992).

212. J. Kruszewski and T. M. Krygowski, *J. Can. Chem.*, **53**, 945 (1975).

213. T. M. Krygowski and J. Kruszewski, *Wiadomosci. Chem.*, **29**, 113 (1975).

214. J. Kruszewski, *Pure Appl. Chem.*, **52**, 1525 (1980).

215. J. Leška and D. Loos, *Coll. Czech. Chem. Commun.*, **49**, 920 (1984).

216. D. Loos and D. Leška, *Coll. Czech. Chem. Commun.*, **45**, 187 (1980).

217. T. M. Krygowski and T. Wieckowski, *Croat. Chem. Acta*, **54**, 193 (1981).

218. T.M. Krygowski, R. Anulewicz, and J. Kruszewski, *Acta Crystallogr.*, **B**, **39**, 732 (1983).

219. T. M. Krygowski and I. Turowska-Tyrk, *Coll. Czech. Chem. Commun.*, **55**, 165 (1990).

220. T. M. Krygowski, Resonance Structure Contributions Derived from the Experimental Geometry of Molecules in J. F. Liebman and A. Greenberg (Eds.), *Stucture and Reactivity*, VCH, New York, 1988, pp. 231–254.

221. T. M. Krygowski, *J. Chem. Res. (S)*, 238 (1984).

222. A. F. Pozharskii and W. Dalnikovskaya, *Usp. Khim.*, **50**, 1559 (1984).

223. A. F. Pozharskii, *Theoretical Principles of Heterocyclic Chemistry*, Khimia, Moscow, 1985.

224. K. Jug, *J. Org. Chem.*, **48**, 1344 (1983).

225. K. Jug, *J. Org. Chem.*, **49**, 4475 (1984).

226. E. J. P. Melar and K. Jug, *Tetrahedron*, **47**, 417 (1986).

227. B. A. Hess, C. S. Ewig, and L. J. Schaad, *J. Org. Chem.*, **50**, 5869 (1985).

228. M. N. Glukhovtsev, B. Ya. Simkin, and V. I. Minkin, *Zh. Org. Khim.*, **23**, 1317 (1987).

229. K. Ewig and J. R. van Wazer, *J. Am. Chem. Soc.*, **111**, 1552 (1989).

230. G. Binsch, E. Heilbronner, and J. N. Murrell, *Mol. Phys.*, **11**, 305 (1966).

231. G. Binsch and E. Heilbronner, *Tetrahedron*, **24**, 1215 (1968).

232. G. Binsch, I. Tamir, and R. D. Hill, *J. Am. Chem. Soc.*, **91**, 2446 (1969).

233. G. Binsch and I. Tamir, *J. Am. Chem. Soc.*, **91**, 2450 (1969).

234. T. Nakajima, *Topics Curr. Chem.*, **32**, 1 (1972).

235. A. Toyota and T. Nakajima, *Bull. Chem. Soc. Jpn.*, **50**, 97 (1977).

236. A. Toyota, T. Nakajima, and S. Koseki, *J. Chem. Soc. Perkin Trans. II*, 85 (1984).

237. T. Nakajima, A. Toyota, and M. Kataoka, *J. Am. Chem. Soc.*, **104**, 5610 (1982).

238. B. Ya. Simkin, M. N. Glukhovtsev, and V. I. Minkin, *Zh. Org. Khim.*, **18**, 1345 (1982).

239. A. Toyota, T. Tanaka, and T. Nakajima, *Int. J. Quant. Chem.*, **10**, 917 (1976).

240. D. Cazes, L. Salem, and C. Tric, *J. Polym. Sci.*, 109 (1970).

241. R. Peierls, *Quantum Theory of Solids*, Clarendon Press, London, 1955.

242. R. C. Haddon and K. Raghavachari, *J. Chem. Phys.*, **79**, 1093 (1983).

243. R. E. Prat, in O. Chalvet, R. Daudel, S. Diner, and J. P. Malrieu (Eds.), *Localization and Delocalization in Quantum Chemistry*, Vol. 1, Reidel, Dordrecht, 1975.

244. G. Chambaud, B. Levy, and P. Millie, *Theor. Chim. Acta*, **48**, 103 (1978).

245. M. M. Mestechkin, The Second-Order Energy Variation in the SCF Theory, in R. Carbo and M. Klobukowski (Eds.), *Self-Consistent Field: Theory and Applications*, Elsevier, Amsterdam, 1990, p. 312.

246. H. Fukutome, *Int. J. Quant. Chem.*, **20**, 955 (1981).

247. K. Yamaguchi, *Chem. Phys.*, **29**, 117 (1981).

248. J. Paldus, Hartree-Fock Stability and Symmetry Breaking in R. Carbo and M. Klobukowski (Eds.), *Self-Consistent Field: Theory and Applications*, Elsevier, Amsterdam, 1990, p. 1.

249. M. Kertesz, J. Koller, and A. Azman, *Int. J. Quant. Chem.*, **18**, 645 (1980).

250. M. M. Mestechkin, *J. Mol. Struct.* (THEOCHEM), **181**, 231 (1988).

251. K. Deguchi, K. Nishikawa, and S. Aono, *J. Chem. Phys.*, **75**, 4165 (1981).

252. J. Cizek and J. Paldus, *J. Chem. Phys.*, **47**, 3976 (1967).

253. R. Seeger and J. A. Pople, *J. Chem. Phys.*, **66**, 3045 (1977).

254. J. Paldus and E. Chin, *Int. J. Quant. Chem.*, **24**, 373 (1983).

255. P. Karadakov and O. Castano, *Theor. Chim. Acta*, **70**, 25 (1986).

256. J. Cizek and J. Paldus, *J. Chem. Phys.*, **53**, 821 (1970).

257. G. G. Dyadyusha and B. I. Lutoshkin, *Teor. Eksp. Khim.* (*USSR*), **9**, 61 (1973).

258. B. Ya. Simkin, M. N. Glukhovtsev, and V. I. Minkin, *Chem. Phys. Lett.*, **71**, 284 (1980).

259. M. N. Glukhovtsev, M. M. Mestechkin, V. I. Minkin, and B. Ya. Simkin, *Zh. Strukt. Khim.* (*USSR*), **23**, 14 (1982).

260. M. N. Glukhovtsev, B. Ya. Simkin, V. I. Minkin, and I. A. Yudilevich, *Zh. Strukt. Khim.* (*USSR*), **23**, 25 (1982).

261. J. A. Jafri and M. D. Newton, *J. Am. Chem. Soc.*, **100**, 5012 (1978).

262. H. Baumann, *J. Am. Chem. Soc.*, **100**, 7196 (1978).

263. R. C. Haddon and K. Raghavachari, *J. Am. Chem. Soc.*, **104**, 3516 (1982).

264. D. J. Raber and W. R. Rodriguez, *J. Am. Chem. Soc.*, **107**, 4146 (1985).

265. B. L. Podlogar, W. A. Glauser, W. R. Rodriguez, and D. J. Raber, *J. Org. Chem.*, **52**, 2126 (1988).

266. T. J. Lee, W. D. Rice, W. D. Allen, R. B. Remington, and H. F. Schaefer, *Chem. Phys.*, **123**, 1 (1988).

267. T. J. Lee, W. D. Rice, R. B. Remington, and H. F. Schaefer, *Chem. Phys. Lett.*, **150**, 63 (1988).

268. L. W. Jenneskens, J. C. Klamer, H. J. R. de Boer, W. H. de Wolf, F. Bickelhaupt, and C. H. Stam, *Angew. Chem. Int. Ed. Engl.*, **23**, 238 (1984).

269. J. E. Gready, T. W. Hambley, K. Kakiuchi, K. Kobiro, S. Sternhell, C. W. Tansey, and Y. Tobe, *J. Am. Chem. Soc.*, **112**, 7537 (1980).

270. F. Pietra, *Acc. Chem. Res.*, **12**, 132 (1979).

271. M. N. Glukhovtsev, B. Ya. Simkin, and V. I. Minkin, *Zh. Org. Khim.*, **20**, 866 (1984).

272. C. J. Finder, D. Chung, and N. L. Allinger, *Tetrahedron Lett.*, 4677 (1972).

273. L. A. Paquette, *Pure Appl. Chem.*, **54**, 987 (1982).

274. R. Gygax, J. Wirz, J. T. Sprague, and N. L. Allinger, *Helv. Chim. Acta*, **60**, 2522 (1977).

275. K. Krogh-Jespersen and P. v. R. Schleyer, *J. Am. Chem. Soc.*, **100**, 4301 (1978).

276. J. Chandrasekhar, P. v. R. Schleyer, and K. Krogh-Jespersen, *J. Comput. Chem.*, **2**, 356 (1981).

277. A. Skancke and I. Agranat, *Nouv. J. Chim.*, **9**, 577 (1985).

278. M. N. Glukhovtsev, P. v. R. Schleyer and C. Maerker, *J. Phys. Chem.*, **97**, 8200 (1993).

279. L. Pauling, *J. Phys. Chem.*, **4**, 673 (1936).

280. J. I. Aihara and T. Horikawa, *Bull. Chem. Soc. Jpn.*, **56**, 1853 (1983).

281. J. A. Pople and K. G. Untch, *J. Am. Chem. Soc.*, **88**, 4811 (1966).

282. Yu. B. Vysotsky and G. T. Klimko, *Zh. Strukt. Khim.*, **27**, 15 (1986).

283. S. Kawajima and Z. G. Soos, *J. Am. Chem. Soc.*, **108**, 1707 (1986).

284. S. Kawajima and Z. G. Soos, *J. Am. Chem. Soc.*, **109**, 107 (1987).

285. J. I. Aihara, *J. Am. Chem. Soc.*, **107**, 298 (1985).

286. J. I. Aihara, *Bull. Chem. Soc. Jpn.*, **53**, 1751 (1980).

287. H. Hosoya, K. Hosoi, and I. Gutman, *Theor. Chim. Acta*, **38**, 37 (1975).

288. J. I. Aihara, *J. Am. Chem. Soc.*, **103**, 5704 (1981).

289. R. B. Mallion, *Pure Appl. Chem.*, **52**, 1541 (1980),

290. J. Hoarau, in O. Chalvet, R. Daudel, S. Diner, and J. P. Malrieu (Eds.), *Localization and Delocalization in Quantum Chemistry*, Vol. 1, Reidel, Dordrecht, 1975.

291. A. K. Burnham, J. Lee, T. G. Schmalz, and W. H. Flygare, *J. Am. Chem. Soc.*, **99**,1836 (1977).

292. W. H. Flygare, *Chem. Rev.*, **74**, 653 (1974).

293. J. A. Pople, *J. Chem. Phys.*, **41**, 2559 (1964).

294. H. J. Dauben, J. D. Wilson, and J.L. Latty, *J. Am. Chem. Soc.*, **91**, 1991 (1969).

295. T. G. Schmalz, T. D. Gierke, P. Beak, and W. H. Flygare, *Tetrahedron Lett.*, 2885 (1974).

296. T. G. Schmalz, C. L. Norris, and W. H. Flygare, *J. Am. Chem. Soc.*, **95**, 7961 (1973).

297. A. A. Bother-By and J. A. Pople, *Rev. Phys. Chem.*, **16**, 43 (1965).

298. K. E. Calderbank, J. A. Calvert, P. B. Lukins, and G. L. D. Ritchie, *Austral. J. Chem.*, **34**, 1835 (1981).

299. P. C. M. Ziji, B. H. Ruessink, J. Balthuis, and C. MacLean, *Acc. Chem. Res.*, **17**, 172 (1984).

300. H. J. Dauben, J. D. Wilson, and J. L. Laity, *J. Am. Chem. Soc.*, **90**, 811 (1968).

301. H. F. Hameka, *J. Chem. Phys.*, **34**, 1996 (1961).

302. A. H. Stollenwerk, B. Kanellakopulos, H. Vogler, A. Juric, and N. Trinajstić, *J. Mol. Struct.* (THEOCHEM), **102**, 377 (1983).

303. P. W. Selwood, *Magnetochemistry*, Interscience, NewYork, 1956.

304. H. Haberditzl, *Angew. Chem. Int. Ed. Engl.*, **5**, 288 (1966).

305. B. Maoche, J. Gayoso, and O. Ouamerali, *Rev. Roum. Chem.*, **29**, 613 (1984).

306. N. Mizoguchi, *Chem. Phys. Lett.*, **106**, 451 (1984).

307. N. Mizoguchi, *Chem. Phys. Lett.*, **134**, 371 (1987).

308. R. F. Childs and I. Pikulic, *Can. J. Chem.*, **55**, 259 (1977).

309. J. I. Aihara, *Bull. Chem. Soc. Jpn.*, **54**, 1245 (1981).

310. B. R. M. de Castro, J. A. N. F. Gomes, and R. B. Mallion, *J. Mol. Struct.* (THEOCHEM), **260**, 123 (1992).

311. V. Elser and R. C. Haddon, *Nature*, **325**, 792 (1987).

312. R. B. Mallion, *Nature*, **325**, 760 (1987).

313. F. Sondheimer, *Acc. Chem. Res.*, **5**, 81 (1972).

314. H. Günther, MNR Spectroscopy, *An Introduction*, Wiley, New York, 1980.

315. B. R. M. de Castro, J. A. N. F. Gomes, and R. B. Mallion, *J. Mol. Struct.* (THEOCHEM), **260**, 133 (1992).

316. M. Barfield, D. M. Grant, and H. Ikenberry, *J. Am. Chem. Soc.*, **97**, 6956 (1975).

317. J. C. Facelli, D. M. Grant, and J. Michl, *Acc. Chem. Res.*, **20**, 152 (1987).

318. N. D. Epiotis, W. R. Cherry, F. Bernardi, and W. J. Hehre, *J. Am. Chem. Soc.*, **98**, 4361 (1976).

319. A. J. Ashe, *Topics Curr. Chem.*, **105**, 125 (1982).

320. A. J. Ashe, T. R. Diephouse, and M. Y. El-Sheikh, *J. Am. Chem. Soc.*, **104**, 5693 (1982).

321. H. Günther and H. Schmickler, *Pure Appl. Chem.*, **44**, 807 (1975).

322. A. J. Jones, P. D. Gardner, D. M. Grand, W. M. Litchman, and V. Boekelheide, *J. Am. Chem. Soc.*, **92**, 2395 (1970).

323. Yu. B. Vysotsky, and M. M. Mestechkin, *Zh. Stukt. Khim.*, **16**, 303 (1975).

324. M. Randic, *J. Magn. Reson.*, **59**, 34 (1984).

325. G. B. Kistiakowsky, J. R. Ruhoff , H. A. Smith, and W. E. Vaughan, *J. Am. Chem. Soc.*, **58**, 146 (1936).

326. M. J. S. Dewar, *Pure Appl. Chem.*, **44**, 767 (1975).

327. G. Winkelhofer, R. Janoschek, F. Fratev, G. W. Spitznagel, J. Chandrasekhar, and P. v. R. Schleyer, *J. Am. Chem. Soc.*, **107**, 332 (1985).

328. M. N. Glukhovtsev and B. Ya. Simkin, *Zh. Strukt. Khim.*, **26**, 168 (1985).

329. F. P. Lossing and J. C. Traeger, *J. Am. Chem. Soc.*, **97**, 1579 (1975).

330. F.-G. Klarner, E. K. G. Schmidt, M. A. A. Rahman, and H. Kolmar, *Agnew. Chem. Int. Ed. Engl.*, **21**, 139 (1982).

331. R. Breslow, *Pure Appl. Chem.*, **28**, 111 (1971).

332. R. Breslow, *Acc. Chem. Res.*, **6**, 393 (1973).

333. M. J. Cook, A. R. Katrizky, and P. Linda, *Adv. Heterocycl. Chem.*, **17**, 255 (1974).

334. R. G. Pearson, *J. Org. Chem.*, **54**, 1423 (1989).

335. R. G. Pearson, *J. Am. Chem. Soc.*, **110**, 2092 (1988).

336. Z. Zhou and R. G. Parr, *J. Am. Chem. Soc.*, **111**, 7371 (1989).

337. Z. Zhou, R. G. Parr, and J. F. Garst, *Tetrahedron Lett.*, **29**, 4843 (1988).

338. R. C. Haddon and T. Fukunaga, *Tetrahedron Lett.*, **21**, 1191 (1980).

339. A. Minsky, A. Y. Meyer, and M. Rabinovitz, *Tetrahedron*, **41**, 785 (1985).

340. F. B. Bramwell, *J. Am. Chem. Soc.*, **101**, 3306 (1979).

341. N. C. Baird, *J. Am. Chem. Soc.*, **94**, 4941 (1972).

342. I. Fischer-Hjalmars, in E. D. Bergman and B. Pullman (Eds.), *Aromaticity, Pseudo-Aromaticity, Anti-Aromaticity*, The Israel Academy of Sciences and Humanities, Jerusalem, 1971, pp. 375–382.

343. W. Kaim, *Angew. Chem. Int. Ed. Engl.*, **20**, 599 (1981).

344. M. N. Glukhovtsev, *The Theoretical Study of the Structure and Valence Isomerisations of Antiaromatic Molecules*, Thesis, Rostov University, Rostov on Don, 1981; M. N. Glukhovtsev and B. Ya. Simkin, *Zh. Strukt. Khim.*, **24**, 31 (1983).

345. G. Hafelinger, *Tetrahedron*, **27**, 4609 (1971).

346. J. M. Bofill, J. Gastels, S. Olivella, and A. Sole, *J. Org. Chem.*, **53**, 5149 (1988).

347. W. T. Dixon, *J. Chem. Soc.*, **B**, 612 (1970).

348. J. I. Aihara, *Bull. Chem. Soc. Jpn.*, **56**, 1935 (1983).

349. R. C. Haddon, M. L. Kaplan, and J. H. Marschall, *J. Am. Chem. Soc.*, **100**, 1235 (1978).

350. K. Dimroth, *Acc. Chem. Res.*, **15**, 58 (1982).

351. R. E. W. Bader, *Acc. Chem. Res.*, **18**, 9 (1985).

352. R. E. W. Bader, *Atoms in Molecules*, Clarendon Press, Oxford, 1990.

353. D. Cremer and E. Kraka, *J. Am. Chem. Soc.*, **107**, 3800 (1985).

354. R. E. W. Bader, T. S. Slee, D. Cremer and E. Kraka, *J. Am. Chem. Soc.*, **105**, 5061 (1983).

355. D. Cremer, in Z. B. Maksic (Ed.), *Modelling of Structure and Properties of Molecules*, Ellis Horwood, Chichester, England, 1988, p. 125.

356. E. Kraka and D. Cremer, Chemical Implications of Local Features of the Electron Density Distribution, in Z. B. Maksic (Ed.), *Theoretical Models of Chemical Bonding,* Part 2, Springer-Verlag, Berlin, 1990, p. 453.

357. D. Cremer, E. Kraka, T. S. Slee, R. E. W. Bader, C. D. H. Lau, T. T. Nguyen-Dang, and O. J. MacDouglall, *J. Am. Chem. Soc.*, **105**, 5069 (1983).

358. M. Barzaghi and C. Gatti, *J. Chim. Phys.*, **84**, 783 (1987).

359. D. Cremer and E. Kraka, in J. F. Liebman and A. Greenberg (Eds.), *Molecular Structure and Energetics*, Vol. 1, VCH Publishers, Deefiel Beach, 1988, p. 65.

360. R. C. Haddon, Nouv. *J. Chim.*, **3**, 719 (1979).

361. C. Edmiston, *J. Mol. Struct. (THEOCHEM)*, **169**, 331 (1988).

362. F. Fratev, D. Bonchev, and V. Enchev, *Croat. Chem. Acta*, **53**, 545 (1980).

363. F. Fratev, V. Enchev, O. E. Polansky, and D. Bonchev, *J. Mol. Struct.*, **88**, 105 (1982).

364. S. B. Bulgarevich, V. S. Bolotnikov, V. N. Scheinker, O. A. Osipov, and A. D. Garnovskii, *Zh. Org. Khim.*, **12**, 197 (1976).

365. S. B. Bulgarevich, T. A. Yusman, and O. A. Osipov, *Zh. Obshch. Khim*, **54**, 1603 (1984).

366. L. W. Jenneskens, F. J. J. de Kanter, P. A. Kraakman, L. A. M. Turkenburg, W. E. Koolhass, W. H. de Wolf, F. Bickelhaupt, Y. Tobe, K. Kakiuchi, and Y. Odaira, *J. Am. Chem. Soc.*, **107**, 3716 (1985).

367. P. G. M. van Ziji, L. W. Jenneskens, E. W. Bastian, C. MacLean, W. H. de Wolf, and F. Bickelhaupt, *J. Am. Chem. Soc.*, **108**, 1415 (1986).

368. Yu. B. Vysotsky, N. A. Kovach, and O. P. Schvaika, *Izv. Sib. Otdel, Akad. Nauk SSSR Ser. Khim. Nauk*, 3 (1980).

369. B. A. Hess, L. J. Schaad, and M. Nakagawa, *J. Org. Chem.*, **42**, 1669 (1977).

370. A. Verbruggen, *Bull. Soc. Chim. Belg.*, **91**, 865 (1982).

371. F.-M. Tao and Y.-K. Pan, *Theor. Chim. Acta*, **83**, 377 (1992).

3

DELOCALIZATION MODES AND ELECTRON-COUNT RULES

One may find in the literature references to about two dozen categories of aromaticity denoted by corresponding terms. This diversity of the aromaticity types originates primarily from the existence of different modes of the electron delocalization, namely, (a) ribbon delocalization of either π- or σ-electrons; (b) surface delocalization of σ-electrons; and (c) volume delocalization of σ-electrons [1] (Fig. 3.1).

3.1 AROMATICITY TYPES STEMMING FROM THE RIBBON DELOCALIZATION IN A CYCLIC SYSTEM WITH PLANAR OR DISTORTED PLANAR GEOMETRY OF A RING

The ribbon delocalization can be realized in acyclic (π- or σ-conjugation) [2] and cyclic systems alike, the latter case being characterized by a variety of linkage fashions between several ribbons (Fig. 3.2) [3]. In the case of a single ribbon only the pericyclic topology is possible (orbital basis sets for a monocyclic A_n system are shown in Fig. 3.3).

The orbital basis sets of a ring may be divided into two main classes, depending on the number of phase inversions. In the case of an even number of the nodes, the basis orbital system is classified as a Hückel type, while with the odd number of nodes it is assigned to the Möbius type (Fig. 3.4). Originally, the aromaticity concept was defined in the course of studies on annulenes, benzenoids, and related compounds [4, 5] in which the basis p_π-orbitals constitute the Hückel system (Chapter 4), such as [6], [8], and [10]annulenes (**1–3**) or naphthalene (**4**):

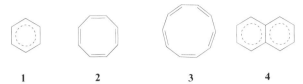

Upon insertion of a saturated CH_2 group (or several groups) into a cyclic system of overlapping p_π-orbitals, an unsaturated so-called homoconjugated system is formed that may have homoaromatic or homoantiaromatic character [6] (Chapter 6):

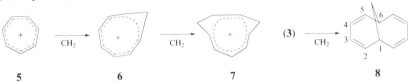

Ribbon Delocalization

Surface Delocalization

Volume Delocalization

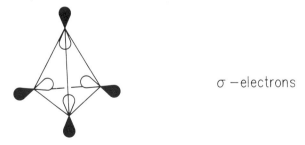

Figure 3.1   Electronic delocalization modes. (From [1] with permission.)

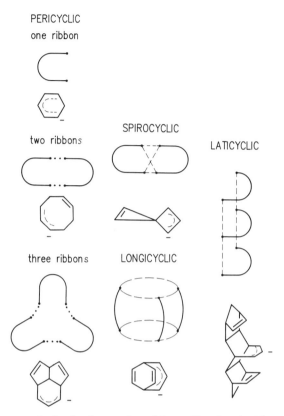

Figure 3.2   Some topologies for interacting ribbons. Reprinted with permission from M. J. Goldstein and R. Hoffmann, *J. Am. Chem. Soc.*, **93**, 6193 (1971).

In addition to the above-mentioned manner of formation of a homo-conjugated system through interruption of the conjugated monocyclic system in one (**6**) or more (**7**) points, it may also be formed through perturbation by additional homoconjugation, as in **8** [7, 8].

Structures **6–8** are nonplanar. In order that the notions of the π-electron delocalization and π-aromaticity might be extended to them, the concepts of the π-orbital and the σ-π separability have to be defined for the three-dimensional case. A convenient tool for establishing a bridge between the σ-π separability in planar conjugated molecules and that in nonplanar ones, as well as for providing a common definition of the π-orbital, is the so called scheme of the π-orbital axis vector (POAV) [9].

The formal peripheral dihedral angle may be used for unambiguous description of the π-orbital alignment when each bonded pair of atoms lies in the same plane as its nearest neighbors [10]. For example, for the fragment shown in Fig. 3.5a the dihedral angle τ provides a meaningful index of the π-orbital alignment

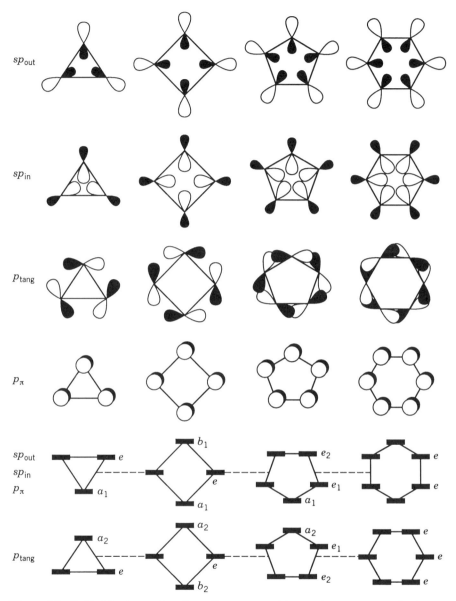

Figure 3.3 Orbital basis sets for A_n cyclic systems. A is a main-group atom in the sp hybridization state, and $n = 3$–6. The p_{tang} orbital systems for $n = 3, 5$ relate to Möbius type, and the remaining to Hückel type. Energy levels for various orbital systems are given in the bottom part of the diagram.

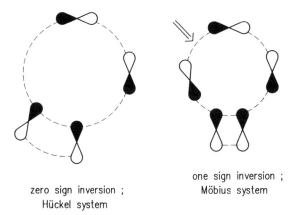

one sign inversion ;
Möbius system

zero sign inversion ;
Hückel system

Figure 3.4 Examples of Hückel and Möbius systems of basis orbitals. Arrow shows the phase inversion.

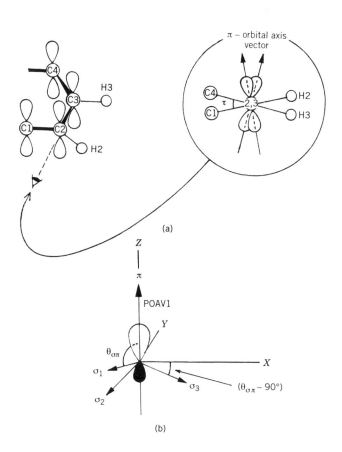

if C1, C2, C3, and H2 are coplanar. In the POAV1 scheme, it is assumed that the π-orbital (and the POAV) makes equal angles with the three σ-bonds in nonplanar structures (Fig. 3.5). In a planar molecule, such as benzene, the angle $\theta_{\pi\sigma}$ equals 90°, while the pyramidalization angle ($\theta_{\pi\sigma} - 90°$) is zero. In nonplanar structures, the values of pyramidalization angles are nonzero, and it is note-worthy that these values and, correspondingly, those of the π-orbital misalign-ments turn out to be substantially smaller than the values of the customarily considered peripheral (skeletal) dihedral angles (Table 3.1). The data in this table show that pyramidalization of even unconstrained carbon atoms may occur. These results suggest that the moving force causing distortion from the planarity is the "striving" to maximize a favorable π-orbital overlap and the aromatic character [9, 10].

A relation between the pyramidalization of a carbon atom and the attendant alterations in the electronic structure of the molecule can be established through introduction of the notion of π-hybridization defined as the fractional s-character m (m, $s^m p$) [9]. Note that in the initial step of the pyramidalization a considerable increase in the ($\theta_{\pi\sigma} - 90°$) angle results in only insignificant rehybridization (Fig. 3.6). Thus, as may be concluded from the results of the POAV1 analysis of hybridization for **8** given in Table 3.1, improvement in the

TABLE 3.1   Peripheral Dihedral Angles α (degrees), Pyramidalization Angles ($\theta - 90°$),[a] POAV1 π- and σ-Hybridizations[b] Calculated from the Geometry for 1,6-Methano [10] annulene (8) [9, 10]

Quantity	C1	C2	C3
α, POAV1	27.3	14.7	0
α, POAV2	26.9	14.1	0
($\theta_{\sigma\pi} - 90°$)	1.9	1.8	3.8
m	0.002	0.002	0.009
$\bar{n}$	2.007	2.006	2.027

[a]For definition of $\theta_{\sigma\pi}$ see Fig. 3.5.
[b]m is the fractional s-character (m, $s^m p$) in the π-orbital; $m = 0.0$ for planarity. $\bar{n}$ is the fractional p-character ($\bar{n}$, sp^n) in the σ-orbital (a group-average σ-hybridization, see Fig. 3.6); for planar geometry $\bar{n} = 2.0$.

◄────

Figure 3.5   (a) Definition of dihedral angles. The formal dihedral (skeletal) angle (τ, C1—C2—C3—C4) is equal to the misalignment angle between the π-orbital axis vectors (POAV) (and all the other three dihedral angles) in the event that C1 and C2 are each in planar geometries. For C2 this would require that C1, C2, C3, and H2 are coplanar (absence of pyramidalization, see Fig. 3.5b) [9,10]. (b) Definitions of $\theta_{\sigma\pi}$ (angle made by the π-orbital with each of the σ-bonds), ($\theta_{\sigma\pi} - 90°$) [pyramidalization angle], and POAV1. The POAV1 is taken to be perpendicular to the local XY plane. Reprinted with permission from R. C. Haddon, Accounts of Chemical Research, 1988, **21**, 243; American Chemical Society.

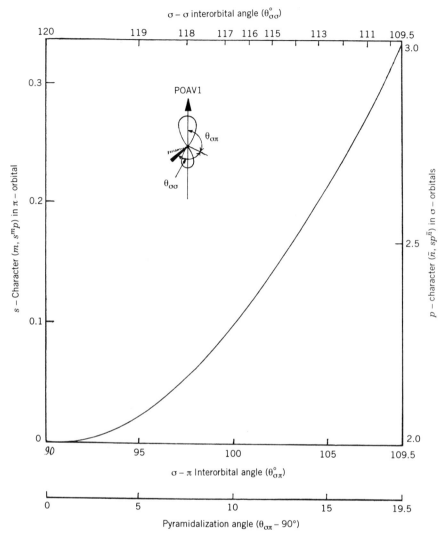

Figure 3.6 Relationship between the σ-σ and σ-π interorbital angles and the hybridization at a carbon atom in the case of the C_{3v} local symmetry (sp^2 hybridizadtion for planar geometry and sp^3 hybridization for tetrahedral geometry). Reprinted with permission from R. C. Haddon, Accounts of Chemical Research, 1988, **21**, 243; American Chemical Society.

π-orbital overlap in conjugated molecules is not associated with any sizeable rehybridization and requires relatively small energetic expenses.

Considering the orthogonality of the σ- and π-orbitals as the basic feature of the σ-π separability (Fig. 3.7), one may suggest a broader definition of the π-orbital applicable to nonplanar systems as well: the π-orbital is the hybrid orbital that is locally orthogonal to the σ-orbitals. The so-called POAV2 analysis is based on the condition of orbital orthogonality [9, 11], with POAV1

ORBITAL ORTHOGONALITY

SYMMETRY element	1–Dimension C_∞ axis	2–Dimension σ_h plane	3–Dimension none
σ–orbital			along internuclear axes (from atom A)
π–orbital			orthogonal to σ–orbitals (of atom A) by construction (POAV)
overlap integral	$S(\pi,\sigma) \equiv 0$ global	$S(\pi,\sigma) \equiv 0$ global	$S(\pi_A,\sigma_A) \equiv 0$ local

Figure 3.7 Orbital orthogonality as the basis for the definition of the π-orbital that is applicable for various dimensions. Reprinted with permission from R. C. Haddon, Accounts of Chemical Research, 1988, **21**, 243; American Chemical Society.

being a particular case of it when the σ-σ bond angles are equal (local C_{3v} symmetry). Within the POAV1 analysis, the π-orbital is defined as a hybrid orbital that makes equal angles with the σ-orbitals. The POAV2 values of the π-orbital misalignment angles (listed for **8** in Table 3.1) lead to the conclusion that the retention of the orbital orthogonality is an important characteristic of the structure of nonplanar conjugated organic molecules. From all appearances, in geometries dictated by the orthogonality condition, the electron–electron repulsions between electrons of hybrid orbitals are reduced to a minimum.

The POAV scheme allows the HMO theory to be extended to nonplanar systems in the form of the 3D-HMO theory (3D standing for three-dimensional) [12]. There is in this case no need for additional parameters. The 3D-HMO theory proved quite useful in examining the problem of homoaromaticity of (Chapter 6) and the aromaticity of spheroidal carbon clusters, such as C_{60} (**9**) [13, 14]:

9

To analyze the π-bonding in nonplanar conjugated molecules by the 3D-HMO theory, the resonance integral is defined as $\beta_{ij} = (S_{ij}/S)\,\beta = \rho_{ij}\,\beta$, where S is the reference overlap integral and β is the standard resonance integral in the HMO method. In a nonplanar distorted structure, the π-orbital overlap integral between atoms i and j is represented as sum of the (s, s), (s, p_σ), (p_σ, p_σ), and (p_π, p_π) components. In the role of the reference overlap integral, one may take either the (p_π, p_π) overlap integral between the nearest neighboring p_π-AOs in benzene (S^B) or the pure (p_π, p_π) overlap integral calculated for the corresponding value of R of the given bond (S^R) [15]. Thus the value of ρ may serve as a measure of strength of the π-bond. The analysis of $\rho_{1,7}^B$ for the equilibrium geometry of the homotropenylium cation **6** and of its change with varying $R_{1,7}$ reveals important features of the electronic structure associated with homoaromaticity (see Chapter 6).

A monocyclic conjugated system can be perturbed by the replacement of a CH group with a heteroatom to give a heterocyclic system, which, in some cases, retains aromatic (antiaromatic) character similar to that of the parent hydrocarbon, as in **10**, or may altogether lose any manifestations of aromaticity, as in **11** [16]:

The term "heteroaromaticity" is often used in reference to such systems [17] (see Chapter 5).

If the heteroatom is represented by an element of the main group of the second or subsequent rows, then, as has already been assumed by Schäfer et al. [18] for λ^5-phosphorine (**12**), there is a possibility of conjugation of the d_{yz} and d_{xz} orbitals of that atom with the π-orbitals of the ring, giving rise to Hückel and Möbius aromatic systems.

When a metal atom is inserted into the conjugated ring, "metalloaromatic" rings can be formed in which the CC bonds are equalized [19], as in **13** and **14**:

The equality of CC bonds is also found in the case of rings containing nontransition metals, as in the 2 π-electron structure (**15**) [20] (see Chapter 5). The terms "metalloaromaticity" [21] and, occasionally, "three–dimensional aromaticity" [22] are employed for structures like **16** and **17** (see Chapter 9):

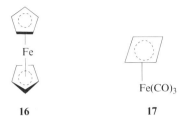

16   **17**

Structures being formed upon insertion of an atom with *d*-orbitals into the center of a cyclic π-system, such as transition metal complexes of unsaturated ligands, are regarded as cruciaromatic [23]. In the case, the relevant *d* AOs should have low enough energy to be capable of effective interaction with the π-MOs of the ligands. In other words, the aromatic stabilization of, for example, **18** is limited by the difference between the iron *d* AO and the carbon 2*p* AO energies.

18   **19**

An example of the structure thought to have aromaticity as a result of the cyclic ribbon delocalization of σ-electrons is given by the dication of hexaiodobenzene (**19**) [24]. Analogous cyclic ribbon systems, but of the Möbius type, formed through σ-overlapping of *p*-orbitals, suggest the notion of the Möbius aromaticity or antiaromaticity. The 3-twist trimethylenemethane (**20**) [25] and its dication (**21**) [26] represent such structures:

20   **21**   **22**   **23**

A Möbius system of carbon *p*-orbitals is formed in the D_{3h} structure **22** with longicyclic topology [3]. An analogous longicyclic array (Fig. 3.2) is made up of π-orbitals of three CH=CH fragments in a barrelene molecule (**23**) [25, 27]. The interaction between the *p*-orbitals lying in one plane suggests the notion of the "in-plane aromaticity" or antiaromaticity (see Chapter 8). The aromatic stabilization due to interaction between the in-plane ethynyl π-orbitals is thought to be likely in hexaethynylbenzene (**24**) [28].

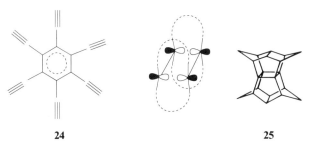

Equally stabilizing is the interaction between in-plane p-orbitals in the pagodane dication (**25**) [29] (see Chapter 8).

3.2 AROMATICITY TYPES DUE TO SURFACE DELOCALIZATION

An example of the molecule with the surface delocalization is given by cyclopropane (**26**) [1]. Consider the skeletal orbitals of the ring (Fig. 3.8). The radial and tangential orbitals lying in the ring plane may belong either to σ-orbitals or

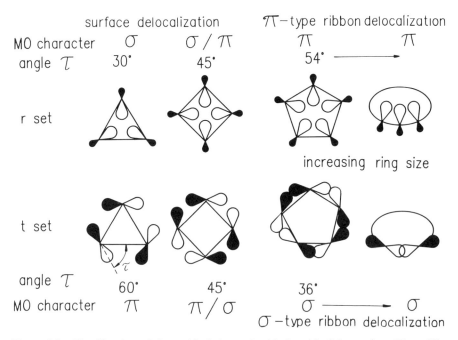

Figure 3.8   Classification of ring orbitals (σ or π) with the aid of the angle τ. (From [1], with permission.)

to π-orbitals, depending on the size of the ring. The radial sp_{in} orbitals or the tangential p-orbitals making angles of $90° \geq \tau > 45°$ with the internuclear connection lines of the ring (Fig. 3.3) form MOs that may be classified as π-orbitals. In the case $45° > \tau \geq 0°$, the corresponding MOs should be regarded as σ-orbitals. In other words, for a three-membered ring, the radial orbitals constitute σ-MOs, while in a four-membered ring they make up π-MOs. As the ring grows, the radial π-MOs become topologically equivalent to the p_π-MOs of a π-conjugated system (Fig. 3.8).

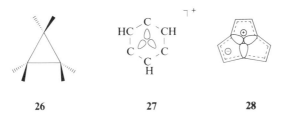

| 26 | 27 | 28 |

An electron pair lying in the MO a_1 formed by radial orbitals takes part in the realization of a three-center two-electron bond ($3c–2e$). This type of bonding has given rise to the notions of σ-aromaticity in cyclopropane (**26**) (see Chapter 7), double aromaticity in the cation (**27**) [30], and trefoil aromaticity in (**28**) [31] (see Chapter 8).

3.3 VOLUME DELOCALIZATION

Systems of this type are represented by the molecule of tetrahedrane (**29**) [1] (see Chapter 9) and the 1,3-dehydro-5,7-adamantanediyl dication (**30**) [32] (see Chapter 8):

$$
\begin{array}{cc}
\text{29} & \text{30}
\end{array}
$$

The inspection of the above-described inventory of aromaticity types invites the inevitable question of whether the introduction of some of these is justified; that is, whether the presence of a corresponding aromatic stabilization has been proved rigorously enough.

This question will be examined in the following chapters in which particular attention will be given to specific features in the electronic structure and geometry of molecules directly attributable to corresponding types of aromaticity (or antiaromaticity). But first, we turn to a subject that is important in all types of aromaticity, namely, the electron-count rules.

3.4 ELECTRON–COUNT RULES: THE HÜCKEL RULE

There exists a definite relationship between the stability of the molecular structure and the filling of the electron shell of a compound. Figure 3.9 depicts energy difference curves as functions of the electron count for two structural problems: (a) relative stabilities of square and rectangular cyclobutadiene structures and (b) cyclobutadiene and its open-chain analog [33]. The cyclobutadiene D_{4h} structure with four π-electrons is seen to be destabilized relative to the rectangular D_{2h} structure with the bond alternation (see Chapter 2); however, when the π-levels are filled with two or six electrons, the relative stability order is reversed. Similarly, the 4 π-electron square structure is less stable, while the 2 and 6 π-electron structures are more stable relative to the open-chain analogs.

To accurately express these relationships, the so-called electron-count rules were proposed (for details see [35, 36] and Chapter 9). Hückel's $4n + 2$ rule was the first among these [37, 38]. This rule being paradigmatic to all others, it is

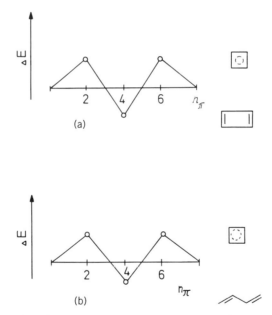

Figure 3.9 Qualitative diagrams energy difference calculated by the Hückel method as a function of the π-electron count. (a) The square structure of singlet cyclobutadiene is unstable towards distortion into the rectangular structure, while the square structures of the cyclobutadiene dication (2 π-electrons) and of the cyclobutadiene dianion (6 π-electrons) are stable to this distortion; (b) 4 π-electron cyclobutadiene is less stable than its open-chain analog, in contrast to the 2 π-electron $(CH)_4^{2+}$ dication and 6 π-electron$(CH)_4^{2-}$ dianion (in the latter case, calculations taking electron repulsion into account show a greater stability of the open-chain structure of the *s-trans*-butadiene dianion (see [34] and Chapter 4). (Adapted from [33].)

deserving of particular attention: it may exemplify both the merits and the limitations of all such rules. Hückel's rule can be formulated within the framework of different approaches.

3.4.1 Formulation of the Hückel Rule

The original formulation of the $(4n + 2)$ Hückel rule was based on the requirement for the closedness of the electron shell as a condition for the stability of a molecule [35]. In the Hückel theory, the energy $e_{(k)}$ of the π-MO of the monocyclic planar conjugated polyene $C_N H_N$ is given by Eq. (3.1) (in β units, on condition that $\alpha = 0$):

$$e_{(k)} = 2 \cos \left(\frac{2\pi k}{N} \right), \quad k = 0, \ldots, N - 1 \qquad (3.1)$$

The lowest π-orbital has the energy $e_0 = 2$ (i.e., $e_0 = \alpha + 2\beta$) at $k = 0$ regardless of N. In all even-membered (alternant) monocyclic polyenes, the energy of the highest π-MO $e_{N-1} = -2$ ($e_{N-1} = \alpha - 2\beta$). For $0 < k < N - 1$, there are pairs of degenerate levels $e_k = e_{N-k}$ (Fig. 3.10). A closed π-electron shell is formed only when the number of π-electrons is $4n + 2$ ($n = 0, 1, 2, 3$).

The highest occupied π-MOs of $[4n + 2]$annulenes, $[4n + 3]$annulenes cations,

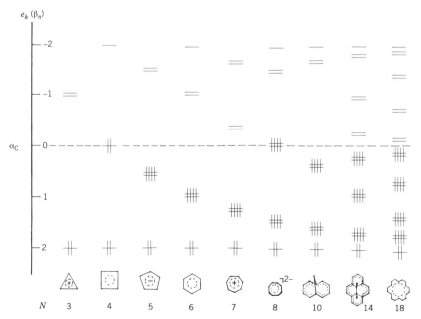

Figure 3.10 Hückel π-electron levels for monocyclic $C_N H_N$ polyenes. (Adapted from [39].)

and [4n + 1]annulene anions are completely filled, while the HOMOs of [4n]annulenes are filled incompletely, which, as Hückel has noted [37], gives rise to the high reactivity of these species. Recently, it has been shown [40] using the formalism of the graph theory that the Hückel rule is applicable to all mono-cyclic conjugated hydrocarbons by virtue of the fact that a monocyclic (4n + 2) π-electron species is indeed always a closed-shell π-system. The expression for the energy of the π-MO of a Möbius monocyclic system [41] is expressed as

$$e_{(k)} = 2\cos\left[\frac{(2k + 1)\pi}{n}\right], \quad k = 0, \ldots, N - 1 \tag{3.2}$$

Thus the electron-count rules for the Möbius system are opposite to those for the Hückel system. The (4n) electron Möbius systems have a closed-shell, while the shell of the (4n + 2) electron Möbius systems is open. The difference between the orbital patterns in the Hückel and the Möbius systems can be visualized by using a mnemonic device [42] (Fig. 3.11).

The above-mentioned condition of the closed π-electronic shell in aromatic systems and the open shell in antiaromatic ones underlies a "narrow" version of the Hückel rule applicable to monocyclic D_{nh} structures. The key antiaromatic molecule of cyclobutadiene has the D_{2h} structure with bond length alternation and a closed electronic shell (see Chapters 2 and 4). The cyclobutadiene hetero-cyclic analogs (e.g., azete) have also a closed π-electronic shell. The problem of classification of these types of structures has prompted a broader interpretation of the (4n + 2) rule based on the requirement of the filling of all bonding orbitals and the vacancy of all nonbonding and antibonding orbitals (e.g., see [43]).

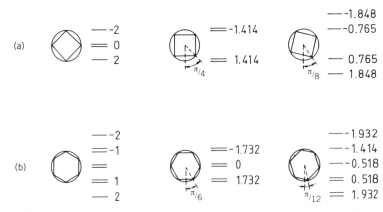

Figure 3.11 Mnemonic device for obtaining energy levels for Hückel, Möbius [N] annulenes and for their reference structures [22, 42]. For calculation of the orbital energy, the general expression $e_{(k)} = 2\cos(2\pi k/N) + A$) is used, where $A = 0$ (Hückel), π/N (Möbius), and $\pi/2N$ (reference) [42] and $k = 0,1,...,N-1$; compare with Eqs (3.1) and (3.2) (a) Cyclobutadiene (b) Benzene.

Evidently this requirement is fulfilled if a given monocyclic conjugated species contains $(4n + 2)$ π-electrons.

The validity of the Hückel rules has been borne out by innumerable amounts of experimental data on those physical properties of the monocyclic conjugated planar molecules that served as a basis for constructing energetic, structural, or magnetic indices of the aromaticity (antiaromaticity) [38, 44, 45]. The $4n + 2$ $(4n)$ rules for determining the aromaticity of Hückel and Möbius systems derived originally on the basis of the simple Hückel theory [42, 43] should not be limited by the conditions of that theory. This has been shown by analysis of these rules in the SCF approximation with electron repulsions and exchange contributions taken into account [46] and with the electron correlation included [47].

Among other approaches to formulating the electron-count rules [35, 36, 48–50], including that of Hückel, of particular interest are those in which there is a direct relationship between "topogeometrical" features of the structure and stability of different types of electronic configuration. Such approaches involve the use of terms that contain combined information about both the electron structure (energy level patterns) and the topology of a given system. An example of such an approach is given by the method of moments [33, 51–54]. For systems with a discrete N-level energy spectrum $\{e_i\}$, the mth moment is defined as follows:

$$\mu_m = e_1^m + e_2^m + \cdots + e_N^m \tag{3.3}$$

A relationship between the moments and the topological features of a molecule is clearly seen in the framework of the Hückel theory. With the use of the Hückel Hamiltonian, the mth moment is presented by Eq. (3.4) where $\mathbf{H}$ is the Hückel Hamiltonian matrix.

$$\mu_m = \mathrm{trace}(\mathbf{H})^m = \sum_j \sum H_{jk} H_{kl} \cdots H_{zj} \tag{3.4}$$

The second summation in Eq. (3.4) is done over all products of order m. Thus the mth moment is, in fact, the weighted sum of all the cyclic (self-returning) walks of length m starting from the j orbital and returning to it in m steps. The weights are the interaction integrals (H_{kl}) between the two orbitals k and l for a given step; in the present case, all H_{kl} are equal to the Hückel resonance integral β. In the expression for the moments, their weights convey the information on electronic structure. The connection of the moments with the topogeometrical characteristics becomes wholly explicit with β set equal to unity, in which case H turns into an adjacency matrix A (see section 2.2.5) and the expression for μ_m takes the form [53]

$$\mu_m = \sum_{i=1}^N (A^m)_{ii} = \sum_{i=1}^N \sum_{\alpha,\beta,\ldots,\delta} A_{i\alpha} A_{\alpha\beta} \cdots A_{\delta i} = \sum_{i=1}^N \sum_{\alpha,\beta,\ldots,\delta} (1) \tag{3.5}$$

In other words, the value of μ_m is the sum of the self-returning walks of length m starting from vertex $1, 2,...,N$ and passing through vertices $\alpha, \beta,...,\delta$.

Owing to its above-mentioned specific features, the method of moments provides for qualitative predictions of the stability of various structural types depending on the orbital filling. The difference between the energies of two structures is expressed in terms of the first disparate moment of their respective energy density of states.

Let us now consider how this method may be helpful in deriving the electron-count rules for determining the aromaticity of a monocyclic system.

Taking as an aromaticity criterion the relative stability of the ring and open chain structures [55], [1] we find that the first moment that will turn out different in the sets $\{\mu_i\}$ for the m-membered ring and the open-chain structures will be the mth moment associated with two sets of complete walks around the ring of length m (clockwise and counterclockwise) since such walks are absent in the open-chain structure. This means that the ring structure possesses the larger moment. Thus for a four-membered ring such as cyclobutadiene, the first moment different from that in the corresponding open-chain structure will be the fourth: μ_4 (ring) $> \mu_4$ (open chain). As may be seen from Fig. 3.12, the ring-structure is unstable with half filled band but becomes more stable than the

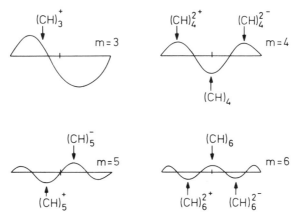

Figure 3.12 Energy difference curves as a function of fractional orbital filling between two structures whose first disparate moment is μ_m. When the curves have positive values, the cyclic structure (with the larger $|\mu_m|$) is stable with respect to the open-chain one. (Adapted from [33].)

[1]It should be noted that this criterion is valid only within the Hückel approach. Beyond it, the ring structures $(CH)_N$, with $N = 4, 6$, and 8, are more stable than the corresponding open-chain structures; this has been shown by calculations using the PPP Hamiltonian with the π-electron correlation taken fully into account [56]. Even so, the difference between their atomization energies per one π-electron still alternates depending on whether $N = 4n$ or $N = 4n + 2$.

open-chain structure for lower and higher orbital fillings. The curve has maxima with the fractional orbital occupancy of $x = \frac{1}{4}$ and $\frac{1}{8}$, and a minimum when $x = \frac{1}{2}$ (by convention when the curves have positive values, the structure with the larger $|\mu_m|$ is more stable [51]). Hence cyclobutadiene will be unstable as a neutral molecule ($n_\pi = 8(\frac{1}{2}) = 4$) but stable in the form of a dication or dianion ($N_\pi = 2$ or 6). In the case of a six-membered ring, the curve has three maxima at $x = \frac{1}{6}, \frac{1}{2}$, and $\frac{5}{6}$ that correspond to $N_\pi = 2, 6$, and 10. In this manner we arrive at the Hückel rule: $N_\pi = 4n_\pi + 2$, $n_\pi = 0, 1, 2, \ldots$. Note that in deriving this rule by means of the method of moments, no features characteristic of only the π-electrons were specially taken into account, so the result is valid for both the π- and the σ-electron monocyclic systems. In the latter case this rule helps classify the σ-aromatic as well as the σ-antiaromatic systems (see Chapter 7). For example, the σ-aromaticity of cyclopropane is seen in terms of the method of moments as a result of a stabilizing effect of the cyclic walk of length 6 linking the six sp^3 hybrid orbitals of the CC bonds [51].

The same approach enables the electron-count rules to be formulated for the Möbius ring system as well. By returning to the condition of $\beta = 1$ employed to obtain Eq. (3.5), we have the situation that all the edges of the monocyclic graph corresponding to the Hückel system have weight 1. As opposed to it, with the monocyclic Möbius graph, one of the edges must have a weight of -1 [57]. Consequently, the absolute value of the mth moment for the Hückel ring is always larger than that for the Möbius ring [33]. As is apparent from the energy difference curves (Fig. 3.12), at a half filled point ($x = \frac{1}{2}$) in the case of four-membered rings, the system with the smaller first disparate (fourth) moment (i.e., the Möbius system) is more stable; by contrast, in the case of the six-membered rings, the system with the larger first disparate (sixth) moment i.e., the Hückel system is more stable [33]. Thus the electron-count rules must be opposite for the Hückel and the Möbius systems.

The method of moments may also yield quantitative estimates of aromaticity, for example, via calculation of resonance energies. The total π-electron energy of a molecule can be expressed in terms of the moments. For acyclic structures this energy is additive. This makes it possible to calculate, after determining the values of the bond energy parameters (five types), the energy of the reference structure by analogy with the Hess–Schaad scheme [53] (see Section 2.24).

3.4.2 Relation of the Hückel Rule with the Energetic and Magnetic Criteria of Aromaticity and with the Reactivity of Cyclic Conjugated Molecules

A direct relation between the Hückel rule and the thermodynamic stability of monocyclic conjugated species has already been shown earlier with the aid of the method of moments. Let us trace some aspects of this relationship, this time in terms of the traditional orbital approach.

From the definition of the topological resonance energy (TRE)—see Section

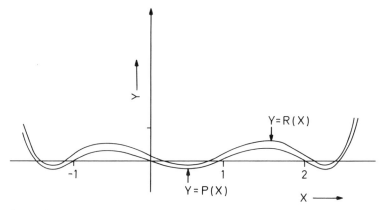

Figure 3.13 Curves $Y = P(X)$ and $Y = R(X)$ for 1,4-dihydropyrazine. (From [40], with permission.)

2.5.5—the term of the mth orbital resonance energy (ORE_m) may be derived [40]:

$$ORE_m = X_m - X_m^{ac} \qquad (3.6)$$

where X_m and X_m^{ac} are the mth roots of the characteristic polynomial $P(X)$ and the reference polynomial $R(X)$.

The roots of $R(X)$ may be obtained using a mnemonic device (see [42]) as has already been pointed out for the $P(X)$ roots for monocyclic Hückel and Möbius systems (Fig. 3.11). As is apparent from Fig. 3.13, where as an example the curves of $Y = P(X)$ and $Y = R(X)$ are given for 1,4-dihydropyrazine, the following inequalities are satisfied [40]:

$$X_m > X_m^{ac} \qquad \text{for } m = 2p + 1 \qquad (3.7)$$

$$X_m < X_m^{ac} \qquad \text{for } m = 2p \qquad (3.8)$$

Hence the ORE_m is positive if m is odd and negative otherwise. Note that, except in the case of $m = 1$ and $m = N$, the absolute values of the ORE_m, while alternating their signs, are fairly close, as the data of Table 3.2 show. Ultimately, the sign of the overall resonance energy is determined in the first place by the sign of ORE_m for the HOMO. From the analysis of the frontier MOs it follows [40] that $ORE(HOMO) > 0$ only for $(4n + 2)$ π-electron monocyclic species. Consequently such species satisfy the Hückel rule that $RE > 0$; in other words, they possess aromatic stability. This is confirmed by the data of Table 2.1.

It is not difficult to reveal a connection between the Hückel rule and the magnetic criteria. As noted in Chapter 2, the ring current induced in a monocyclic system by an external field is approximately proportional to the RE multiplied by the area squared of the ring—Eq. (2.86). Since in $(4n + 2)$ π-electron mono

TABLE 3.2   Orbital Resonance Energies (ORE_m) for Antiaromatic 1,4-Dihydropyrazine (TRE = − 0.132) [40] and Aromatic Borepin (TRE = 0.154) [61]

m	X_m	X_m^{ac}	ORE_m, β units
		1, 4-Dihydropyrazine	
1	2.547	2.494	0.053
2	2.034	2.146	− 0.112
3	1.000	0.833	0.167
4	− 0.047	0.127	− 0.174
5	− 1.000	− 1.121	0.121
6	− 1.534	− 1.479	− 0.055
		Borepin	
1	1.847	1.826	0.021
2	1.247	1.336	− 0.089
3	0.760	0.615	0.145
4	− 0.445	− 0.211	− 0.234
5	− 0.767	− 1.000	0.233
6	− 1.802	− 1.615	− 0.187
7	− 1.840	− 1.951	0.111

cyclic species RE > 0, they will be diatropic, whereas the (4n) π-electron compounds where RE < 0, will be paratropic [40].

The Hückel (4n + 2) rule allows certain conclusions to be drawn about the reactivity of monocyclic conjugated systems. Indeed as follows from inequalities (3.7) and (3.8), in a (4n + 2) π-electron monocyclic conjugated system, the HOMO has always a lower energy than in the reference acyclic olefinic-type structure; and, vice versa, the energy of the LUMO of the former is higher than that of the latter. Thus monocyclic (4n + 2) π-species must have much smaller superdelocalizability than that for olefinic reference structures. This means that such monocyclic systems are characterized by a smaller reactivity in electrophilic and nucleophilic reactions in comparison with the reference structures of the same geometry [40]. As for the (4n) π-electron monocyclic species and their olefinic reference systems, the relation between their reactivities is reversed; that is, the HOMO of the former has a higher energy and the LUMO has a lower one than the corresponding energies of the frontier MOs of the latter.

3.4.3   Limits of Applicability of the Hückel Rule

This rule loses its effectiveness in the case of highly charged cations and anions (e.g., the (4n + 2) π-electron benzene tetracation or the cycloheptatrienyl trianion cannot be assigned to aromatic species [58, 59]) because these species are characterized by the predominance of electrostatic interactions. Some comments are also in order regarding the application of the Hückel rule to monocyclic heteroatomic systems. In this case four situations may be singled out, depending on the number of π-electrons in the system (N_π) and the values of h and k parameters that serve to describe the heteroatom in the Hückel MO theory (the Coulomb integral of heteroatom X $\alpha_x = \alpha + h\beta$ and the resonance integral for CX bond $\beta_{CX} = k\beta$ [38]) [60]:

1. The number of π-electrons $N_\pi = 4m + 2$: the Hückel rule is applicable irrespective of the value of the h and k parameters,
2. $N_\pi = 4m + 1$: the rule cannot be applied for $h < -2k^2$,
3. $N_\pi = 4m + 3$: the rule fails for $h \geq 2k^2$,
4. $N_\pi = 4m$: the rule is violated for $h < 0$ and satisfied at $h > 0$ regardless of the magnitude of the parameter k.

For example, in the case of 4 π-electron borol (**31**) at $h_B = -1$ [38] there are only two bonding MOs. Since in **31** all bonding levels are filled and all nonbonding and antibonding MOs are vacant, this molecule should be assigned to the aromatic class. However, such an assignment is inconsistent with the negative value of the TRE (-0.321), the values of the structural indices [61], and the very high reactivity of borol [62]. By contrast, for borepin (**32**) (situation 1) the Hückel rule is valid since the number of the bonding π-MOs is the same as in the analogous hydrocarbon analog $(CH)_7^+$:

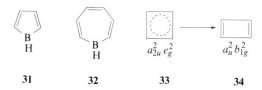

Strictly speaking, the use of the Hückel rule, based on the open/closed electron shell condition, becomes purely conventional when applied not to a high-symmetry D_{nh} structure of the $4n$ π-electron species but rather to lower-symmetry structures with bond alternation. For example, in the D_{4h} cyclobutadiene structure (**33**) the nonbonding e_g MOs are not vacant and it belongs to an antiaromatic species, but in the D_{2h} structure of cyclobutadiene (**34**) all antibonding π-MOs are vacant, the nonbonding MOs are absent, and all bonding MOs are completely filled. Thus, the π-electron shell is closed, but species **34** is still considered to be antiaromatic because the RE is negative in this structure.

Even more interesting is the role of bond alternation when the electron-count rule is applied to Möbius annulenes. For all such systems, the Hückel rule is reversed [22, 41, 63]. Indeed, whereas for the total π-energy per electron of Hückel annulenes relation (3.9) is always satisfied, in the case of Möbius annulenes without bond alternation relation (3.10) is valid [64]:

$$E^{4n+2}_{\pi,\text{Hückel}} \leq E^{4n}_{\pi,\text{Hückel}} \quad (n = 1, 2, \ldots) \tag{3.9}$$

$$E_{\pi,\text{Möbius}}^{4n} \leq E_{\pi,\text{Möbius}}^{4n+2} \quad (n = 1, 2, \ldots) \tag{3.10}$$

where E_π is the total π-energy per electron. However, when the bond alternation (BA) is large enough (e.g., $R_a = 1.35$ Å, $R_b = 1.45$ Å), inequality (3.10) turns around to (3.11). In other words, when the $(4n + 2)$ and the $(4n)$ π-electron Möbius systems with strong bond alternation are compared, the former, though formally classified as antiaromatic, may turn out to be more stable [64].

$$E_{\pi,\text{Möbius}}^{4n}(\text{BA}) \geq E_{\pi,\text{Möbius}}^{4n+2}(\text{BA}) \quad (n = 1, 2, \ldots) \tag{3.11}$$

The Hückel rule should be confined to monocyclic systems. Attempts to extend it without modifications to polycyclic systems are not justified since in the case the rule may fail. Generally, the presence of an odd number of bonding π-orbitals (i.e., $4n + 2$ carbon atoms) in polycyclic hydrocarbons, in contrast to annulenes, is not a sufficient condition for the closedness of a π-electron shell (i.e., complete filling of all bonding π-orbitals). Indeed, the Hückel rule may be violated when applied to such systems [65, 66]. Examples of the failure of the $(4n + 2)$ rule are not confined to rare topologies with the open electronic shell [67]: another case in point is represented by a redundant number of bonding π-orbitals, as in the 14 π-electron molecule of pentalenopentalene (**35**), where there are 8 bonding π-MOs [68]:

The isomeric systems **36** and **37** consist of condensed five-membered rings; while having equal numbers of these rings, they possess π-electron shells of different types. The **37**-type systems with the number of five-membered rings equal to $4l$ (hence, with $12l + 2$ π-electrons) have a closed π-electron shell, indicating that the $(4n + 2)$ rule is valid in this case. In contrast to **37**, the **36**-type systems with $(4n + 2)$ carbon atoms have vacant bonding orbitals whose number grows with the growing number of five-membered rings [69].

An example of the systems that satisfy the Hückel rule but are characterized by the filling of the NBMOs is given by certain cata-condensed polycyclic hydrocarbons, such as **38** [70]:

In closed loop odd-alternant polycyclic polyenes **39** and **40**, the electronic structure, as has been shown by MNDO-calculated bond lengths and bond orders [71], is independent of whether the system as a whole has $(4n)$ or $(4n + 2)$ π-electrons.

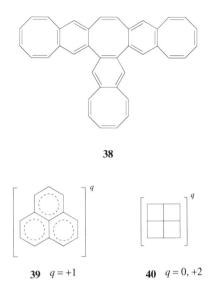

38

39 $q = +1$ **40** $q = 0, +2$

One characteristic difference between the polycyclic systems and the monocyclic ones is that in the former the aromatic character manifesting itself in neutral species may be retained also in their doubly charged derivatives, cf **41** and **42** [72]. This behavior contrasts with that of the dianions or dications of aromatic [4n + 2] annulenes in which there occurs inversion of aromaticity.

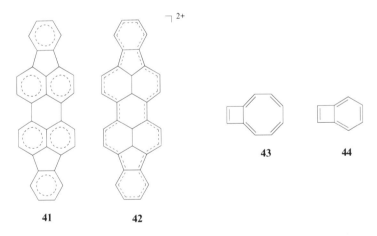

41   **42**   **43**   **44**

After ascertaining whether all π-electrons of individual rings of a polycyclic species are included in the overall conjugated system or whether for a given polycyclic structure fragments with localized π-bonds are characteristic, a kind of "arithmetic of polycyclic aromaticity" [73] may be worked out, which, in the former case, will coincide with the conventional "arithmetic" while in the latter case it is not necessarily so. Thus benzocyclooctatetraene (**43**) is a system with peripheral 10 π-electron delocalization (see Section 4.3.4), hence $2 + 8 \Rightarrow 10$. On the other hand, for **44** bond fixation in the four-membered ring is characteristic, so $2 + 6 \neq 8$.

The Hückel rule can be modified to the so-called perimeter rule, according to which the $(4n + 2)$-membered perimeter ensures the stabilization of a polycyclic conjugated hydrocarbon, while the $(4n)$-membered one provides, on the contrary, for its destabilization, but the rule does not provide reliable predictions. For example, molecule **45** has a 10-membered perimeter; however, the condition of the closed shell is not met in terms of the Hückel rule [74, 75]:

45

3.4.4 The Generalized Hückel Rule

Considering the success of the conjugated circuit model in calculating REs of polycyclic hydrocarbons (see Section 2.2.6), one may assume that an approach to the generalization of the Hückel rule based on the examination of various circuits in polycyclic systems should be useful. Such an approach involving the use of directed polycyclic graphs with weighted edges has in effect produced unified rules for polycyclic systems containing the Hückel- and/or Möbius-type circuits [76]. Table 3.3 presents such rules for bicyclic systems.

TABLE 3.3   Rule for Stability of Generalized Graphs: Effects of a Circuit and Two Disjoint Circuits on Stability

Rule	Type of Circuit	Effect of Circuit(s)
$N(C_j)^a = 4n$	Hückel	Destabilize
	Möbius	Stabilize
$N(C_j) = 4n + 2$	Hückel	Stabilize
	Möbius	Destabilize
$N(C_j) + N(C_k)^b = 4n$	Hückel Hückel	Stabilize
	Möbius Möbius	Stabilize
	Hückel Möbius	Destabilize
$N(C_j) + N(C_k) = 4n + 2$	Hückel Hückel	Destabilize
	Möbius Möbius	Destabilize
	Hückel Möbius	Stabilize

From [76], with permission.
[a]Number of vertices in a circuit.
[b]Number of vertices in a pair of disjoint circuits.

The validity of the generalized Hückel rule, according to which a $(4n + 2)$-membered ring of the Hückel type makes a stabilizing contribution to the energy of a π-electron polycyclic system while the contribution from a $(4n)$-membered ring is destabilizing, has been proved by Hosoya et al. [77]. The proof is based on the examination of the relationship between the total π-electron energy E_π and the molecular topology. The so-called modified topological index $\tilde{Z}_G$ proposed by Hosoya et al. [77] correlates with E_π:

$$E_\pi = 6.092 \log \tilde{Z}_{\bar{N}} + 0.129 \quad \text{(linear polyene)} \tag{3.12}$$

$$E_\pi = 6.092 \log \tilde{Z}_{N^\circ} \quad \text{(annulene)} \tag{3.12a}$$

where $\bar{N}$ and N° denote, respectively, linear polyene and annulene with N carbon atoms. The $\tilde{Z}_G$ index is a function of most coefficients of the characteristic polynomial $P(X)$ [77]

$$\tilde{Z}_G = \sum_{k=0}^{m} (-1)^k a_{2k} \tag{3.13}$$

where $m = N/2$ is the largest integer not exceeding $N/2$. For a chain hydrocarbon the modified index $\tilde{Z}_G$ is identical to the Z_G index, which, in turn, is defined [78, 79] as

$$Z_G = \sum_{k=0}^{m} p(G, k) \tag{3.14}$$

where $p(G, K)$ is the number of ways in which k bonds are so chosen from the G graph that no two of them are connected ($p(G, 1)$ equals the number of bonds in the molecule); m is the maximum number of k for G. Thus the difference between the values of $\tilde{Z}_G$ and Z_G may be used as an aromaticity index, which will be nonzero in the case of molecules whose structures contains rings (except the molecules with only one odd-membered ring) [77]:

$$\Delta Z_G = \tilde{Z}_G - Z_G \tag{3.15}$$

A $(4n + 2)$-membered ring makes a positive contribution to ΔZ_G, while the contribution of a $(4n)$-membered ring is negative. This may be visualized by writing the definition of ΔZ_G for corresponding monocyclic systems:

$$\Delta Z_G = 2 \sum_{\substack{r=1 \\ n=4k+2}}^{R} Z_{G \ominus R} \qquad \bigcirc \,, \quad \bigcirc\!\!\!\bigcirc \,, \quad \bigcirc\!\!\!\bigcirc\!\!\!\bigcirc \,, \cdots \tag{3.16}$$

$$\Delta Z_G = -2 \sum_{\substack{r=1 \\ n=4k}}^{R} Z_{G \ominus R} \qquad \square \,, \quad \bigcirc \,, \quad \bigcirc\!\!\!\bigcirc \,, \cdots \tag{3.17}$$

The summations in Eqs. (3.16) and (3.17) are performed over the cyclic systems that have an even number of carbon atoms in an independent ring or a set of independent rings (e.g., a 4-membered ring or two 3- and 5-membered rings); the graph $G \ominus R$ is the subgraph of graph G derived from thorough removal of the R ring together with all the edges incident to R.

We have examined the key rule for counting electrons, that of Hückel for planar conjugated molecules, concentrating on the problems that are common to most such rules. Many of these serve to ascertain qualitatively various types of aromaticity in systems with different topology (three-dimensional, σ-, in-plane, bicycloaromaticity). We will return to these rules in the following chapters. For the present, however, we wish to draw the reader's attention to one more approach that emphasizes the common idea behind different rules since it enables such rules to be formulated, from the number of nonbonding MOs, for quite diverse compounds, from alternant hydrocarbons to polyhedral molecules. We have in mind the method of the alternant operator A and parity operator P [80]. In the alternant hydrocarbon, the bonding MOs ϕ_μ^+ belong to the irreducible representation Γ_+ and the paired antibonding MOs ϕ_μ to Γ_-, with Γ_+ and Γ_- being related as follows:

$$\Gamma_+ = \Gamma_A \times \Gamma_- \tag{3.18}$$

$$\Gamma_- = \Gamma_A \times \Gamma_+ \tag{3.19}$$

where Γ_A is one-dimensional representation that interconverts the starred and unstarred carbon atoms in an alternant system. In the case where a symmetry operation changes the positions of the starred atoms, its character in Γ_A equals 1; when interconversion of the starred and unstarred atoms occurs, the character of such an operation equals -1. For cyclobutadiene (D_{4h}) the alternant operator A corresponds to the b_{1g} representation. Hence, for the bonding a_{2u} π-MO and antibonding b_{2u} π-MO, one may write $b_{2u} = b_{1g} \times a_{2u}$. Noteworthy is that the nonbonding MOs remain unaffected: $e_g = b_{1g} \times e_g$. The MOs b_{2u} and a_{2u} are called conjugate, while the e_g MOs are referred to as self-conjugate [80]. The latter are nonbonding MOs.

For benzene, Γ_A is a b_{1u} representation and there are no self-conjugate MOs-hence no nonbonding MOs: $b_{2g} = b_{1g} \times a_{2u}$, $e_{2u} = b_{1u} \times e_{1u}$. Analogously, for Möbius cyclobutadiene and Möbius benzene, it may be shown, making use of representations of the double group, that in the former case self-conjugate representations are absent, while in the latter they are present, thereby indicating the presence of two nonbonding MOs. This approach may be applied to other molecules, for example, the three-dimensional polyhedral ones.

The following chapters will examine molecules of quite diverse types in which various forms of aromaticity are realized. We shall try to verify, by invoking the above-described criteria, whether any suggested type of aromaticity may be rightfully considered as valid; furthermore, specific features of the electronic structure and geometry of the key molecules representing a given type of aromaticity will be identified, and, finally, we will examine how the effects of the aromatic stabilization or antiaromatic destabilization manifest themselves.

REFERENCES

1. D. Cremer, *Tetrahedron*, **44**, 7427 (1988).
2. M. J. S. Dewar, *J. Am. Chem. Soc.*, **106**, 669 (1984).
3. M. J. Goldstein and R. Hoffman, *J. Am. Chem. Soc.*, **93**, 6193 (1971).
4. A. T. Balaban, *Pure Appl. Chem.*, **52**, 1409 (1980).
5. J. R. Partington, *A History of Chemistry*, Vol. 4, Macmillan, New York, 1964.
6. R. E. Childs, *Acc. Chem. Res.*, **17**, 347 (1984).
7. A. Jurić, N. Trinajstić, and G. Jashari, *Croat. Chem. Acta,* **59**, 617 (1986).
8. L. T. Scott, *Pure Appl. Chem.*, **58**, 105 (1986).
9. R. C. Haddon, *Acc. Chem. Res.*, **21**, 243 (1988).
10. R. C. Haddon and L. T. Scott, *Pure Appl. Chem.*, **58**, 137 (1986).
11. R. C. Haddon, *J. Phys. Chem.*, **91**, 3719 (1987).
12. R. C. Haddon, *J. Am. Chem. Soc.*, **109**, 1676 (1987).
13. R. C. Haddon, *J. Am. Chem. Soc.*, **112**, 3385 (1990).
14. R. C. Haddon, L. E. Brus, and K. Raghavachari, *Chem. Phys. Lett.*, **131**, 165 (1986).
15. R. C. Haddon, *J. Am. Chem. Soc.*, **110**, 1108 (1988).
16. M. N. Glukhovtsev, *Zh. Org.Khim. USSR (Engl. Transl)*, **27**, 363 (1991).
17. A. F. Pozharskii, *Khim. Geterotsikl. Soed.*, 867 (1985).
18. W. Schäfer, A. Schweig, K. Dimroth, and H. Kanter, *J. Am. Chem. Soc.*, **98**, 4410 (1976).
19. D.L. Thorn and R. Hoffmann, *Nouv. J. Chim.*, **3**, 39 (1979).
20. P. v. R. Schleyer, E. Kaufmann, G. W. Spitznagel, R. Janoschek, and G. Winkelhofer, *Organometallics*, **5**, 79 (1986).
21. B. E. Bursten and R. F. Fenske, *Inorg. Chem.*, **18**, 1760 (1979).
22. J. I. Aihara, Bull. Chem. *Soc. Jpn.*, **51**, 1541 (1978); **58**, 266 (1985).
23. M. J. S. Dewar, E. F. Healy, and J. Ruiz, *Pure Appl. Chem.*, **58**, 67 (1986).
24. D. J. Sagl and J. C. Martin, *J. Am. Chem. Soc.*, **110**, 5827 (1988).
25. H. E. Zimmerman, *Acc. Chem. Res.*, **4**, 69 (1971).
26. I. Agranat and A. Skancke, *New J. Chem.*, **12**, 87 (1988).
27. L. P. Schmitz, N. L. Allinger, and K. M. Flurchick, *J. Comput. Chem.*, **9**, 281 (1988).
28. M. S. El–Shall and K. P. C. Vollhardt, *J. Mol. Struct. (THEOCHEM)*, **183**, 175 (1989).
29. R. Herges, P. v. R. Schleyer, M. Schindler, and W.–D. Fessner, *J. Am. Chem. Soc.*, **113**, 3649 (1991).
30. J. Chandrasekhar, E. D. Jemmis, and P. v. R. Schleyer, *Tetrahedron Lett.*, 3707 (1979).
31. T. Fukunaga, H. E. Simmons, J.J. Wendoloski, and M. D. Gordon, *J. Am. Chem. Soc.*, **105**, 2729 (1983).
32. M. Bremer, P. v. R. Schleyer, K. Scholz, M. Kausch, and M. Schindler, *Angew. Chem. Int. Ed. Engl.*, **26**, 761 (1987).
33. J. K. Burdett, *Acc. Chem. Res.*, **21**, 189 (1988).
34. A. Skancke and I. Agranat, *Nouv. J. Chim.*, **9**, 577 (1985).

35. D. M. P. Mingos and J. C. Hawes, *Struct. Bonding (Berlin)*, **63**, 1 (1985).

36. D. J. Wales, D. M. P. Mingos, T. Slee, and L. Zhenyang, *Acc. Chem. Res.*, **23**, 17 (1990).

37. E. Hückel, *Z. Phys.*, **70**, 204 (1931).

38. A. Streitweiser, *Molecular Orbital Theory for Organic Chemists*, Wiley, New York, 1961.

39. H. Bock, *Agnew. Chem. Int. Ed. Engl.*, **16**, 613 (1977).

40. J. I. Aihara and H. Ichikawa, *Bull. Chem. Soc. Jpn.*, **61**, 223 (1988).

41. E. Heilbronner, *Tetrahedron Lett.*, 1923 (1964).

42. N. Mizoguchi, *J. Am. Chem. Soc.*, **107**, 4419 (1985)

43. D. J. Klein and A. T. Balaban, *J. Mol. Stuct., (THEOCHEM)*, **259**, 307 (1992).

44. D. Lewis and D. Peters, *Facts and Theories of Aromaticity*, Macmillan, London, 1975.

45. P.J. Garrat, *Aromaticity*, Wiley, New York, 1986.

46. H. E. Zimmerman, *Tetrahedron*, **38**, 753 (1982).

47. D. J. Klein and N. Trinajstić, *J. Am. Chem. Soc.*, **106**, 8050 (1984).

48. R. B. King and D. H. Rouvray, *J. Am. Chem. Soc.*, **99**, 7834 (1977).

49. D. M. P. Mingos, *Acc. Chem. Res.*, **17**, 311 (1984).

50. B.K. Teo, *Inorg. Chem.*, **23**, 1251 (1984).

51. J. K. Burdett and S. Lee, *J. Am. Chem. Soc.*, **107**, 3063 (1985).

52. J. K. Burdett, S. Lee, and W. C. Sha, *Croat. Chem. Acta*, **57**, 1193 (1984).

53. Y. Jiang, A. Tang, and R. Hoffmann, *Theor. Chim. Acta*, **66**, 183 (1984).

54. S. Pick, *Coll. Czech. Chem. Commun.*, **53**, 1607 (1988).

55. R. Breslow, *Angew. Chem. Int. Ed. Engl.*, **7**, 565 (1968).

56. A. V. Luzanov and I. I. Ivanov, unpublished results.

57. A. Groovac and N. Trinajstić, *Croat. Chem. Acta*, **47**, 95 (1975).

58. L. Radom and H. F. Schaefer, *J. Am. Chem. Soc.*, **99**, 7522 (1977).

59. P. v. R. Schleyer, D. Willhelm, and T. Clark, *J. Organomet. Chem.*, **281**, C17 (1985).

60. I. Gutman and A. K. Mukherjee, *Indian J. Chem.*, **27a**, 847 (1988).

61. M. N. Glukhovtsev, M. E. Kletsky, B. Ya. Simkin, and V. I. Minkin, *Zh. Org. Khim. (USSR)*, (Engl. Transl.), **27**, 1583 (1991).

62. J. J. Eisch, J. E. Galle, and S. Kizima, *J. Am. Chem. Soc.*, **108**, 379 (1986).

63. I. Gutman, *Z. Naturforsch.*, **33a**, 214 (1978).

64. P. Kardakov, V. Enchev, F. Fratev, and O. Castaño, *Chem. Phys. Lett.*, **83**, 529 (1981).

65. D.A. Botchvar and I.V. Stankevich, *Zh. Strukt. Khim.*, **10**, 680 (1969).

66. D.A. Botchvar and I.V. Stankevich, *Zh. Strukt. Khim.*, **12**, 142 (1971).

67. R. V. Mallion and D. H. Rouvray, *Mol. Phys.*, **36**, 125 (1978).

68. D.A. Botchvar, I.V. Stankevich, and A. V. Tyutkevich, *Izv. Akad. Nauk SSSR Ser. Khim.*, 1185 (1969).

69. D.A. Botchvar and I.V. Stankevich, *Zh. Strukt. Khim.*, **13**, 1123 (1972).

70. A. T. Balaban, *Rev. Roum. Chem.*, **17**, 1513 (1972).

71. C. Glidewell and D. Lloyd, *J. Chem. Res.*, **(S)**, 106 (1986).

72. M. Rabinovitz, I. Willner, and A. Minsky, *Acc. Chem. Res.*, **16**, 298 (1983).

73. C. Glidwell and D. Lloyd, *J. Chem. Educ.*, **63**, 306 (1986).

74. I. M. Gutman and S. Bosonac, *Bull. Soc. Chim. Beograd*, **42**, 499 (1977).

75. I. M. Gutman and S. Bosonac, *Tetrahedron*, **33**, 1809 (1977).

76. N. Mizoguchi, *J. Phys. Chem.*, **92**, 2754 (1988).

77. H. Hosoya, K. Hosoi, and I. Gutman, *Theor. Chim. Acta*, **38**, 37 (1975).

78. H. Hosoya, *Bull. Chem. Soc. Jpn.*, **44**, 2332 (1971).

79. H. Hosoya, *Theor. Chim. Acta*, **25**, 215 (1972).

80. D. M. P. Mingos and L. Zhenyang, *New J. Chem.*, **12**, 787 (1988).

4

ANNULENES, MONOCYCLIC CONJUGATED IONS, AND ANNULENOANNULENES

The present chapter is concerned with the simplest monocyclic and bicyclic conjugated hydrocarbons, such as annulenes, monocyclic conjugated ions, and annulenoannulenes. These compounds may be regarded as a kind of testing ground in the process of developing notions about the principal type of aromaticity, the π-aromaticity (including its antipode, the π-antiaromaticity). This and the following chapters will attempt to show how the criteria of aromaticity (discussed earlier) actually work in organic chemistry. We will try to demonstrate how the aromaticity is "encoded" in electronic and structural characteristics of a given compound. The body of data on synthesis and reactions will be drawn up only when absolutely necessary. Should the reader wish to obtain information on subjects lying on the fringe of our discussion, relevant material may be found in recently published books [1, 2] or in review articles [3–6].

4.1 ANNULENES

4.1.1 Benzene and Cyclobutadiene

4.1.1.1 Ground-State Structures and Their Thermal Automerizations The aromaticity is manifested in numerous properties of compounds. Comparison between such manifestations in the quintessential contrasting cases of aromatic benzene and antiaromatic cyclobutadiene, which in effect represent the reference structures, throws into sharp relief those effects in which the opposite sides of the aromaticity concept, namely aromatic stabilization and antiaromatic destabilization, are reflected.

If the reader is prepared to accept the analogy, one may liken each property of the aromatic and antiaromatic compounds to a definite sort of wine. Now every such property, like every wine's bouquet, is a combination of various effects. One such effect is associated with aromaticity, and, after the generalities of Chapters 2 and 3, we now move on to tasting the "wines" with the view of finding out precisely these effects. The properties we shall check up on include ground-state structures and their thermal automerizations, excited-state structures, the stability of benzene and cyclobutadiene with respect to their valence isomers, the stability and geometry of dication and dianion structures of benzene and cyclobutadiene. We shall stick to this scheme for the pairs of cyclooctatetraene and [10]annulene, cyclopropenyl cation and cyclopropenide anion, butalene and naphthalene, and so on.

The general ideas considered in Chapter 2 suggest that benzene is a stable molecule with a structure without bond length alternation and not prone to thermal automerization, while cyclobutadiene is a species of very high reactivity with considerable bond length alternation undergoing automerization at a low activation barrier.

Benzene was first isolated by M. Faraday in 1825 in the condensed gases of pyrolyzed whale oil. The first unsuccessful attempts to obtain cyclobutadiene (by elimination of two HBr molecules from 1,2-dibromocyclobutane) were undertaken by Willstätter in the early 20th century. Since then, it has been a coveted target for chemists. It was only in 1973 that cyclobutadiene was actu-

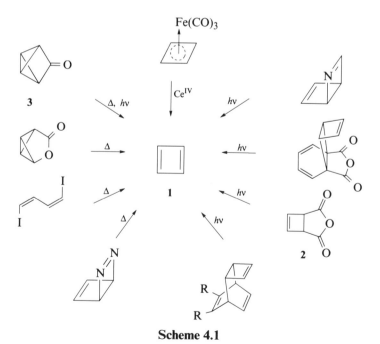

Scheme 4.1

ally isolated in an argon matrix [7, 8]. Further research into its structure occasionally reminds one of a story of adventure (for a review see [3, 4, 9]). Various
currently known approaches to obtaining cyclobutadiene are illustrated by
Scheme 4.1 [1–4].

The most convenient method for the matrix isolation of cyclobutadiene (**1**) is
the photochemical cleavage of the anhydride (**2**); and in its purest form it can be
obtained from tricyclopentanone (**3**). Benzene has in its ground state a D_{6h} structure (for data on geometry see Table 2.4), and two Kekule formulas **4** and **5**
represent actually one structure, that is, 1,2-disubstituted benzene (**6**). In contrast, cyclobutadiene has in its ground state a D_{2h} structure (**1**) with bond alternation (Table 4.1) and two isomers of 1,2-disubstituted cyclobutadiene are conceivable, that is, cyclobutadiene-1,2-d_2 (**7**) and cyclobutadiene-1,4-d_2 (**8**). The
existence of these isomers has been confirmed by experiments with the trapping
of **7** and **8** with an acrylic acid derivative [10].

The interpretation of the experimental data with IR spectra to the effect that
7 and **8** are distinct species has been supported by *ab initio* calculations of their
vibrational frequencies and infrared intensities [11–13]. Findings reported in
[10, 11] have led one to revise a previous report [14] according to which the

TABLE 4.1 *Ab Initio* Calculated Relative Energies ΔE (in kcal/mol) and Geometries of the D_{4h} and D_{2h} Structures of the Cyclobutadiene Lowest Electronic States

States	Computational Level	STO-3G Full π-CI [18]	DZ Basis Set, IEPA [19]	6-31G*, Limited (π + σ)CI [20]	TZ Basis: C(6,4,1) H(3,1) MC SCF [21]	[3s2p1d2s] GVB-CISD [22][f]	6-31G MC SCF [23][g]
$^1B_{1g}$ (D_{4h})	R(CC)	1.453	$(1.44)^{a,b}$	1.428^d	1.446	1.448	1.4522
	R(CH)	(1.100)	(1.10)	(1.075)	(1.085)	1.093	1.0673
	E(au)	−151.8409	−153.4801	−153.73717	−153.7365	−154.09822	−153.6430
$^3A_{2g}$ (D_{4h})	R(CC)	1.451	$(1.44)^b$	1.425	1.440		1.4493
	R(CH)	(1.100)	(1.10)	(1.075)	(1.085)		1.0673
	ΔE	18.0	9.8^c	11	9.71		12.6
$^1A_{1g}$ (D_{4h})	R(CC)			(1.424)	1.447		1.4582
	R(CH)			(1.075)	(1.085)		1.0665
	ΔE			45	55.4		55.8
$^1B_{2g}$ (D_{4h})	R(CC)			(1.424)	1.423		1.4308
	R(CH)			(1.075)	(1.085)		1.0661
	ΔE			74	80.7		95.5

$^1A_{1g}$ (D_{2h}) rectangular						
R(C=C)	1.369	$(1.34)^b$	1.334^d	1.375	1.346	1.3657
R(C—C)	1.539	(1.54)	1.564	1.548	1.567	1.5532
R(C—C)	(1.100)	(1.10)	(1.075)	(1.085)	1.084	1.0679
∠CCH	(135)	(135)		(135)	134.9	134.8
ΔE	-4.2^e	-13.4	-12.0	-6.25	-9.0	-3.2^h

^1A_g (D_{2h}) rhombus						
R(CC)						1.4503
R(CH1)						1.0731
R(CH2)						1.0611
∠CCC						85.8
∠CCH						137.1
ΔE						52.5

[a] The geometry parameters indicated in parentheses have not been optimized; all ΔE values are relative to the $^1B_{1g}$ state structure.

[b] According to calculation with inclusion of the valence electron correlation energy by CEPA approach for D_{2h} structure geometry R(C=C) = 1.336, R(C—C) = 1.571 Å. The geometry with this basis set for the $^1B_{1g}$ and $^3A_{2g}$ states of D_{4h} structure leads to R(CC) = 1.444 and 1.441 Å [19].

[c] With the basis set augmented with d-type functions $\Delta E = 7.3$ kcal/mol [18].

[d] The comparison of the energies of the square and rectangular structure was based on a D_{4h} square geometry with R(CC) = 1.424 Å, while for rectangular structure the geometry is optimized in the two-configuration SCF calculation (TCSCF) using the 4-31G basis set [20].

[e] With $\sigma + \pi$ CI allowed for $\Delta E = -8.3$ kcal/mol [19].

[f] GVB-CISD calculation of the geometry of $^1B_{1g}$ state is optimized in the 6-31G* basis set; the optimum geometry of D_{2h} rectangular structure was obtained by MBPT(2)/6-31G* calculation [25].

[g] The CASSCF (4,4)-CISD calculations [24] give the D_{2h} structure to be 6.7 kcal/mol lower in energy than the $^1B_{1g}$ (D_{4h}) singlet. The $^3A_{2g} - {}^1B_{1g}$ energy differences is 9.4 kcal/mol [24].

photolysis of α-pyrone-6,6-d_2 (**9**) and α-pyrone-3,6-d_2 (**10**) in an argon matrix should result in a square structure (**11**). It has turned out that in this case an equilibrium 1:1 mixture of two isomers (**7** and **8**) is obtained. The dynamic equilibrium of the rectangular structures of cyclobutadiene is observed also at low temperatures. For example, the equilibrium of the tri-*tert*-butylcyclobutadiene structures **12** and **12a** cannot be frozen out in the ^{13}C NMR experiment even at 88 K (see [15] and the literature cited therein). This suggests that the activation energy for this process is no more than 2.5 kcal/mol.

$$\text{(4.1)}$$

12   **12a**

The use of the dynamic ^{13}C NMR spectroscopy [15] with the isotopic perturbation method of Saunders et al. [16] (R = *t*-C_4D_9) leads one to conclude from the temperature dependence of the splitting of the signal for C1/C3 that there is a dynamic equilibrium (Eq. (4.1)) rather than a single structure with equalized bond lengths in the ring. The same conclusion was arrived at after comparing the observed ^{13}C NMR spectrum of vicinally ^{13}C-dilabeled cyclobutadiene with simulated spectra for a static noninterconverting 1:1 mixture of bond alternating structures and for a case of rapid interconversion [17]. At 25 K the rate of interconversion exceeds $10^3 \, s^{-1}$.

According to *ab initio* calculations [18–23], the D_{4h} square structure (**13**) of the $^1B_{1g}$ state of cyclobutadiene (Fig. 4.1) corresponds to the transition state of automerization of the D_{2h} rectangular structures **1** and **1a**:

$$\text{(4.2)}$$

1   **13**   **1a**

The activation barrier of this reaction is estimated to be from 3.2 to 13.4 kcal/mol (Table 4.1). The activation parameters for the cyclobutadiene automerization **7** ⇌ **8** measured in the 223–263 K range are $1.6 \le \Delta H^{\neq} \le 10$ kcal/mol and $-32 \le \Delta S^{\neq} \le -17$ cal/(mol·K) [26]. Carpenter [27] suggested that the large negative value of activation entropy may be accounted for by the tunneling process. He modeled cyclobutadiene in one-dimensional tunneling calculations by a homonuclear pseudodiatomic molecule (Fig. 4.2). A model for the barrier form of the cyclobutadiene automerization was provided by the truncated inverted parabola whose width at the base $\Delta R = R(C-C) - R(C=C)$ and whose height was equal to the magnitude of the potential-energy barrier ΔE of the automerization. A calculation has shown that the tunneling may account for over 97% of the value of the total rate constant of automerization below 273K

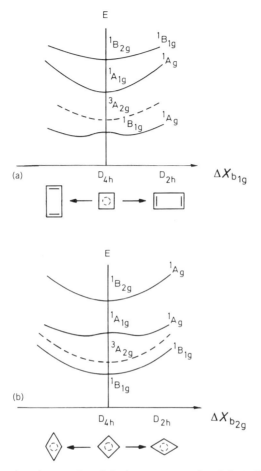

Figure 4.1   Alternations in energies of the lowest states of cyclobutadiene in the case of the b_{1g} and b_{2g} distortions of the D_{4h} structure. (Adapted from [23].)

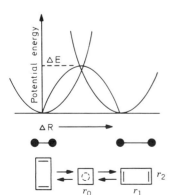

Figure 4.2   Model potential energy functions for the automerization of cyclobutadiene. $\Delta R = R(C\!-\!C) - R(C\!=\!C)$; in [22] the "reference" coordinates $S_1 = r_1 - r_2$ and $S_2 = r_1 + r_2 - 2r_0$ were used. (Adapted from [27].)

[27]. This overriding role of the heavy-atom tunneling can be explained by the fact that in reaction (4.2) the displacement of carbon atoms (0.2 Å) is comparable to the de Broglie wavelength of carbon [28].

In subsequent calculations [22, 28, 29], this conclusion was confirmed by the use of more sophisticated models. For example, using the model with the symmetrical double-well potential and calculating the tunneling rate constants from Eq. (4.3), with the same values of the barrier height and the distance between two minima ΔR as given by Carpenter [27], the splitting of the vibrational ground state ΔE_0 and the tunneling rate constant k_0 were found to be, respectively, 0.009 cm^{-1} and $3 \cdot 10^{10}$ s^{-1} (263 K) [29]:

$$k_i = 2\Delta E_i/h \quad i = 0, 1, \ldots \tag{4.3}$$

where the subscript i refers to the ith doublet of vibration levels with the energy splitting of ΔE_i.

The MINDO/3 method, which reproduces fairly well the difference between the energies of the D_{4h} and D_{2h} structures of cyclobutadiene (5.8 kcal/mol [30], the 3×3 CI taken into account raises it to 8.1 kcal/mol [28]), yielded $\Delta E_0 = 3.80$ cm^{-1} (with 10% correction of the calculated vibrational frequency) and $k_0 = 2.28 \cdot 10^{11}$ s^{-1} [27]. The calculation of the classical rate constants has shown that the rate of tunneling in the cyclobutadiene automerization is greater than the classical rate by a factor of 1000 even at 350 K [28].

Čarsky et al. [22] used a two dimensional model which was based on the so-called reference coordinates S_1 (equaling $r_1 - r_2$, antisymmetrical CC stretch of D_{4h} structure) and S_2 (equaling $r_1 + r_2 - 2r_0$, symmetrical CC stretch); see Fig. 4.2 for 39 combinations of these two coordinates. *Ab initio* GVB/4-31G calculations were performed with the aim of constructing the two dimensional potential surface of the automerization. The data obtained were used to determine the form of the potential energy function. The calculated value of ΔE_0 turned out to be 4.2 cm^{-1}, which, according to Eq. (4.3), corresponds to $k_0 = 2.5 \cdot 10^{11}$ s^{-1}.

According to the dynamic ^{13}C NMR spectroscopy data [31], the automerization barrier for silylsubstituted cyclobutadiene (**12 b**) is $\Delta G^{\neq} = 5.8 \pm 0.2$ kcal/mol. For **12c**, the barrier is even less, only 4.5 kcal/mol [31]. If the tunneling process for the automerizations of **12a** and **12b** took place, the ring C atoms had been moved along with their large substituents. These results [31] bring up a question as to whether the role of heavy-atom tunneling for cyclobutadiene is as considerable as suggested [27–29].

12b R = SiMe$_2$(OCHMe$_2$)
12c R = SiMe$_2$Ph

Along with the above-discussed distortion of D_{4h} structure of cyclobutadiene into a D_{2h} rectangular structure, one might assume a possibility of distortions into a D_{2d} structure (**14**) or a rhombic D_{2h} structure (**15**). However, according to MINDO/3 calculations [32], the D_{2d} structure is 45.8 kcal/mol higher in energy than the rectangular one and does not correspond to a minimum on the PES but rather to a transition state of the valence isomerization of bicyclo [2.2.0] butan-2,4-diyl (**16**). The activation barrier of this isomerization is a mere 4.4 kcal/mol [32].

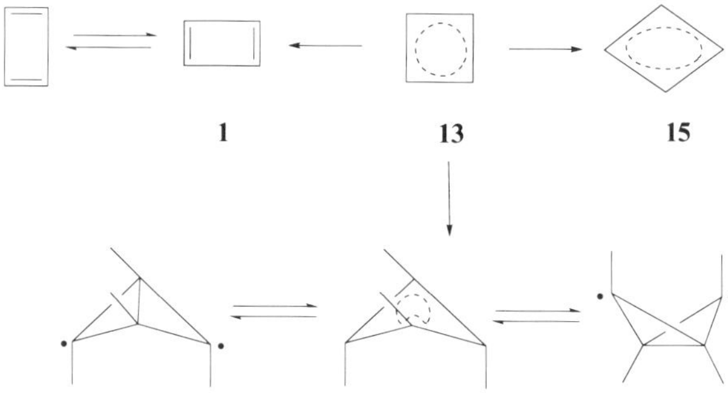

As for the rhombic D_{2h} structure (**15**), the assumption was that it might represent an alternative transition state of the cyclobutadiene automerization [33]. But *ab initio* [20, 23] calculations alike indicate that this structure has a higher energy than that of the D_{4h} $^1B_{1g}$ singlet (Table 4.1) and is neither a minimum nor a saddle point on the PES [23].

Unlike cyclobutadiene, for benzene and its derivatives thermal automerizations associated with a redistribution of the labels (^{13}C, substituents) are not typical. Only a few cases of these are known, for example, the interconversions of difluorobenzenes and di-^{13}C-labeled benzene in the gas phase [34–36]. According to experimental estimates, the $\Delta G^{\neq}$ barrier for the thermal automerization (e.g., of difluorobenzene $C_6H_4F_2$) amounts to no less than 90 kcal/mol [34]. The thermal (1100° C) automerization of benzene has been found experimentally to occur as a result of successive 1,2-shifts of the labeled center [36]:

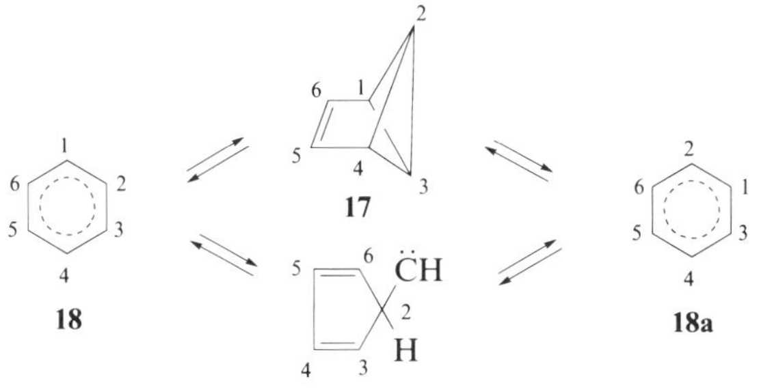

This reaction is assumed to develop via benzvalene (**17**) as an intermediate [36]. MINDO/3 calculations suggest the possibility of direct automerization,

18 ⇌ 18a, via a transition-state structure of 2,4-cyclopentadienylcarbene (**19**) [37]. This reaction channel with the activation barrier of 101 kcal/mol can compete with the automerization developing via intermediate benzvalene (activation energy of the benzene–benzvalene isomerization is ~ 100 kcal/mol, MINDO/3 [38, 39]; experimental value of $\Delta H^{\neq} = 93.4$ kcal/mol [36]).

As is apparent from calculated and experimental data on energy barriers [34–36], "archaromatic" benzene, as opposed to "archantiaromatic" cyclobutadiene, is "disinclined" to take part in automerization reactions [34].

4.1.1.2   The Excited States: Structures and Aromaticity   The material presented in Chapter 2 leads one to expect inversion of the aromatic (antiaromatic) character in the lowest excited states of conjugated hydrocarbons relative to the ground state. In other words, one may, for example, expect antiaromatic destabilization or, at least, substantial reduction of the aromatic stabilization for the lowest singlet and triplet states of benzene and, conversely, aromatic stabilization in the case of such states of cyclobutadiene. Thus the difference between the energies of the ground state and the presumed excited states must be quite sizable for benzene, but much less so for cyclobutadiene. The lowest excited states of benzene can, in accordance with the structural criteria of aromaticity (Section 2.3), have structures with bond length alternation or even nonplanar structures, while a high-symmetry structure (D_{4h}) may be characteristic of the lowest excited states of cyclobutadiene.

Next, we consider π-level systems and the corresponding electronic states of benzene and cyclobutadiene. In conjugated hydrocarbons the HOMOs and LUMOs are the π-orbitals. It should be noted that the π-orbitals considered as a whole do not necessarily form a unified system of levels distinct in energy from the body of the σ-MOs. For example, the levels of the HOMOs and LUMOs of benzene depicted in Fig. 4.3 show that the bonding π-MOs $1a_{2u}$ and $1e_{1g}$ lie within the scale of orbital energies of the σ-MOs. In this case, a pair of degenerate $3e_{2g}$ σ-MOs forms a layer between $1a_{2u}$ and $1e_{1g}$ π-MOs [40]. Moreover, such an important property as the geometry of the ground-state structure is determined by the σ-system. As has been pointed out in Chapter 2, the distortion into a D_{3h} structure is energetically advantageous for the π-system, in contrast to the σ-system. *Ab initio* calculations have shown [42] that in this case the energies of the $1a_{2u}$ π-MO and $3e_{2g}$ σ-MO are increased, while that of the e_{1g} π-MO gets lower. An integral group of the π-MOs is formed in hexafluorobenzene since in this molecule the perfluoro effect stabilizes the uppermost σ-MO [43] (*ab initio* calculations [44]).

The direct products of the irreducible representations for the occupied and lowest unoccupied π-orbitals of benzene give the $^1B_{1u}$, B_{2u}, E_{1u}, and E_{2g} valence excited states, which are expected to be the low-lying states ($E_{1g} \times E_{2u} = B_{1u} + B_{2u} + E_{1u}$, $A_{2u} \times E_{2u} = E_{2g}$). The experimental excitation energies show (Table 4.2) that the energy order of the singlet states is $^1B_{2u} < {}^1B_{1u} < {}^1E_{1u}$, while for the triplet states it is $^3B_{1u} < {}^3E_{1u} < {}^3B_{2u}$. The valence excited states are classified into two groups. One of these is comprised of the

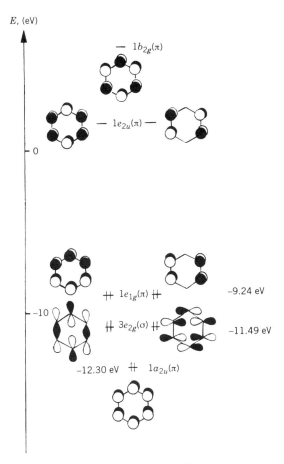

Figure 4.3 Energies of the highest occupied MOs of benzene according to data of photoelectron spectroscopy [40]. The experimental electron affinities of the unoccupied π-MOs are -1.1 eV ($1e_{2u}$) and -4.8 eV ($1b_{2g}$) [41].

states $^1B_{2u}$, $^3B_{1u}$, $^3E_{1u}$, $^1E_{2g}$, and $^3E_{2g}$, which have mainly covalent character. The other group consists of the states with predominantly ionic structure—$^1B_{1u}$, $^3B_{2u}$, and $^1E_{1u}$ [45, 47]. According to experiment [48], the structure of the lowest excited singlet, the $^1B_{2u}$ state, has D_{6h} symmetry (Table 4.3), similarly to the ground-state structure (cf. Fig. 4.4). *Ab initio* calculations show [49] that this structure corresponds to a minimum on the PES and its energy is higher by 133–135 kcal/mol relative to the D_{6h} structure of the ground $^1A_{1g}$ state. Unlike the $^1B_{2u}$ state, the singlet $^1B_{1u}$ state (S_2) has a biradicaloid structure (**20**) of C_s symmetry ($^1A'$ state). Note that this finding based on the *ab initio* calculation with the 4-31 G basis set and limited CI [50] is opposite to the results of calculations by the SINDO1 method [51] (see Table 4.3).

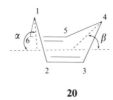

20

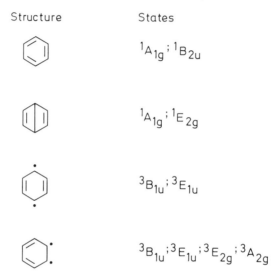

Figure 4.4   Representation of the π-electron states arising from nonpolar valence-bond structure of benzene. (Adapted from [47].)

TABLE 4.2   Experimental and Calculated Excitation Energies (in eV) for the Lowest Valence Electronic Excited States of Benzene

State	Experiment[a]	CASSCF[b] [45]	MRDCI[c] [46]
$^1B_{2u}$	4.9 –5.0	4.9	4.73
$^1B_{1u}$	6.2 –6.3	7.4	6.87
$^1E_{1u}$	6.98–7.0	7.8	7.80
$^1E_{2g}$	7.3	8.1	8.39
$^3B_{1u}$	3.9	3.9	
$^3E_{1u}$	4.9	4.7	
$^3B_{2u}$	5.6	6.7	
$^3E_{2g}$	6.55	7.2	

[a]Taken from [45].
[b]Complete-active-space SCF (CASSCF) calculations.
[c]Basis set $4s2p/2s$, multireference CI calculations.

TABLE 4.3   Relative Energies (in kcal/mol) and Geometries (Bond Lengths are in Å) of the Structures of the Benzene Lowest Electronic States According to Semiempirical and Nonempirical Calculations

State		SINDO-CI [51, 55]	MC SCF[a] 6-31G [54]	MRDCI[b] [53]	MC SCF-CI[c] MIDI [49]
		Singlet States			
$^1A_{1g}$	E_{rel}	0			0
	$R(CC)^d$	1.419			1.403
	$R(CH)$				1.082
	RCI^e	1.75			
$^1B_{2u}$	E_{rel}	134.2			144.8
	$R(CC)$	1.442			1.449
	$R(CH)$	1.078			1.082
	RCI	1.52			
$^1B_{1u}^f$	E_{rel}	147.3			
	$R(CC)$	1.438			
	$R(CH)$	1.079			
		Triplet States			
Quinoid D_{2h} structure	E_{rel}	82.1	0	0	
	$R(C1C2)$	1.368	1.371		
	$R(C2C3)$	1.442	1.466		
	RCI	1.40			
Antiquinoid D_{2h} structure	E_{rel}		0.4[g]		
	$R(C1C2)$		1.494		
	$R(C2C3)$		1.404		
D_{6h} structure	E_{rel}		0.4[g]	2.4[g]	
	$R(CC)$		1.432	1.440	
	$R(CH)$		1.072	1.086	

[a]In the CASSCF calculations [54], active space consisted of six π-electrons and of the six π- orbitals, CASSCF (6,6).
[b]Including σ–σ and σ–π correlations with the basis set close to [4s2p/2s].
[c]The geometry is optimized with the use of the STO-3G basis set.
[d]Experimental and computational data on the geometry of the ground state of benzene are listed in Table 2.5.
[e]Ring-current index of aromaticity; see Section 2.4.
[f]According to *ab initio* calculations [50], this state has C_s structure (**20**), $R(C1C2) = 1.46$, $R(C3C4) = 1.45$, $R(C2C3) = 1.37$, $R(C1C4) = 2.21$ Å, $\alpha = 44°$, $\beta = 67°$.
[g]The energy is given relative to that of structure **21**.

Both calculational [51–55] and experimental (ESR, ENDOR experiments on C_6H_6 in the C_6D_6 host crystal) [53, 56] results show that the structure of the lowest triplet state of benzene is of D_{2h} symmetry. The instability of the D_{6h} structure of the $^3B_{1u}$ state is due to the second-order Jahn-Teller effect (vibronic coupling between the $^3B_{1u}$ and $^3E_{1u}$ states through e_{2g} modes [53]). According to *ab initio* calculations [54], the quinoid structure **(21)** (cf. Fig. 4.4) corresponds to a minimum on the PES, while the antiquinoid structure **(22)** represents a transition state of the topomerization of **(21)**; see Table 4.3.

21   **22**

For the electronic configuration $(e_g)^2$ of the D_{4h} structure of cyclobutadiene, there are three singlet ($^1B_{1g}$, $^1B_{2g}$, $^1A_{1g}$) states and one triplet ($^3A_{2g}$) state; $(E_g \times E_g)^+ = A_{1g} + B_{1g} + B_{2g}$ and $(E_g \times E_g)^- = A_{2g}$. The description of these low-lying states with the aid of the MO theory is depicted in Fig. 4.5.

When rhomboidal e_g (A and B) orbitals are used [20, 57], the $^1B_{1g}$ state is of covalent "biradical" character, while the $^1B_{2g}$ and $^1A_{1g}$ states are zwitterionic (Fig. 4.5) [57]. The exchange integral K_{AB} is vanishingly small (equal to zero in the ZDO approximation). The Coulomb integral J_{AA} is larger than J_{AB}.

$$K_{AB} = \; < A(1)B(2)\left|\frac{1}{r_{12}}\right|A(2)B(1) > \tag{4.4}$$

$$J_{AB} = \; < A(1)B(2)\left|\frac{1}{r_{12}}\right|A(1)B(2) > \tag{4.5}$$

Hence in calculations that do not take CI into account, the energy order of the states should be as follows: $^1A_{1g} \gtrsim {}^1B_{2g} > {}^1B_{1g} \gtrsim {}^3A_{2g}$ (see Fig. 4.5). The role of the CI in calculating the lowest electronic states of the D_{4h} cyclobutadiene structure is fundamentally important. When the CI is taken into account, the energy order of both the states $^3A_{2g}$, $^1B_{2g}$, and $^1B_{2g}$, $^1A_{1g}$ is reversed (Table 4.1). The lower energy of the $^1B_{1g}$ state can be accounted for by the dynamical spin polarization of the doubly occupied a_{2u} π-MO [19]. This effect operates as follows. The e_g MOs are localized at different carbon atoms, and the electrons lying in them and having in the case of the singlet state opposite spins will determine the spatial positions of the remaining electrons in the preceding a_{2u} MO. The spin polarization of the a_{2u} orbital is achieved by the mixing into it of a virtual antibonding b_{1u} orbital (Fig. 4.6). As a result, the a_{2u} orbital "splits" into two, one of which accommodates an electron with spin α and the other is for the electron with the opposite spin β. This reduces somewhat the Coulomb repulsion between the electrons with antiparallel spins in the e_g π-MO and those in the a_{2u} MO. In the triplet $^3A_{2g}$ state, radial spin-polarization occurs to give rise to a

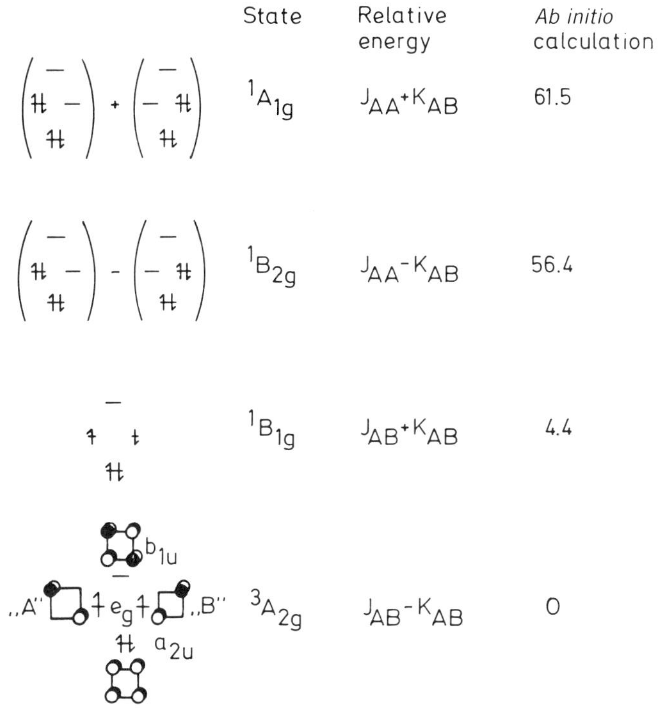

State	Relative energy	Ab initio calculation

$\left(\begin{array}{c}\text{--} \\ \text{⥮ --} \\ \text{⥮}\end{array}\right) + \left(\begin{array}{c}\text{--} \\ \text{-- ⥮} \\ \text{⥮}\end{array}\right)$ $^1A_{1g}$ $J_{AA} + K_{AB}$ 61.5

$\left(\begin{array}{c}\text{--} \\ \text{⥮ --} \\ \text{⥮}\end{array}\right) - \left(\begin{array}{c}\text{--} \\ \text{-- ⥮} \\ \text{⥮}\end{array}\right)$ $^1B_{2g}$ $J_{AA} - K_{AB}$ 56.4

$^1B_{1g}$ $J_{AB} + K_{AB}$ 4.4

$^3A_{2g}$ $J_{AB} - K_{AB}$ 0

Figure 4.5 MO theory description of the four low-lying states of cyclobutadiene [20, 57–59]. Rhomboid e_g (A and B) π-MOs. Expressions for the relative energies correspond to the condition according to which the form of the orbital is the same for all states. J and K are the usual two-electron Coulomb (Eq. (4.5)) and exchange (Eq. (4.4)) integrals. For the real A and B orbitals, J and K are positive. The values of the relative energies (in kcal/mol) are given, derived through *ab initio* SCF calculations without CI [58]. Dunning's DZ basis set was used; the geometry was taken from the calculations [18] (see Table 4.1).

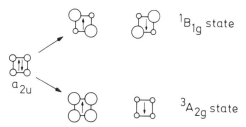

$^1B_{1g}$ state

$^3A_{2g}$ state

Figure 4.6 Schematic representation of the spin-polarization of the a_{2u} π-MO of cyclobutadiene. (Adapted from [19,57].)

large amplitude for the electron with spin α at all four carbon atoms and to a small one for the electron with spin β (Fig. 4.6) [49]. The stabilization of the singlet state proves sufficient for the energy of this state to grow lower than that of the $^3A_{2g}$ state (Table 4.1). Since an e_g MO accommodates two electrons and their spin states can interconvert, the spin-polarization has dynamic character, unlike the similar spin-polarization in the radicals, which is static.

Owing to the interaction of the $^1A_{1g}$ state with higher-lying states, its energy falls below the energy of the $^1B_{2g}$ state (Table 4.1). According to *ab initio* calculations [18–23], the D_{4h} structure of the $^1B_{1g}$ state corresponds to a transition state of the automerization of cyclobutadiene, and the D_{4h} structure of the $^1A_{1g}$ state to a transition state of the interconversion of rhombic D_{2h} structures (Fig. 4.1). The rhombic structure (**15**) does not correspond to a minimum on the PES [23].

As is apparent from the data in Tables 4.1 and 4.3, geometries of the lowest triplet states of cyclobutadiene and benzene differ in principle from that of the ground state in that the bond length alternation in one state occurs despite its absence in the other. According to structural criteria of aromaticity (Chapter 2), this leads one to expect changes in the aromatic (antiaromatic) character relative to the ground state. The absence of the bond length alternation in the $^3A_{2g}$ state of cyclobutadiene suggests a certain degree of the aromatic character [60]. A quantitative evaluation may be made with the aid of the ring-current index (RCI) [55]. This index is based on the fact that a necessary condition for sustaining the ring current is a considerable degree of delocalization of the π-electrons over all ring bonds (see Section 2.3.2) and, consequently, the degree of aromaticity is indicated by the value of the weakest bond order. Whereas for the ^1A_g state of cyclobutadiene (D_{2h} structure) the RCI = 0.98, which is typical of the antiaromatic molecules (RCI < 1.20), in the case of the $^3A_{2g}$ state the RCI = 1.51 (SINDO 1) [51, 55]. The RCI values (Table 4.3) show a considerable lessening of aromaticity in the ^3A_g state of benzene as compared to the ground state. A decrease in the RCI value for the $^1B_{2u}$ state of benzene (Table 4.3) is associated with the lengthening of the CC bonds [51].

A change in the aromaticity in the lowest triplet $^3\pi\pi^*$ state of conjugated cyclic hydrocarbons relative to the ground-state aromaticity may equally be demonstrated with the aid of an energy criterion. Even when the topological resonance energy (TRE) is calculated by the simple Hückel method, it indicates the inversion of the aromatic (antiaromatic) character when the ground and the lowest excited states are contrasted in benzene and in cyclobutadiene (Table 2.1) [61]. The general conclusion drawn by Baird [62] summarizes these and similar facts: the aromatic (antiaromatic) character of the ground state of conjugated cyclic hydrocarbons is reversed in the lowest triplet state. The electron-count rules are reversed accordingly. Thus the lowest triplet state of [4n]annulene will have aromatic character, while that of [4n + 2]annulene will be antiaromatic. In this case, the lowest $^3\pi\pi^*$ state of the open-chain polyene, which contains the same number of carbon atoms, is taken as the reference structure. Since one of the internal C=C bonds in triplet polyene is twisted by 90°, its bending energy is in effect that of two corresponding radical chains linked by a purely single $C(sp^2)$—$C(sp^2)$ bond. Thus for calculating the Dewar resonance energy (DRE) (Section 2.2.3), modified in accordance with this model, for the triplet state of cyclobutadiene, the total calculated carbon–carbon bonding energy of the $^3A_{2g}$ state has to be compared with the carbon–carbon energy of an allyl radical (with the CCC angle of 90°) plus the energy of two purely single CC bonds (88 kcal/mol each).

According to NNDO calculations [62], the DRE for the $^3A_{2g}$ state of cyclobutadiene is equal to 14.1 kcal/mol, while for the ^3A_g state of benzene it is -12.3 kcal/mol (structure **21**) and -16.4 kcal/mol (structure **22**).

From the aromatic stabilization of the lowest triplet state of [4n]annulenes that are antiaromatic in the ground state and the antiaromatic destabilization of this state in the case of [4n + 2]annulenes, one may conclude that in the former case the corresponding excitation energy must be a good deal lower. This is indeed confirmed by the data in Tables 4.1 and 4.3. It has been found using the flash-photolysis technique that for peralkylated cyclobutadiene (**23**) the adiabatic triplet excitation energy is 12 kcal/mol and the vertical excitation energy is ≥ 28.7 kcal/mol [63] (cf. corresponding values for benzene in Tables 4.2 and 4.3).

23

4.1.1.3 *Stability Relative to Valence Isomers*

Since the aromatic compounds are characterized by the retention of their structural type in chemical transformations, benzene, as contrasted with cyclobutadiene, should be thermodynamically the most stable of the six valence isomers $(CH)_6$. Indeed, Table 4.4 shows that the rest have higher energy. (The other $(CH)_6$ isomers (e.g., see [64]) are Dewar benzene (**24**), benzvalene (**17**), prismane (**25**), bicycloprop-2-enyl (**26**), and the rather exotic "Möbius stripane" (*cis-cis-trans*-cyclohexatriene (**27**) that is assumed to form [65] in the reaction of photoisomerization of benzene—Eq. (4.6).[1]). Dewar benzene (**24**) was synthesized in 1963, benzvalene (**17**) and prismane (**25**) were prepared in the early 1970s, and the unsubstituted bicycloprop-2-enyl (**26**), the least stable of the above isomers, is quite a recent arrival, having been synthesized in 1989 [67]

Whereas in the aromatic ground state of benzene its valence isomerization (automerization) can occur under extreme conditions only, for the lowest excited states of benzene the photoisomerization reactions are characteristically aided by decreased aromaticity in these states (Table 4.3). Some of the latter

---

[1] An isomer-computation program generates altogether 217 structural formulas for the C_6H_6 system, provided that the coordination numbers for carbon are ≤ 4 and that for hydrogen this number is unity; see [66].

TABLE 4.4   Experimental and *Ab Initio* Calculated ΔH_f (298 K) Values (in kcal/mol) for the (CH)$_6$ Valence Isomers and for the C$_4$H$_4$ Isomers

Isomers	ΔH_f(298 K)^a, Experimental	E_{rel}	ΔH_f(298 K)^b, Calculated	E_{rel}^d
Benzene (**18**)	19.8	0	20.8	0
Dewar benzene (**24**)	79.3	59.5	94.0	73.2
Benzvalene (**17**)	87.3	67.5	90.2	69.4
Prismane (**25**)	111.0	91.2	136.4	115.6
3,3′-Bicyclopropenyl (**26**)	130–140	110–120	137.6	116.8
Cyclobutadiene(**1**)	—	—	103.7 [69]^c (105.1 [72])	0
Tetrahedrane (**40**)	—	—	132.3	27.8
Methylenecyclopropene (**48**)	98.0 [69] (91.9 [72])	—	—	– 13.5
Vinylacetylene (**49**)	72.9	—	—	– 37.0
Bicyclobutadiene (**50**)	—	—	—	23.0
Cyclobutyne (**51**)	—	—	—	27.0
Butatriene (**52**)	83.4	—		

^{a}Experimental values for the (CH)$_6$ isomers are taken from [68], for the C$_4$H$_4$ isomers from [69].
^{b}Determined from the heats of homodesmotic reactions calculated by an *ab initio* method (MP2/6-31G*) [70]. For benzene, ΔH_f is calculated by group equivalents (6-31G*).
^{c}Calculated [69] by using ΔH_f of **48**.
^{d}For C$_4$H$_4$ isomers, MP2/6-31G* relative energies [73] are given. For **51**, the data of DZ + P TCSCF calculations are given [75]. At MP4SDTQ/6-31G**//MP2/6-31G**, tetrahedrane is 29.4 kcal/mol higher in energy than cyclobutadiene [74]. At HF/6-31G*, the relative energies of the (CH)$_6$ isomers with respect to benzene are as follows: 81.0 (**24**), 74.8 (**17**), 117.5 (**25**), 126.4 (**26**) kcal/mol [71].

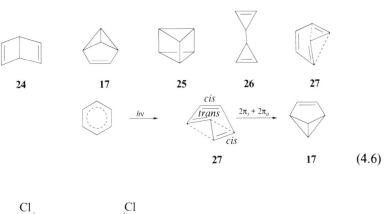

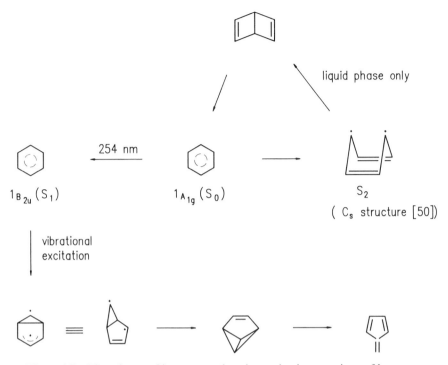

Figure 4.7   The scheme of important photoisomerization reactions of benzene.

reactions are shown in Fig. 4.7. The fact that the valence isomers **17**, **24**, **25**, and **26** are stable with respect to rearomatization into benzene (**18**), even though they have much higher energies than **18**, is explained by the relatively large activation barriers of the rearrangements [76–78]; see also [79]. For example, according to MINDO/3 calculations, the activation energy of the thermal rearomatization of benzvalene into benzene amounts to 21.5 kcal/mol (for the reverse reaction it is ~ 100 kcal/mol) [38, 39].

Clearly, tautomeric processes associated with the loss of the aromatic cyclic electron delocalization cannot be characteristic of benzene, while the reverse transformations will be facilitated by the thermodynamic factor. Thus 3-methylene-1,4-cyclohexadiene (**28**) and 5-methylene-1,3-cyclohexadiene (**29**) fairly readily undergo thermal rearrangements into toluene (**30**) [80–82].

| 28 | 30 | 29 |

At room temperature, **28** and **29** can be handled for up to 1 h. There are experimental data [82] showing that the value of ΔH of the tautomerizations **28** → **30** and **29** → **30** defined as the difference in the gas-phase acidity between two corresponding tautomeric forms (ΔH_{acid} (**30**) − ΔH_{acid} (**28**) or (**29**)) equals 24.0 kcal/mol in both cases. The MNDO calculations [82] yield for **28** and **29** the value of the ΔH_{taut} = 21.9 kcal/mol. According to the same calculations, structures **31** and **32** have energies higher by 42.1 and 55.1 kcal/mol than for toluene.

A lowering of ΔH_{taut} may be effected through perfluoration (for $C_6H_5CF_2H$, ΔH_{taut} = 17.2 kcal/mol [82]), benzoannelation (ΔH_{taut} for **33** → **34** is 3.4 ± 4.0 kcal/mol [82]) or else via introduction of "push–pull" substituents [81, 82]. Thus the nonbenzenoid form of dialdehyde (**35**) is stable in the crystalline state at 293 K for several weeks [83].

According to MNDO calculations [84], the difference between the energies of nonbenzenoid structure **35** and benzenoid structure **36** is 13.4 kcal/mol, which is 1.5 times less than that between the energies of toluene (**30**) and structure **28**. The X-ray analysis of **37** has shown the cyclohexadiene ring to be in a boat conformation [84].

As for methylenecyclobutene (**38**), a study of its reactivity has shown that it does not rearrange into methylcyclobutadiene (**39**) under usual conditions [85].

| **38** | **39** | **1** | **40** |

A MNDO calculation on methylcyclobutadiene (**39**) [86] has indicated that its structure is 31.8 kcal/mol higher in energy than **38**. Tetrahedrane (**40**) is the only valence isomer of cyclobutadiene (**1**) [66]. Owing to its substantial strain energy of 148.8 kcal/mol (STO-3G) [87] (for details see [68, 88]), this molecule is 29.4 kcal/mol [74] destabilized even relative to cyclobutadiene (Table 4.4). Unsubstituted tetrahedrane has not been isolated so far. One known molecule is that of tetra-*tert*-butyltetrahedrane (**41**), stabilized through the "corset effect" of the *tert*-butyl group [89]; its synthesis was performed [90] by a photochemical procedure starting from tetra-*tert*-butylcyclopentadienone (**42**). The excitation of **43**, isolated in an argon matrix, with 254 nm light results in the tricyclic valence isomer (**44**) whose continued irradiation gives rise to photochemical decarbonylation:

| **43** | **44** |

| **42** | **41** |

Tetra-*tert*-butyltetrahedrane (**41**) exhibits a high kinetic stability and isomerizes to the corresponding cyclobutadiene (**42**) only at 135°C [90]. This reaction can be reversed photochemically. According to MNDO calculations [91, 92], tetra-*tert*-butylcyclobutadiene (**42**) is more stable than **41** by a mere 6.7 kcal/mol, while for unsubstituted species **1** and **40**, the difference between the energies calculated by the same method amounts to 45.9 kcal/mol (see also [93]). At the MP4SDTQ/6-31G**//MP2(full)/6-31G** computational level, the **1** – **40** energy difference is 29.4 kcal/mol [74]. At the Hartree–Fock level, the result is 27.1 kcal/mol(6-31G*) [94]. The geometries of **1** and **40** optimized at MP2/6-31G*//MP2/6-31G* are shown below [72, 94]:

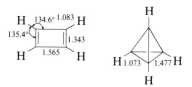

Compare isomerizations of benzene and cyclobutadiene to, respectively, benzvalene (**17**) and tetrahedrane (**40**):

Reaction (4.7) proceeds in the singlet state (for the D_{6h} ground-state structure of benzene RCI = 1.75 [51, 55]) with the activation barrier of ~100 kcal/mol

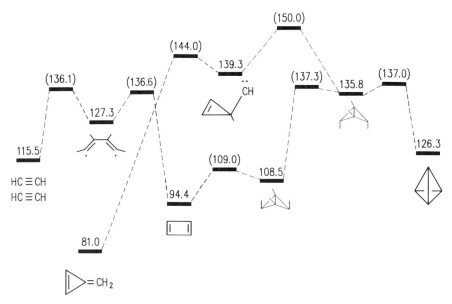

Figure 4.8   The MINDO/3 calculated values of heats of formation ΔH_f (kcal/mol) for some C_4H_4 species [95]. Values of ΔH_f of transition states are given in parentheses.

(MINDO/3) [39]. In the lowest triplet state, the intermediate in the course of the isomerization (4.8) is represented by the bicyclic structure of prefulvene (**45**) (for the ^3A_g state of D_{2h} benzene structure RCI = 1.40; see Table 4.3). The activation barrier of the isomerization (4.8) to **45** is ~ 55 kcal/mol (MINDO/3) [39]. On the other hand, in the case of cyclobutadiene the barrier of the isomerization (4.9) (ground singlet state, RCI = 0.98 [55]) to an analogous bicyclic structure (**46**) is calculated by the same method to be as low as 14.6 kcal/mol [95] (Fig. 4.8).

The thermal concerted isomerization **1** → **40** is forbidden by the orbital symmetry conservation rules [96], and this reaction develops with the formation of intermediate biradicaloids **46** and **16** [24, 95, 96] (Fig. 4.8).

Relative values of the energies of the C_4H_4 isomers **40** and **48**–**52** show that cyclobutadiene (**1**) is destabilized with respect to (**48**) and (**49**).

| 48 | 49 | 50 | 51 | 52 |

A number of approaches have been suggested that enable this destabilization to be reduced. One of these consists in the introduction of conjugated electron-donor ("push") and electron-acceptor ("pull") substituents [97, 98]. This "push–pull" stabilization is realized in the most effective manner in tetrasubstituted cyclobutadienes (for review see [99, 100]).

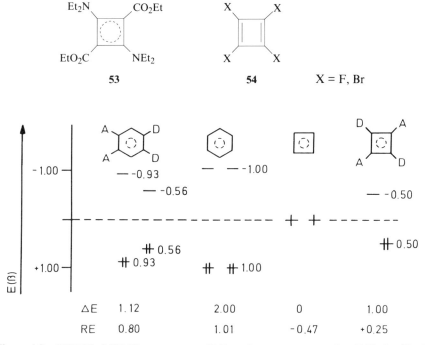

As a result of such a substitution, the degeneracy of the e_g π-MO of the D_{4h} structure of cyclobutadiene is removed and the stabilization occurs of the rhomboidal MO (Fig. 4.5) that spans the two carbon atoms where the acceptor groups are located. Since this MO is antibonding between these two atoms, the distortion into a rhombic structure is energetically advantageous, in which the atoms in question lie on the greater diagonal. In the case where the donor and acceptor properties of the substituents are strongly pronounced, the zwitterionic singlet state (corresponding to the $^1A_{1g}$ state in the D_{4h} structure, see Fig. 4.5) may turn out lower than the nonionic singlet state ($^1B_{1g}$ in the D_{4h} structure). The probability of a rhombic structure undergoing a bond alternation distortion [101] is lessened as the HOMO–LUMO energy gap increases due to the "push–pull" effect. Thus, unlike the highly unstable cyclobutadiene, whose fixation is possible only in a matrix at low temperatures, diethyl-2,4-bis(diethylamino)cyclobutadiene-1,3-dicarboxylate (**53**) (yellow crystals, mp 56°C) has a stable rhombic structure of the ring and is apt to get dimerized at temperatures not lower than 120°C [100]:

Figure 4.9   HOMO–LUMO energy gaps (ΔE) and resonance energies (RE) (in β) of donor–acceptor-substituted benzene and cyclobutadiene, calculated by HMO method [100]. RE $= E_\pi$ (ring) $- E_\pi$ (acyclic polyene). $\alpha_X = \alpha_C + h_X\beta_{CC}$, $h_{X=A} = -1.5$, $h_{X=D} = +1.5$.

HMO calculations of the resonance energy show that the impact of substituents on a molecule may prove so strong that it can reliably be described as not belonging to the antiaromatic class (Fig. 4.9). In contrast, benzene undergoes destabilization upon such substitution (Fig. 4.9) [100]. CNDO/2 calculations indicate that the stabilization with respect to a dimerization can be achieved in perfluoro- and perbromo-substituted cyclobutadienes (54) [102].

Another approach to stabilization connected with the introduction of heteroatoms into a ring [97] will be discussed in Chapter 5.

4.1.1.4 *Dications and Dianions of Benzene and Cyclobutadiene*   Following the Hückel rule (Chapter 3), the dication (55) and dianion (56) of benzene should be assigned to antiaromatic molecules, while the dication (57) and the dianion (58) of cyclobutadiene are to be classified as aromatic. Such an assignment is consistent with the RE values (Table 2.1). The quite large endothermicity of the formation of the salt (56) from benzene and sodium (+96 kcal/mol, approximation from experimental data for polyacene dianion salts) is attributed to its antiaromaticity [103]:

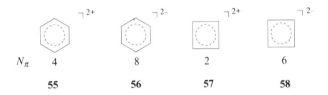

N_π	4	8	2	6
	55	**56**	**57**	**58**

The two-electron oxidation of benzene, for example, through the action of SbF_5/SO_2ClF, failed to produce the unsubstituted antiaromatic dication (55) in solution (for review see [104, 105]), which is additional evidence for the stability of benzene against oxidation. This dication could, however, be obtained in the gas phase by electron-impact ionization[106]. The perchlorobenzene dication (59) was isolated in solution upon treatment of perchlorobenzene at room temperature with a viscous solution of SbF_5 saturated with Cl_2 [107]. The deep-purple solution of the radical cation is cooled to ≤ 77 K and subjected to photolysis ($\lambda \geq 310$ nm), which leads to the loss of the second electron to form **59**:

According to ESR studies, the ground state of **59** is triplet ($^3A_{2g}$ state of D_{6h} structure) [107]. MINDO/3 calculations indicate the chair conformation (60) of the triplet state to be more stable than **55** and **59**—by 7.8 and 2 kcal/mol, respectively [108].

TABLE 4.5 **Relative Energies for Valence Isomers of Dication $(CH)_6^{2+}$ (in kcal/mol)**

Structure	3-21G [109]	MNDO/3 [108]	MNDO [109]	Ab Initio MP3/6-31G*//HF/6-31G* +ZPE(HF/6-31G*)[110]^a
55 ($^3A_{2g}$)	0	0	0	0
60	—	− 7.8	—	
61	1.2	1.0	—	—
62	—	− 10.3	7.2	6.6
68	11.3	− 10.9	—	− 10.0

^{a}At this level, the fulvenyl dication (**71**) is 13.9 kcal/mol higher in energy than **68** [110].

According to the structural criteria of aromaticity (Section 2.3), the high-symmetry D_{6h} structure of the singlet state of the antiaromatic dication $(CH)_6^{2+}$ must be unstable to distortion into a lower-symmetry structure with bond length alternation. Indeed, the singlet $^1E_{2g}$ state of D_{6h} structures **55** and **59** is subject to the first-order Jahn-Teller effect. Both MINDO/3 [108] and MNDO [109] calculations show that the planar structure **61** of the singlet state undergoes a Jahn-Teller distortion. The energetically preferred chair like C_{2h} structures **60** and **62** possess a higher energy relative to structure **55** of the triplet state (Table 4.5). At MP3/6-31G*//HF/6-31G* + ZPE(HF/6-31G*) **62** (C_{2h}) is 6.6 kcal/mol higher in energy than **55** [110].

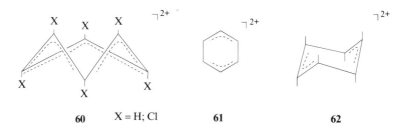

60 X = H; Cl   **61**   **62**

The MNDO calculations on the periodobenzene dication $(CI)_6^{2+}$ equally bear witness to its triplet ground state [111].

The interest in the dication **55** and in some other antiaromatic species is explained, in part, by the possibility of obtaining stable molecules with low reactivity whose ground state is triplet, which may serve as components of ferromagnetic organic compounds [112, 113]. The hexakis (dimethylamino)benzene dication (**63**) had been assumed as a possible representative of such a stable system [111], which was obtained as the bis(triodate) salt (**65**) in the reaction of **64** with iodine in acetonitrile/ether [114]. According to X-ray data, structure **65** has a twisted benzene ring with unequal bond lengths and approximate D_2 molecular symmetry [114] rather than a planar or chair ring conformation characteristic of the triplet state of the $(CR)_6^{2+}$ structures.

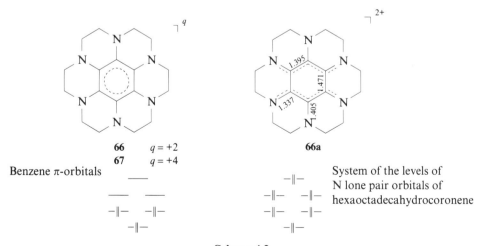

Another candidate for this role had been the hexaazaoctadecahydro-coronene dication (**66**), which has a triplet ground state in certain frozen solutions [115]. However, solid-state magnetic susceptibility and ESR measurements have shown that in the solid state the dication **66** possesses a singlet ground state [112, 116], while the X-ray analysis of some of its salts indicates a distorted structure (**66a**) of C_{2h} symmetry [112] that corresponds to coupled cyanine fragments (bond lengths in angstroms are given according to data of the single-crystal X-ray diffraction study [112]).

Since the π-orbitals of **66** stem from the π-orbitals of benzene (Scheme 4.2, left) and the orbitals of N lone pairs (orbital levels, Scheme 4.2, right), one might assume that, after the removal of four electrons and remixing of the orbitals, the tetracation **67** would acquire the electronic configuration e^2 and the triplet ground state.

However, the tetracation **67**, even though it has, according to X-ray data on its salt, a structure that is only slightly distorted from the idealized D_{3d} symmetry, does not possess a triplet ground state, as has been demonstrated by the magnetic susceptibility measurements [112, 117].

Unlike benzene, its dication, retaining the structure of the six-membered ring, is not the most stable valence isomer of $(CH)_6^{2+}$. According to MINDO/3 calculations [108], structures **55** and **68**, similar to their hexamethyl derivatives **69** and **70**, have nearly equal energies. Dication **70**, first observed by Hogeveen and Kwant [118], possesses nonclassical pyramidal structure, as has been demonstrated by ^{1}H and ^{13}C NMR studies [118] and by Saunders' deuterium-induced isotopic perturbation method [119] (see also Chapter 9).

Ab initio calculations [110] show that the most stable $C_6H_6^{2+}$ C_{5v} structure (**68**) is the global $C_6H_6^{2+}$ minimum. The cyclobutadiene dication that, unlike the benzene dication, obeys the $(4n + 2)$ rule, is , accordingly, to be assigned to aromatic species. An unsubstituted dication of cyclobutadiene has not been obtained so far; however, a number of its substituted dications (**72**) were generated and then characterized by NMR spectra; for review see [4, 104, 105].

68   **69**   **70**

71   **72**   **73**

$R_1 = R_2 = CH_3; R_1 = R_2 = C_6H_5;$
$R_1 = F, R_2 = C_6H_5; R_1 = H, R_2 = C_6H_5$

Quite surprising have been the results of *ab initio* calculations on **72** ($R_1 = R_2 = H$) [120–122]. It has turned out that this dication possesses a nonplanar D_{2d} structure (**73**), rather than the D_{4h} one as might be expected for an aromatic species [123]. According to 6-31G* calculations [122], the planar D_{4h}

structure is a transition state for the ring inversion of the D_{2d} structure (**73**) with the activation barrier of 9.6 kcal/mol (MP4SDTQ/6-31G*//6-31G*) [122]. A convincing proof of the D_{2d} structure of the cyclobutadiene dication has been supplied by the direct comparison of the IGLO-calculated ^{13}C chemical shifts with the experimental data for (CMe)$_4^{2+}$ [122]. In the case of the D_{2d} structure for the ring and methyl carbons, the δ ^{13}C values (IGLO, DZ) are 209 and 18.7 ppm and for the D_{4h} structure they are 263 and 25.2 ppm, respectively. The experimental values of the chemical shifts rule out unequivocally the D_{4h} structure.

Thus the cyclobutadiene dication that obeys the $(4n + 2)$ rule has, nevertheless, the nonplanar structure **73**, similar to the benzene dication **60**, belonging to the $4n$ π-electron species (see calculation results in [108, 109]). Does this situation result from the absence of the π-stabilization in (CH)$_4^{2+}$? As has been pointed out in Chapter 3, nonplanar distortions of the geometry of conjugated cyclic molecules may facilitate interaction between the hybrid π-orbitals (POAV2). Note that in the $D_{4h} - D_{2d}$ transition, the energies of the MOs that correspond to the e_g π-MOs in the (CH)$_4^{2+}$ D_{4h} structure get lower [120] and the CC bond lengths are shortened. In the D_{2d} structure **73**, stabilizing 1,3-and 1,2-interactions are operative, and the π-system in **73**, as it were, strives to achieve the three-dimensional aromaticity (see [119] and Chapter 9). An illustrative example of such a situation is given by the 1,3-dehydroadamantane-5,7-diyl dication (see species **30** in Chapter 3).

The stabilization of a planar structure is possible, according to the *ab initio* calculations [120], in the perfluorocyclobutadiene dication (CF)$_4^{2+}$. Unlike cyclobutadiene (**1**), the cyclobutadiene dication (**73**) is more stable than the corresponding C$_4$H$_4^{2+}$ isomers, **74** and **75** (cf. Tables 4.4 and 4.6).

74   **75**

The assignment following the Hückel rule, of the benzene dianion (**56**) to antiaromatic or, at least, nonaromatic species has been confirmed by RE calcu-

TABLE 4.6   Relative Energies (in kcal/mol) of C$_4$H$_4^{2+}$ Isomers [121]

Structures	6-31*//4-31G	MP2/6-31G* (Estimate)[a]
73	0	0
72	7.5	9.1[b]
74	6.4	9.4
75	14.2	19.7

[a]These values were obtained by adding the MP2/4-31G corrections to the 6-31G*//4-31G values.
[b]According to MP4SDTQ/6-31G* calculations [122], 9.6 kcal/mol; in the D_{2h} structure the puckering angle is 42.6°; for (CMe)$_4^{2+}$ $\Delta E(D_{4h} - D_{2d}) = 5.0$ kcal/mol (HF/6-31G*), inclusion of the electron correlation increases this value to 7 kcal/mol according to estimates in [122].

TABLE 4.7 Relative Energies (in kcal/mol) for Benzene Dianion Structures (56), (76), and (77). Calculated by MNDO [124] and *Ab Initio* [125, 126] Methods

Structure	MNDO[124]	Ab Initio		
		CI, STO-3G [125]	4-31G [125]	MP2/6-31 + G*//HF/6-31G [126]
56 ($^3A_{2g}$)	0	0	0	0 (5)[a]
56 ($^1E_{2g}$)	19.1	19.7	20.4	—
76	12.7	10.8	9.2	3.6 (4)
77	11.5	5.9	8.3	4.6 (4)
76a	—	—	—	− 3.7 (0)
76b	—	—	—	− 4.1 (0)

[a]Number of imaginary frequencies (HF/6-31G*, [126]) is given in parenthesis.

lations (Table 2.1) as well as by the calculated value of RCI = 1.31 [55]. MNDO [124] and *ab initio* [125] calculations alike show the ground state of **56** to be the $^3A_{2g}$ triplet: the $^1E_{2g}$ singlet undergoes the first-order Jahn-Teller effect. However, at HF/6-31G* the D_{6h} triplet of **56** has 5 (!) imaginary frequencies [126]. The planar quinoid (**76**) and antiquinoid (**77**) singlet structures distorted from the D_{6h} symmetry are not minima as well (Table 4.7). The nonplanar C_{2h} and C_{2v} singlet structures **76a** and **76b**, are 3.7 and 4.1 kcal/mol (MP2/6-31+G*//HF/6-31G*) lower in energy than **56** and correspond to minima (the HF/6-31G* bond lengths are given):

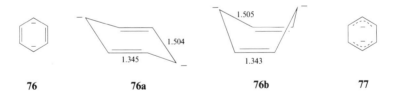

The CC bond lengths in **56** ($^3A_{2g}$) are 1.434 Å [126].

The presence of counter ions may appreciably affect the geometry of the hydrocarbon fragment, giving rise, for example, to nonplanar distortions [127]. Indeed, the X-ray data on the bis [(tetrahydrofuran)lithium(I)]hexakis (trimethylsilyl) benzenide (**78**) [128] show that the six-membered ring in **78** is significantly folded. Both the lithium cations are located on the same side of the ring. *Ab initio* calculations of $C_6H_6Li_2$ [126] indicate the nonplanar C_{2v} structure (**78a**), which resembles the X-ray structure (**78**).

According to RE calculations (Table 2.1) and judging from the value of RCI = 1.51 [55], the cyclobutadiene dianion (**58**) should be regarded as an aromatic species. However, the unsubstituted dianion (**58**) has not yet been isolated and the available evidence concerning its formation indicates its very high

78   78a

reactivity [129]. Only its tetraphenyl derivative (**79**) is known [130]. Measurements of pK_a for **80** suggest that the dianion formed from **80** is unstable [131].

79   80   81

The instability of **58** follows, as has been shown by MNDO [132] and *ab initio* [133] calculations, from the Coulombic repulsions in the four-membered ring. *Ab initio* calculation of the π-RE for **58** that has taken into account the difference between the energies of **58** and 1,3-butadiene dianion (**81**) corrected for the presence of two extra hydrogens in the latter and for strain energy in the four-membered ring in the former, has yielded the value of – 40 kcal/mol at HF/6-31G) [133]. MINDO/3 [134] and *ab initio* [135] calculations show that the planar D_{4h} structure (**58**) is not a minimum. Structures **82** and **83** have lower energies [134].

82   58   83

As is apparent from *ab initio* calculations [135], the cyclobutadiene dianion has the C_s structure (**84**) in which the negative charge is delocalized at the allylic anion fragment and localized at the C4 atom (bond lengths for **84** and **85** are given as calculated at HF/6-31G*).

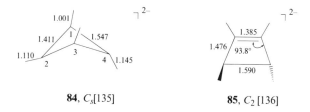

84, C_s[135]   **85**, C_2 [136]

In **84** the dihedral angle made by the $C_2C_1C_4$ and $C_2C_3C_4$ planes is 167.6°. Subsequent 6–31G* calculations [136] have shown, however, that the C_s structure (**84**) corresponds to a saddle point on the potential energy surface of $C_4H_4^{2-}$ (Scheme 4.3). The geometry optimization, when the C_s constraint is lifted, leads to the C_2 structure (**85**), which is the only minimum on the PES from among the four-membered ring structures $(CH)_4^{2-}$.

Thus, for the small, four-membered ring, the Coulombic interactions prove much stronger than the effects of the aromatic stabilization. As will be shown in the following section, for the cyclooctatetraene dianion $(CH)_8^{2-}$ (eight-membered ring), the Hückel rule is fulfilled.

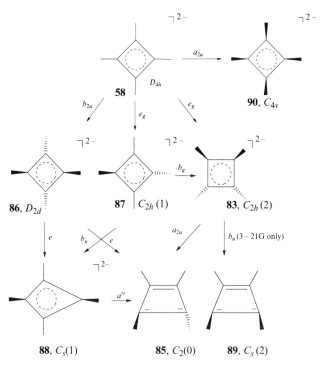

Scheme 4.3 The number of the 6-31G* calculated imaginary frequencies are shown in brackets [136]. The C_{4v} structure (**90**) has an open-shell π-electron configuration $((a_1)^2 (b_1)^2 (e)^2)$ [134].

4.1.2   Cyclooctatetraene and [10]Annulene

4.1.2.1 Structures of the Ground and Lowest Excited States   With the growing ring size the aromaticity (antiaromaticity) effects are diminished, as follows, for example, from the REPE values (Table 2.1). RE values calculated for planar structures of [n]annulenes using the scheme of the homodesmotic reactions [131] (see Section 2.2.9) as well as the ratios between REs for [n]annulenes and that for benzene indicate that planar cyclooctatetraene (COT) has a very small negative resonance energy (Table 4.8) and will hardly exhibit any substantial manifestations of antiaromatic destabilization.

Indeed, both COT synthesized by Willstätter in 1911 and its derivatives are, in contrast to cyclobutadiene, stable substances studied in considerable detail [138, 139]. There are fairly reliable experimental data on their geometry, the type of the electron ground state, and characteristic chemical properties. In its ground state the COT molecule possesses a D_{2d} tub structure (**91**) with alternating single and double bonds.

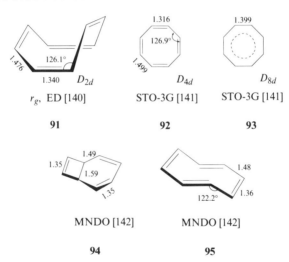

TABLE 4.8   *Ab Initio* Resonance Energies for [n]Annulenes Calculated by Means of Homodesmotic Reaction Schemes as Well as the Ratio Between REs Calculated by Various Schemes for [n]Annulenes (Section 2.2) and Those for Benzene [137]

[n]Annulene	HSE[a], kcal/mol	Ratio to the Benzene Value			
		HSE	DRE	TRE	HSRE
[4], D_{2h}	− 48	− 1.85			
[6], D_{6h}	27	1	1	1	1
[8], D_{4h}	− 2.5	− 0.09	− 0.16	− 2.18	− 1.23
[10], D_{10h}	26	0.96	0.45	0.58	0.67
[18], D_{3h}	12	0.15	0.15	0.33	0.56

[a]MP4/6-31G**//6-31G.

No antiaromatic destabilization occurs in the nonplanar D_{2d} structure of COT, and this structure is the most stable among the chemically reasonable $(CH)_8$ isomers (Table 4.9)[2].

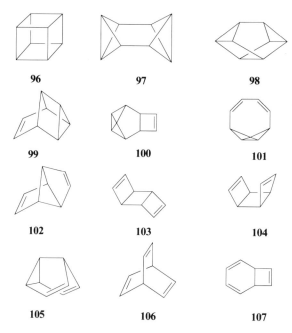

At first sight, the treatment of the nonplanarity of COT as the corollary of antiaromaticity appears a reasonable enough approach. However, as has already been pointed out in Section 2.3.4, the relationship between the trend toward nonplanar distortions and antiaromaticity is by no means straightforward. For example, the determinative role in the distortion of planarity may belong to strains in the σ-system. Thus, in the case of COT, the nonplanarity of the carbon ring is caused by angular strains and related phenomena characteristic of the planar structure (their contribution to the enthalpy of the $D_{4h} \rightarrow D_{2h}$ transition may be as high as 85%) [143–145]. The π-electron energy of COT, on the contrary, is minimal precisely for the planar ring geometry and any nonplanar distortion leads to its increase [145].

According to calculations (Table 4.9), the $(CH)_8$ D_{2d} structure is 10.6 kcal/mol more stable than the D_{4h} structure [149]. The nonplanar chair structure (**95**), alternative to the D_{2d} tub structure (**91**), is 55.2 kcal/mol (MNDO; 59.3 MINDO/3 [142]) higher in energy than **91**. The activation barrier for the **95**→**94** rearrangement is rather insignificant (15.3 kcal/mol, MNDO; 12.9 kcal/mol, MINDO/3) [142]. Structure **94** rearranges readily into structure **91**, which is the

---

[2]Twenty-one valence isomers of $(CH)_8$ (including four pairs of stereoisomers) are conceivable [150].

TABLE 4.9   Relative Energies (in kcal/mol) of $(CH)_8$ Isomers Calculated by Semiempirical Methods (MINDO/3, MNDO, and AM1) and Experimental Values of ΔH_f

Isomers	MINDO/3 [142, 145]	MNDO [142, 145, 147]	MNDO 3 × 3 CI [146]	AM1 [147]	ΔH_f(exptl) Review [147]	E_{rel} (exptl)
91, D_{2h}	0^a	0^b	0^c	0^d	71.1^e	0
92, D_{4h}	1.1	7.7	8.8	13.0^f		
93, $D_{8h}(T)$	17.1	20.8	17.6			
93, $D_{8h}(S)$	17.1	20.8	17.6			
94	20.1	7.7			76.6	5.5
95	20.1	7.9				
96	83.2	42.8		87.7	148.7	77.6
97		86.7		108.9		
98		44.6		70.4		
99		42.6		66.0		
100		62.6		86.2		
101		39.8		52.2		
102	75.7	47.2		61.8		
103	63.6	46.3		74.7		
104	66.5	48.8		78.2		
105	34.7	31.6		35.6	73.6	2.5
106		11.3		3.8	72.5	1.4
107		36.5		54.5	-	-

$^a\Delta H_f$ (**91**) = 56.6 kcal/mol [142]. The MINDO/3 calculated value of ΔH_f (**91**) given in [148] and quoted in [147] requires correction: see [145].
$^b\Delta H_f$(**91**) = 56.2 kcal/mol (MNDO) [142].
$^c\Delta H_f$(**91**) = 60.0 kcal/mol (MNDO, 3 × 3 CI) [146].
$^d\Delta H_f$(**91**) = 63.5 kcal/mol (AM1) [147].
^{e}MP2/6-31G* calculations by means of the homodesmotic reaction scheme gave the following values of ΔH_f = 69.1 kcal/mol (**91**) and 148.7 (**96**) [71].
^{f}At CASSCF (8, 8)/6-31G*//CASSCF(8, 8)/3-21G + ΔZPE(CASSCF(8, 8)/3-21G), ΔE
(**91**–**92**) = 10.6 kcal/mol and ΔE (**93**–**92**) = 4.1 kcal/mol [149].

final product of the transformation of **95**. If, however, the **95** → **94** barrier could be raised by appropriate substituents, one might expect a substituted **95** to be stable [142].

The flattening of the COT ring may be achieved through annulation via CC single bonds by small rings or suitable rigid bicyclic frameworks [151, 152], as is indicated by the MINDO/3 calculations [153–155] and the experimental data [152, 156–158] (Fig. 4.10). The planarity of the COT ring structure can also be affected through fusion of four fluorinated cyclobutane rings (X-ray analysis [158]). An eight-membered ring is flattened in dehydro[8]annulenes annulated by benzene rings; for a review see [159].

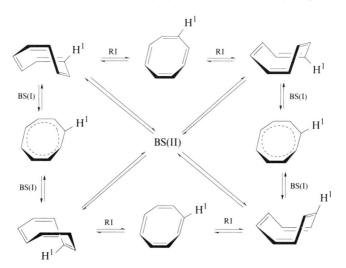

Figure 4.10 (a) Flattening of the COT ring with the decreasing size of the 1,2-annulated ring, according to MM2 calculations [152]. (b) Flattening of the COT ring in several COT derivatives according to X-ray date [157, 158].

The D_{2d} structure (**91**) is subject to transformations of three principal types. Two of these are the ring inversion (RI) and the bond shift (BS) [147, 160]:

The RI process develops via the intermediate D_{4h} structure. The activation barrier for the RI of COT derivatives amounts to ~10 kcal/mol; for a review see [144, 160]. The barrier for the BS reaction analogous to the automerization of the preceding [4n]annulene (i.e., cyclobutadiene) is higher by roughly 4 kcal/mol. For the BS a greater amount of the activation energy, compared to the RI process, can be attributed to the antiaromatic destabilization of the D_{8h}

structure (**93**). Hence the energy of this destabilization may be estimated from the following equation [161]:

$$RE = \Delta H^{\ddagger}_{(RI)} - \Delta H^{\ddagger}_{(BS)} \tag{4.10}$$

For 1,3-di-*tert*-butylcyclooctatetraene this value is 23.3–19.9 = 3.4 kcal/mol [141], which is consistent with other calculations of the RE (Tables 2.1 and 4.8). After the introduction of alkyl substituents, the difference $\Delta G^{\ddagger}(BS) - \Delta G^{\ddagger}(RI)$ is reduced [144] and for 1,2,3,4-tetramethylcyclooctateraene these values become equal. For the structure of 1,3-octamethylene-bridged cyclooctatetraene (**108**), the bond shift is the kinetically most accessible dynamic process ($\Delta G^{\ddagger}(BS) = 9.3$ kcal/mol at 202 K, while $\Delta G^{\ddagger}(RI) = 16.0$ kcal/mol at 334K) [162].

108

The transition state of the bond shifting in **108** is probably a nonplanar flattened saddle-like structure without the antiaromatic destabilization characteristic of the D_{8h} COT structure (**93**) [151, 163].

As has been shown by the complete line shape analysis of the dynamic NMR spectra of unsubstituted COT dissolved in nematic solvents, for the bond shift $\Delta H^{\ddagger} = 10.0$ kcal/mol and $\Delta S^{\ddagger} = -9.7$ eu [164]. According to MNDO 3 × 3 CI calculations [146], the difference between the energies of the D_{8h} structure (**93**) and the D_{4h} structure (**92**) is 8.8 kcal/mol. The calculation of the tunneling rate constant for the automerization of **92** via **93** shows that, even at 398 K, k(tunneling) is 10^3 times as great as the classical rate constant.

The CASSCF(8,8)/6-31G*//CASSCF(8,8)/3-21G calculations (in these calculations of COT, CASSCF(8,8)—all possible occupancies of the eight π-orbitals by the eight electrons) [149] show the RI barrier in COT to be 10.6 kcal/mol. The energy difference between the RI barrier and the BS barrier is 4.1 kcal/mol, in excellent agreement with the experimental data for monosubstituted derivatives of COT [160, 161].

The third transformation of **91** may be the valence isomerization into bicyclo[4.2.0]octa-2,4,7-triene (**94**). The equilibrium of the monocyclic form (**91**) ⇌ bicyclic form (**94**) process is shifted toward the first of these forms, which is more stable (Table 4.9). Experimental thermodynamic parameters for this equilibrium obtained with the use of a high-temperature trapping technique are

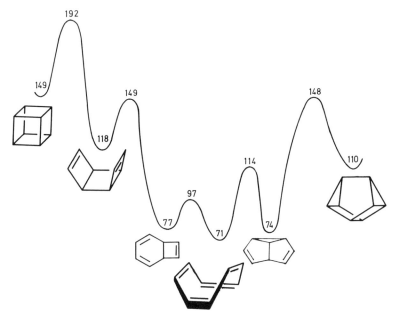

Figure 4.11   Energy profiles for selected rearrangements of $(CH)_8$ hydrocarbons. The activation energies (in kcal/mol) were estimated from kinetic studies. Experimental heats of formation for structures **91**, **94**, and **96** are given. ΔH_f of **104** was calculated using the additive scheme; ΔH_f of **98** was obtained by means of molecular mechanics (MMP1) calculations. (Adapted from [147].)

as follows: $\Delta H° = 5.5 \pm 0.6$ kcal/mol and $\Delta S° = -4.3 \pm 0.7$ eu [165]. The experimental value of the activation barrier of the **94 → 91** rearrangement is $\Delta G^{\ddagger} = 18.7$ kcal/mol [166] (cf. 22.7 kcal/mol, MINDO/3 [142]). The energy profiles for the rearrangements of the $(CH)_8$ valence isomers are shown in Fig. 4.11. The triplet state of COT has the D_{8h} structure ($R(CC) = 1.398$ Å, STO-3G [141]) whose energy is higher than that of the D_{4h} (**92**) and D_{2d} (**91**) singlet state structures (Table 4.9). As has been pointed out in Section 4.1.1, in the case of the lowest $\pi\pi^*$ state of cyclic conjugated hydrocarbons, the Hückel rule is reversed [62]. Thus for the lowest triplet state of COT, RCI = 1.62, whereas for the D_{2d} ground-state structure (**91**), it is 1.29 [55] (SINDO1). The energy of the aromatic stabilization is in the former case 17.7 kcal/mol [62].

For the next member of the [n] annulene series, [10]annulene, the planar configurations **109** and **110** as well as the nonplanar ones **111–114**, are conceivable:

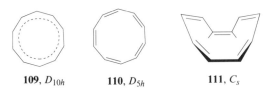

109, D_{10h} **110**, D_{5h} **111**, C_s

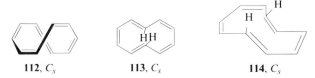

112, C_s 113, C_s 114, C_s

Following the Hückel rule, [10]annulene should be classified as aromatic, and the structural criteria of aromaticity (Section 2.3) suggest that the lowest energy will be possessed by the planar D_{10h} structure with no alternation of bond lengths. However, MINDO/2 [167] and *ab initio* [168] calculations indicate that the most stable are the nonplanar structures, **111** and **112** (Table 4.10).

Even though the HHSE value for **109** is rather high (Table 4.8), numerous attempts to synthesize [10]annulene had been unsuccessful until two decades ago when it was at last isolated. Its precursors were *trans-* and *cis*-9,10-dihydronaphthalenes, **115** and **116**. The irradiation of **115** at –70°C produced all-*cis*-[10]annulene (**111**), which at – 10°C underwent thermal rearrangement to **116** [171, 172]. The irradiation of **116** at – 60°C gives rise to **112**, which at – 25°C rearranges to **115**. Upon irradiation of **116** at – 60°C, a mixture of isomers is formed. Following the separation of the tetracyclic isomer **117**, the isomers **111** and **112** were separated as crystalline products by chromatography on alumina at – 80°C [171].

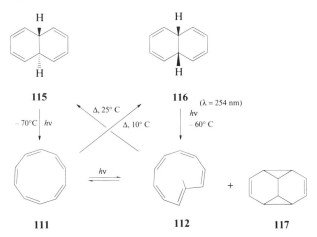

TABLE 4.10   *Ab Initio* Calculated Relative Energies of Some $(CH)_{10}$ Isomers

Structures	109	110	111	112	113	114
4-31G//STO-2G[168]	30.7[a]	32.7	2.1	0	2.8	11.5

[a]At MP2/6-31G//HF/6-31G, structure **109** ($R(CC)$ = 1.329 Å) is 1.2 kcal/mol more stable than **110** ($R(CC)$ = 1.376 Å and 1.408 Å) [169]. MP4/6-31G* calculation at the standard geometries (for **109** $R(CC)$ = 1.40 Å, for **110** $R(CC)$ = 1.34 and 1.46 Å) gives the energy difference as 11 kcal/mol [170]. Geometry optimization at MP2/6-31G leads to D_{10h} structure **109** ($R(CC)$ = 1.417 Å) [169].

Thus two isomers of [10]annulene have been isolated, namely, the all-*cis* (**111**) and mono-*trans* (**112**) structures, which, similar to the D_{2d} structure of COT, are nonplanar. They are stable at low temperature only; their NMR spectra resemble those of polyenes and there are no signs of aromaticity. Pseudorotations (4.11) and (4.12) are characteristic of structures **111** and **112** respectively.

(4.11)

111   **111a**

(4.12)

112   **112a**

The instability of structure **111** with respect to cyclization into **116** is explained by effects of the bond angle strain in the all-*cis*-isomer **111**, where the angles differ considerably from 120°—a value that is characteristic of sp^2 hybridized carbon atoms (e.g., the internal angle is 144°). In the di-*trans* structure **113**, the 1,6-hydrogen nonbonding interactions prove decisive in determining instability. The mono-*trans* isomer **112**, in which both effects, albeit on a smaller scale, are operative, is also unstable.

The ease with which **111** and **112** undergo thermal rearrangements[3] is in stark contrast to the stability of the preceding member of the [4n + 2] annulene series, that is, benzene (Section 4.1.1). For the thermally allowed isomerization of **111** into the bicyclic structure **116** that lessens the strain, $\Delta H^{\neq} = 20$ kcal/mol, and for the analogous **112** → **115** isomerization, $\Delta H^{\neq} = 17$ kcal/mol [172].

The replacement in the di-*trans* structure of two spatially close hydrogen atoms by a methylene bridge stops nonbonding interactions in **118**. Molecule **118** (1,6-methano [10]annulene) is a representative of the bridged annulenes; it has been studied in a detailed fashion both experimentally and theoretically (e.g., see [1, 2, 5, 170, 173, 174]). Pertinent results of these studies indicate the aromaticity of **118**: its TREPE is 0.029β [175] and its SRTRE = 1.04 eV [176]. Specific features of the geometry of **118** were dealt with in Chapter 3. According to MP2/6-31G calculations [170], structure **118** is 1.5 kcal/mol more stable than structure **119**, which is characterized by bond alternation, and it is more stable than the norcaradienic isomer **120** by 15.6 kcal/mol.

---

[3] General schemes of the thermal and photochemical interconversions of $(CH)_{10}$ valence isomers are given in [150]

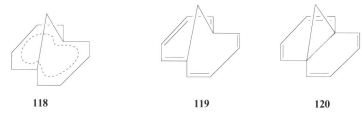

| 118 | 119 | 120 |

Thus, in the planar structure **109**, the strain effects overshadow the aromatic stabilization. However, the aromatic character of the planar structure of [10] annulene may still be revealed through comparison between the energies of a structure without bond length alternation (**109**) and one with alternation (**110**). *Ab initio* calculations with the electron correlation taken into account [136, 169, 170] indicate greater stability of the former (Table 4.10).

According to NNDO calculations [62], the lowest triplet state of [10]annulene is weakly antiaromatic (the energy of antiaromatic destabilization is a mere − 0.8 kcal/mol).

4.1.2.2   COT's Dication and Dianion   The COT dianion (**121**) may be obtained in various ways, for example, by treatment of solutions of COT in ether or tetrahydrofuran with alkali metals (e.g., see [1, 2, 177]). Judging from experimental [178] and calculational [141] data, this 10 π-electron species has planar D_{8h} structure (R(CC) = 1.399 Å, STO-3G [141]). The detailed study of the physical and chemical properties of **121** has enabled it to be assigned to aromatic systems [177]. It will be recalled that the cyclobutadiene dianion possesses a nonplanar structure.

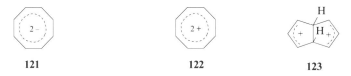

| 121 | 122 | 123 |

An unsubstituted COT dianion (**122**) is not known so far; however, its methyl and phenyl derivatives have been obtained by way of two-electron oxidation of the corresponding cyclooctatetraenes in SbF$_5$/SO$_2$ClF at − 78°C [179] (for reviews see [105, 106]).

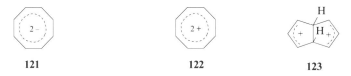

| 124 | 125 |

Experimental data obtained using the Spiesecke–Schneider correlation [180] and comparing the total ^{13}C deshielding with that of suitable model compounds (see Section 2.4.3) show that **122** has a planar structure (6 π-electrons, $R(CC) = 1.415$ Å at HF/STO-3G [123, 141]). According to *ab initio* calculations [123], the COT dication is unstable with respect to the following fragmentation reactions (HF/STO-3G, ΔE are given in kcal/mol):

$$C_6H_6^+ + C_2H_2^+ \longleftarrow C_8H_8^{2+} \longrightarrow C_7H_7^+ + CH^+$$

$$\Delta E = 30 \qquad\qquad\qquad\qquad \Delta E = 31$$

$$C_6H_5^+ + C_2H_3^+ \qquad\qquad 2C_3H_3^+ + C_2H_2$$

$$\Delta E = 29 \qquad\qquad\qquad \Delta E = 14$$

Upon warming **124** up to $-20°C$, electrocyclic ring closure occurs to give *cis*-2,3a,5,6a-tetramethyldihydropentalene dication (**125**) [179]. MNDO calculations indicate [181] that **122** and **123** are very close in energy (ΔH_f is 537.7 for the former and 537.6 kcal/mol for the latter). The loss of aromaticity on account of the **122** → **123** rearrangement is offset by the diminished angular strain in **123** and the hyperconjugative stabilization resulting from the interaction between the allyl system and the carbon–carbon sigma framework [181].

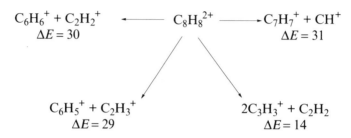

126

For the dianion (**126**) this interaction proves destabilizing; thus the COT dianion (**121**) is more stable than **126** by 27.7 kcal/mol (MNDO, [181]). This is consistent with experimental data, which bear witness to the ease with which the ring opening occurs in the bicyclic structure (**126**); the final product of reaction is **121** [181]:

$$(4.13)$$

According to MINDO/3 calculations [182], the 6 π-electron dication of COT (**122**) is the most stable among the $(CH)_8^{2+}$ isomers (**123**–**123e**) (in parentheses MINDO/3 calculated heats of formation are given in kcal/mol; ΔH_f (**122**) = 506 kcal/mol:

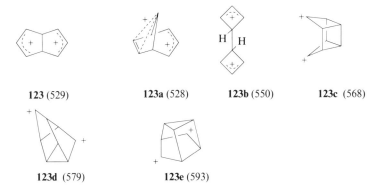

123 (529)   123a (528)   123b (550)   123c (568)

123d (579)   123e (593)

4.1.3   [18]Annulene

As the size of the ring grows, the aromaticity effects become weaker (Table 4.8) so the annulenes that follow the [10]annulene species considered in the preceding section, namely, the [14] and [18] species, should apparently, have less pronounced manifestations of aromaticity.

In [14] annulene the four inner hydrogens are repulsed from one another and thereby move away from the carbon ring plane so that the structure becomes nonplanar. The above conclusion is based on an X-ray study [183] as well as on CNDO/2 [184], SINDO 1 [185], MINDO/2 [186], and AM1 [187] calculations. On the other hand, [18]annulene, unlike the [10] species, is almost strainless and, judging from experimental data [188, 189], has a planar structure. In other words, in [18]annulene, there are no effects that might obscure the energetic and structural manifestations of aromaticity, as indeed occurs in the case of the [10] and [14]annulenes.

Thus, among the annulenes following the [10] species, [18]annulene is of particular interest. Its comparison with benzene makes it possible to examine the changes in aromaticity, unobscured by strain effects, that result from the growth of the ring size [177]. Furthermore, the comparison of [18]annulene with [10]annulene, together with the results of the preceding section, can show more clearly how the strain affects the properties of the latter. So, for the reasons stated, [18]annulene will be examined next.

Like [10]annulene, [18]annulene is a $(4n + 2)$ π-electron system, but, unlike it, [18]annulene is much more stable (incidentally, it was synthesized a decade

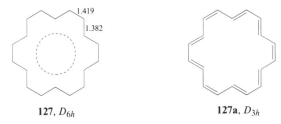

127, D_{6h} 127a, D_{3h}

earlier [190]). According to X-ray data [188], it has a planar structure (**127**) of D_{6h} symmetry in which "inner" (1.382 Å) and "outer" (1.419 Å) bonds may be singled out. ^{1}H and ^{13}C NMR spectra show diatropicity of this molecule (Table 2.9) and again the D_{6h} structure. (For a historical review of studies on [18]annulene see [189]).

[18]Annulene possesses conformational mobility associated with interchange of positions of the inner and outer hydrogens during interconversion of three equivalent forms [160]:

On the assumption that the aromatic stabilization is lost in the transition state of this process, this stabilization of [18]annulene will have been not more than 16 kcal/mol [160]. This value is close to HSE (Table 4.8)

The aromaticity of the D_{6h} structure (**127**) is also evidenced by the quite high value of the $HOMA_W$ index (see Section 2.3.2), equal to 0.985 (CNDO/2) [184] (for the geometry obtained by the X-ray study in [188], $HOMA_W$ is high too, namely, 0.978).

The RCI index (see Section 2.3.2) of 1.72 (SINDO 1) is equally high for **127** [185]. However, for the D_{3h} structure (**127a**) this index is 1.35, thus relegating it to the class of weakly aromatic or even nonaromatic molecules. Thus the D_{6h} structure of aromatic [18]annulene must be more stable than the D_{3h} structure with bond alternation, as is indeed corroborated by experimental evidence [188, 189].

At the same time, both semiempirical (MINDO/3, MNDO) [191, 192] and *ab initio* calculations without inclusion of the electron correlation [137,170,193] indicate greater stability of the D_{3h} structure (**127a**) with bond alternation compared to the D_{6h} structure (**127**). As has been demonstrated in [186], agreement with the experimental data showing greater stability of the D_{6h} structure of [18]annulene can be achieved only when the energy of the electron correlation is appropriately taken into account, the way it has been done in the case of [10] annulene. The HF solution for [18]annulene is triplet unstable (see Section 2.3) so the UHF approximation should be used. Indeed, the UMNDO calculation shows the preference of the D_{6h} structure [192] in accordance with experiment. The importance of the inclusion of electron correlation for the problem in hand is confirmed by the MNDOC calculations [194] as well as the SINDO1 calculations with the CI (60 configurations) and the Langhoff–Davidson correction for quadruple excitations taken into account. They show greater stability of the D_{6h} compared with the D_{3h} structure (by 1.1 and 11.5 kcal/mol, respectively).

When a more general formula for estimating the contributions from quadruple and higher excitations is applied in the MNDOC calculations [195], the D_{6h} structure (**127**) is found to have 4.8 kcal/mol lower energy than that of **127a**.

4.2 MONOCYCLIC AROMATIC AND ANTIAROMATIC IONS

4.2.1 Cyclopropenyl Cation and Cyclopropenide Anion

The cyclopropenyl cation (**128**) and cyclopropenide anion (**129**) are 2 π- and 4 π-electron species, respectively. Hence the former should be classified as aromatic while the latter as antiaromatic. The RE values support this assignment (see Table 2.1). The unsubstituted cation (**128**) was isolated in the late 1960s [1], and it has recently been detected in the tail of Halley's comet [196]. In contrast to **128**, anion **129** is known only in the form of derivatives formed in solution as a result of electrochemical reduction [197] or by reaction of fluorodesilylation [198].

TABLE 4.11 Experimental Heats of Formation and *Ab Initio* Calculated Relative Energies (in kcal/mol) of Cyclopropenyl Cation and of Its Open-Chain Isomers

	Experiment		*Ab Initio*					
			E_{rel}^a	E_{rel}^c	E_{rel}^c	E_{rel}^d	E_{rel}^e	E_{rel}^f
Structures	ΔH_f	E_{rel}	[203]	[204]	[205]	[206]	[207]	[208]
128	255 ± 1 [209]	0	0	0	0	0	0	0
131	281 ± 2 [206]	26 ± 3	15.3	28.9[g]	35.1	31.1	25.8	27.5
132	—	—	—	117.1	—	—	117.4	—
133	325	70	55.1	72.4	69.3	—	—	69.8
134	~368	113	110.0	100.4				—

[a]4-31G/STO-3G.
[b]MP4/6-31G**//6-31G*.
[c]6-31G*//6-31G*.
[d]MP4/6-31G**//6-31G*.
[e]MP4/6-311G**//MP2/6-31G* [207].
[f]According to MP2/6-31G* calculations of **128** [207,210], $R(CC) = 1.368$ Å, $R(CH) = 1.083$ Å. The MP2 calculations (basis set ($5s2p/3s$), augmented with polarization functions, lead to $R(CC) = 1.3647$ Å, $R(CC) = 1.0753$ Å [211]; CI-SD/DZP calculations give $R(CC) = 1.3705$ Å, $R(CH) = 1.0795$ [211].
[g]27.7 kcal/mol with ZPVE correction.

The triphenylcyclopropenide anion (**130**) was detected in the gas phase using the ion-cyclotron resonance technique; it is stable for the time scale of the experiment (i.e., 1–2 s) [199].

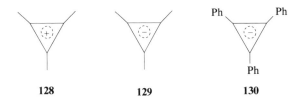

128   **129**   **130**

For the 2 π-electron $(CH)_3^+$ ion, structural criteria of aromaticity predict a structure lacking bond alternation. The X-ray data show that cation **128** has, indeed, the D_{3h} structure for which the mean ring-bond lengths vary from 1.363 to 1.384 Å, depending on the type of substituent, with the overall mean being 1.373 Å [200]. This value is in agreement with the results of *ab initio* calculations (Table 4.11). The data of this table as well as those from collisional activation mass spectra [201] indicate that structure **128** is the global minimum on the PES while its isomers **131–134** possess higher energies. In contrast, the open-chain singlet valence isomers **135–137** of the antiaromatic cyclopropenide anion (**129**) have lower energies compared to the cyclic anion itself [202] (Table 4.12).

For **138–141**, the HF/6-31G* geometry parameters are given.

Noteworthy is the fact that the CC bonds in **128** are shorter (1.373 Å) than that in benzene (see Table 2.4). The experimental value of the force constant for the CC stretching in **128** determined from the IR and Raman spectra is 7.89 mdyn/Å [213] (*ab initio* value is 7.92 [214]). This value is larger than the CC stretching constant (6.578 mdyn/Å [215]). Thus, from the well-known correla-

TABLE 4.12 Relative Energies (in kcal/mol) of the $C_3H_3^-$ Isomers According to *Ab Initio* Calculations

Isomer	MP/4-31 + G//4-31 + G [202]	MP3/6-31 ++ G [212][d]
140	0^a	0
129 ($^3A_2'$)	28.3	
138	$(55.6)^b$	
139	35.3	63.3
141	1.3	4.3
142	9.5	
135	– 5.1	
136	– 5.0	
137	0.8^c	

[a]The heat of formation of **140**, estimated from energies of isodesmic reactions, amounts to 110–120 kcal/mol [202].
[b]At HF/3-21 + G.
[c]At HF/6-31 + G*//HF/4-31 + G, E_{rel} (**137**) = – 14.3 kcal/mol [202].
[d]Including ZPVE correction.

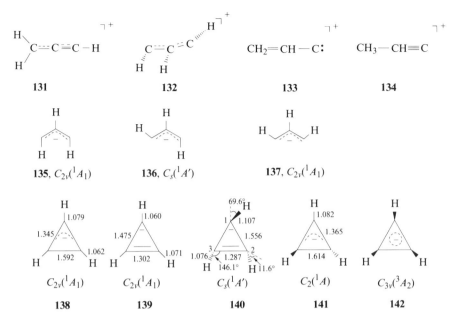

tion between the magnitude of the force constant and the bond strength, the conclusion may be drawn that the CC bond strength is greater in **128** than in benzene [213]. Moreover, the value of the TREPE is also greater in **128** than in benzene (Table 2.1).

Thus there is some substance in the assertion that **128** represent a "superaromatic" system [213]. However, it should be taken into consideration that the shortening of the CC bonds in **128** may be accounted for by primarily the effects of σ-aromaticity (Chapter 7). Indeed, we see from Table 4.13 that the ISE value

TABLE 4.13   ISE (in kcal/mol) of Some Monocyclic Conjugated Ions with ISEs of Benzene and Cyclobutadiene Given for Comparison

Isodesmic Reactions	3-21G//3-21G [217–219]	ΔE_{exptl} [220]	R^a
$(CH)_3^+ + 2CH_4 + CH_3^+ \rightarrow 2CH_3CH_2^+ + CH_2{=}CH_2$	-34	-30	1.20
$(CH)_5^+ + 4CH_4 + CH_3^+ \rightarrow CH_3CH_3 + 2CH_3CH_2^+ + 2CH_2{=}CH_2$	-5	-4	-0.40
$(CH)_6 + 6CH_4 \rightarrow 3CH_3CH_3 + 3CH_2{=}CH_2$	60	64^b	1
$(CH)_4 + 4CH_4 \rightarrow 2CH_3CH_3 + 2CH_2{=}CH_2$	-70		-1.05
$(CH)_5^- + 4CH_4 + CH_3^- \rightarrow CH_3CH_3 + 2CH_3CH_2^- + 2CH_2{=}CH_2$	87		1.27
$(CH)_7^+ + 6CH_4 + CH_3^+ \rightarrow 2CH_3CH_3 + 2CH_3CH_2^+ + 3CH_2{=}CH_2$	80	73	1.33

$^a R$ is the ISE/ISE(benzene) ratio, with ISE normalized per one π-electron. For evaluating strain energies in a three-membered ring and a five-membered ring, the ISEs of cyclopropene (-60 kcal/mol, exptl.) and cyclopentadiene (11 kcal/mol, 3-21G//3-21G [217]; 22 kcal/mol., exptl.) have been taken. Thus the corrected values are ISE($C_3H_3^+$) = 24 kcal/mol, ISE($C_5H_5^+$) = -16 kcal/mol, ISE(C_5H_5) = 76 kcal/mol. For cyclobutadiene the corrected value is ISE = -42 kcal/mol. The corrected ISE values were used in calculations of R.
^{b}Calculated from the data in [221].

(see Section 2.2.9) for the cyclopropenyl cation corrected for the strain energy even exceeds that of benzene (calculated per one π-electron).

Additional stabilization of the cyclopropenyl cation can be achieved by the introduction of such π-donor substituents as the amino and hydroxy groups—see Eq. (4.14) [205]; by contrast, F, CN, and NC groups lead to its destabilization (HF/6-31G*) [205].

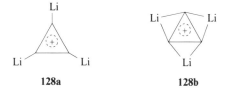

$$(4.14)$$

Here ΔE (Eq. (4.14)) = 31.8 (X = NH$_2$), 13.3 (X = OH), -10 (X = F), -24.7 (X = CN), and -12.6 kcal/mol (X = NC). With all the above substituents, the corresponding derivative of the cyclopropenyl cation is lower in energy than the γ- and α-substituted propargyl cations.

Both classical and bridged structures of trilithiocyclopropenium cation, $C_3Li_3^+$, **128a** and **128b**, are minima at MP2/6-31G* and appear to be very stable ions [216].

Li
△
Li   Li
128a

Li—△—Li
△
Li
128b

Indeed, the σ-electron donating ability of lithium may stabilize carbocations nearly as effectively as π-donation from an amino group [217]. This is confirmed by the energies of reactions (4.14a) and (4.14b) (MP2/6-31G*//HF/6-31G*) [216]:

$$C_3H_3^+\ (\mathbf{128}) + 3\ CH_3Li \rightarrow C_3Li_3^+\ (\mathbf{128b}) + 3\ CH_4\quad \Delta E = -167.2\ \text{kcal/mol}\quad (4.14a)$$

$$C_3H_3^+\ (\mathbf{128}) + 3\ CH_3NH_2 \rightarrow C_3(NH_2)_3^+ + 3\ CH_4\quad \Delta E = -97.5\ \text{kcal/mol}\quad (4.14b)$$

The energy of stabilization of **128** found from the calculated energy of the isodesmic reactions (4.15) is 70 kcal/mol while an analogous estimate for **140**— see Eq. (4.16)—comes out at -3 kcal/mol [202]. At MP2/6-31G**//MP2/6-31G* the stabilization energy of (**128**) (Eq. (4.15)) is 62.4 kcal/mol [208].

△ + △ ⟶ △ + △⁺

$$(4.15)$$

△ + △ ⟶ △ + △⁻

$$(4.16)$$

The unexpected low energy of destabilization of the cyclopropenide anion is explained by geometry distortions from the planarity, practically reducing the antiaromatic destabilization effects to nil. According to structural criteria of aromaticity (Section 2.3.3), the planar $(CH)_3^- D_{3h}$ structure of the lowest singlet state does not correspond to a minimum [146].

Ab initio calculations [202, 212, 222] showed that the ground electronic state of the cyclopropenide anion is the singlet ($^1A'$). The corresponding structure (**140**) is, as opposed to the D_{3h} structure of the aromatic cyclopropenyl cation, nonplanar with bond length alternation. The singlet $^1E'$ state of the D_{3h} structure undergoes the first-order Jahn-Teller effect, which leads to lower energy structures with C_{2v} symmetry; they correspond to the following states: $^1A_1(\cdots b_1^2 a_2^0)$ (**139**), $^1A_1(a_2^2 b_1^0)$ (**138**) and $^1B_2$(**143**) [59, 223, 224].

$$C_{2v}(^1B_2)$$

143

As a result of the second-order Jahn-Teller effect, the $^1A_1$ state acquires a lower energy than the $^1B_2$ state. Although the "allyl-type" planar structure (**138**) is more stable than the planar structure of "ethylenic type" (**139**) [223], the pyramidalization of the carbon C1 atom at which the electron pair is localized as well as the pyramidalization (but to a lesser degree) of the C2 and C3 atoms result in structure **140** of C_s symmetry becoming the ground-state structure of the cyclopropenide anion [202, 211]. Structure **140** possesses lower energy than the triplet-state structures **129**, ($^3A_2'$) and **142** ($^3A_2$) and of the nonplanar singlet-state structure **141** (Table 4.12). *Ab initio* calculations [212] show **141** to be transition-state structure for the pseudorotation of **140**:

$$\cdots \rightleftharpoons \triangle \rightleftharpoons \triangle \rightleftharpoons \triangle \rightleftharpoons \cdots \qquad (4.17)$$

The realization of this rapid pseudorotation is evidenced by the experimental data indicating a nearly statistical distribution of the ^{13}C label in triphenycyclopropene that is formed in a reaction developing via the intermediate, **130** [198]:

$$(4.18)$$

The activation barrier of reaction (4.17) is a mere 4.3 kcal/mol [212] (Table 4.12). Calculations of the classical and tunneling rates of this interconversion (k^{class} = 2.42·10^7 s^{-1}, k^{class} = 7.13·10^8 s^{-1}, 298 K, MINDO 3 × 3 Cl [146]) show that at room temperature the latter is faster than the former by one order of magnitude. The structure of the triplet **142** is equally nonplanar [202].

Since experimental studies are conducted on anions stabilized by counterions, which may appreciably affect the geometry of the hydrocarbon fragment [127], data of *ab initio* calculations on (CH)$_3$Li isomers [225] are important. It has turned out that the cyclopropenyllithium, similar to the cyclopropenide anion, has nonplanar ground singlet state structure (**144**):

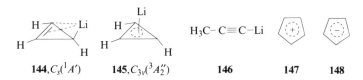

$144, C_s(^1A')$ $145, C_{3v}(^3A_2'')$ **146**   **147**   **148**

Triplet **145** is 14.3 kcal/mol (MP2/6-31G*) higher in energy than **144**.

The aromatic cation **128** is thermodynamically and kinetically stable against rearrangements into C$_3$H$_3^+$ isomers [208]. This may be illustrated by the energy profile (see Fig. 4.12) of the ring-opening rearrangement of the cyclopropenyl cation into the propargyl cation (**131**). In contrast to the (CH)$_3^+$ isomers, the open-chain (CH)$_3^-$ isomers have lower energies than that of **140**. However, one should not expect a smooth rearrangement of an analogous type for the cyclo-

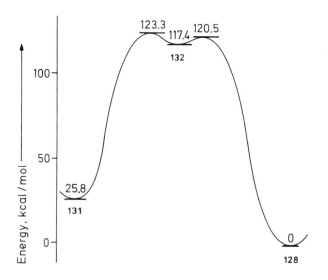

Figure 4.12   Reaction profile (MP3/6-311G**//MP2/6-31G* + ZPVE) for rearrangement of the cyclopropenyl cation (**128**) into the propargyl cation (**131**) [219].

propenide anion, since such a rearrangement is symmetry forbidden [202]. The open-chain isomers of C_3H_3Li have lower energy than the cyclic structure (**144**). For example, **146** is 62.3 kcal/mol more stable (MP2/6-31G*) than **144** [225].

4.2.2   Cyclopentadienyl Cation and Cyclopentadienide Anion

Following the Hückel rule and judging from the RE values (Table 2.1) and ISE values (Table 4.13), **147** should be classified as antiaromatic while anion **148** as aromatic. Accordingly, one may expect that the high-symmetry D_{5h} structure of $(CH)_5^+$ has an open-shell electronic configuration and **148** has the closed-shell configuration. Since, as also in the case of the cyclopropenide anion, the degenerate highest occupied e_1'' π-MOs of **147** share carbon atoms (cf. rhombic e_g π-MOs of cyclobutadiene), Hund's rule is not violated. Both the experimental data (e.g., ESR spectra of the cyclopentadienyl cation obtained by treatment of 5-bromocyclopentadiene with SbF_5 in di-n-butylphthalate at 78 K) [226] and calculation results [227–229] indicate that the ground state of **147** is triplet ($^3A_2'$, structure of D_{5h} symmetry).

147, D_{5h} **149**

One should, however, keep in mind that the adiabatic singlet–triplet energy difference for **147** is quite small (7.3 kcal/mol, STO-3G + CI [229]) and at the equilibrium geometry of the singlet state (see below) the energy of the triplet state is higher. Consequently, substituents may change the multiplicity of the ground state. For example, for the pentaphenylcyclopentadienyl cation (**149**), the triplet state has 0.35–1.15 kcal/mol more higher energy than the singlet state (see [111]).

The first-order Jahn-Teller effect is present in the lowest singlet $^1E_2'$ state of D_{5h} structure. Also, the second-order Jahn-Teller effect is well pronounced in this state, which, as a result of the e_2' distortion, is mixed with the low-lying $^1A_1'$ state [229]. In this case, the energy of stabilization amounts to 13.4 kcal/mol (STO-3G + CI) [229]. According to *ab initio* calculations, structures **150** and **151** of the lower C_{2v} symmetry possess nearly equal energies (Table 4.12). There are reasons to assume [230] that calculations at a higher level would show structure **151** to be a transition state of the pseudorotation—Eq. (4.19):

(4.19)

150   **151**

The D_{5h} structure cannot be a transition state of the above isomerization as this would contradict the Murrell–Laidler theorem [231].

According to MNDO calculations [232], structure **151** of the "ethylene" type may be stabilized in the derivatives **152** and **153** of the cyclopentadienyl cation. These derivatives have been synthesized; they are much more stable than the parent species $(CH)_5^+$ (**147**); for a review see [100].

152   **153**

The aromatic cyclopentadienide anion (**148**) was isolated in 1901 by reaction of cyclopentadiene with sodium dispersed in benzene (e.g., see [1,177]). The antiaromatic cation (**147**) was obtained much later. The IR, Raman, and ^{1}H NMR spectra indicate a planar D_{5h} structure of this anion. Turning to the planar structure (**150**) of the cyclopentadienyl cation, the following is to be noted. Even though its antiaromaticity is, compared to the D_{5h} structure (**147**), reduced on account of the bond length alternation, this reduction is still insufficient for stabilizing this planar structure. The energy of the nonplanar C_s structure (**154**) is, as has been shown by MINDO/3 calculations [227, 233], lower by 0.7 kcal/mol than that of **150**. At HF/6-31G*, **150** is a minimum, but **151** corresponds to a transition state [234]. However, CASSCF/3-21G calculations give **151** to be a minimum. At MP2(fc)/6-31G*, **150** turns out to be a transition state and **151** is a minimum [234]. The C_s structure **154** seems to be an artificial minimum in the MINDO/3 calculations. However, **150** and **151** are very close in energy, the energy difference being less than 1 kcal/mol at MP4SDTQ/6-31G(2d, p)//MP2 (full)/6-31G* [234]. The C_{4v} pyramidal structure is 3.5 kcal/mol higher in energy (MP2(full)/6-31G*//MP2(full)/6-31G*) than **151** [234].

154

Whereas the antiaromatic cyclopentadienyl cation $(CH)_5^+$ is destabilized relative to benzene (see Table 4.13), the aromatic anion (**148**) possesses even extra stabilization, as it may be judged from the value of ΔE of reaction (4.20), which is the difference between the energies of the corresponding reactions given in Table 4.13.

$$\text{(benzene)} + 2CH_4 + 2CH_3CH_2^- \longrightarrow \text{(cyclopentadienide)} + CH_3^- + 2C_2H_6 + C_2H_4 \quad (4.20)$$

$$\Delta E = -28 \text{ kcal/mol } (3-21G) \text{ [218]}$$

However, this conclusion on the greater aromaticity of **148** compared to benzene does not agree with the estimations obtained using the magnetic criterion (Chapter 2), namely, the ^{1}H NMR shifts [235]. Aromaticity of the cyclopentadienide anion (**148**) and of benzene can be directly compared using the ^{1}H NMR data for the benzannulene (**148a**) and for the anion (**148b**). The large difference between the chemical shift (corrected for delocalized charge) of the methyl protons of the anion (**148b**) ($\sigma_{Me} = -2.6$ ppm) and of the benzannulene (**148a**) ($\sigma_{Me} = -1.6$ ppm) indicates **148** to have an aromaticity substantially less effective than that of benzene [235].

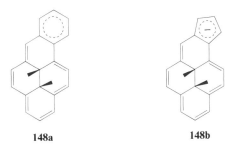

148a   148b

The pyramidal cation (**155**) characterized by the so-called three-dimensional aromaticity (see Chapter 9) is less stable than **150** by as little as 3.7 kcal/mol [230].

$$\text{(4.21)}$$

150   154, C_s *allowed*   155 $\uparrow C_{4v}$ *forbidden*   151

156, $D_{3h}(^1E')$

The activation barrier of the **150** → **155** isomerization is 43 kcal/mol (MINDO/3) [227]. Structure **151** is a lumomer of **150**, that is, a species with HOMO and LUMO inversion, and the **151** → **155** isomerization is forbidden for symmetry reasons. Structure **155** may be regarded as being the result of a Jahn-Teller distortion of the Möbius antiaromatic structure **156** [236], whose energy is higher by 69.5 kcal/mol compared to **155** (MINDO/3) [237]. An analogous pyramidal C_{5v} structure of the $(CH)_5$ anion does not satisfy the electron-count rule [88] and cannot correspond to a minimum on the PES. The $C_5H_5^+$ isomer **157**, which is a derivative of the aromatic cyclopropenyl cation, is more stable compared to the antiaromatic cyclopentadienyl cation (**150**) and other $C_5H_5^+$ isomers (Table 4.14).

TABLE 4.14 Relative Energies (in kcal/mol) of the $C_5H_5^+$ Isomers According to MP2/6-31G**//6-31G* Calculations (Including ZPVE Corrections) [230]

Isomer	150	151	155	157	158	159	160	161
E_{rel}	0^a	~0	3.7^b	-9.2^c	12.4	11.8	8.8	17.1

^{a}At MP2(fc)/6-31G*, **150** and **151** are a transition state and a minimum, respectively [234].
^{b}The D_{5h} triplet **147** is 2.6 kcal/mol lower in energy than **157** at MP4SDTQ/6-31G**//MP2(full)/6-31G* [234]. At QCISD(T)/6-31G**//MP2-31G*, this energy difference is 2.0 kcal/mol [234].
^{c}The pyradimidal C_{4v} structure **155** has 12.8 kcal/mol higher energy than that of **151** at MP4SDTQ/6-31G**//MP(full)/6-31G* [234].

| 157 | 158 | 159 | 160 | 161 |

It is assumed that in the gas phase the carbon atom scrambling in the $(CH)_5^+$ proceeds via the pyramidal structure (**155**) (for a review see [238]). A competitive process for the elimination of acetylene—see Eq. (4.22)—is an endothermic reaction (*ab initio* calculations [230], for experimental data see [238]), which apparently may be attributed to the strain energy of the three-membered ring (**128**):

$$\Delta E = -44.5 \text{ kcal/mol (MP2/6–31G}^{**}//6–31G^* \text{ including ZPVE [230])}$$

4.2.3 Tropylium Cation and Cycloheptatrienide Anion

When Doering and Knox obtained the cycloheptatrienium (tropylium) cation (**162**) in 1954, it was a vivid demonstration of the prognostic power of the Hückel rule. According to both experimental (IR, Raman, and ^1H NMR spectra) [1, 2, 177] and calculation data [209, 239, 240], the tropylium cation possesses a planar D_{7h} structure ($R(CC) = 1.400 \pm 0.002$ Å (for the $C_7H_7^+$ moiety in π-cycloheptatrienylium molybdenum (0) tetrafluoroborate, X-ray [241], 1.405 Å (MINDO/3[239]). The ISE value of **162** is even greater than that of benzene (Table 4.13). The unsubstituted cycloheptatrienide anion (**163**), which the Hückel rule assigns to the antiaromatic species, was registered in the early 1960s [1, 2]; its reactivity is fairly high.

The heavily substituted cycloheptatrienide anions are more stable. An NMR spectroscopic study of the monosubstituted anions $C_7H_6X^-$ ($X = CO_2R$, SO_2R, $CONR_1R_2$) [242] has shown that they have nonplanar structure (**164**), as distinct from the planar D_{7h} structure of the aromatic cation (**162**)

For the singlet $^1E_3'$ state of the D_{7h} structure (**163**), one may expect the first- and second-order Jahn-Teller effects. It is the latter effect due to the interaction of the $^1E_3'$ state with the energetically close $^1A_1'$ excited state that results in distortion into structures **165** and **166** with bond alternation [242]. According to MINDO/3 calculations [242], both these planar C_{2v} structures apparently correspond to minima on the PES, as opposed to the analogous structures **150** and **151** of $C_5 H_5^+$. However, this conclusion cannot be regarded as final, seeing that the MINDO/3 method overestimates the flatness of cyclic structures.

The norcaradiene-type anion structure (**167**) is also a minimum on the PES [243]. The analogous cation structure (**168**) is unstable with respect to the ring opening to the tropylium cation (MINDO/3 [239]).

The pyramidal structure **169** may, in contrast to the analogous structure **155**, be regarded as one possessing a three-dimensional antiaromaticity. According to the electron-count rule [88], it cannot correspond to a minimum on the PES, as has indeed been confirmed by MINDO/3 calculations [239]. Unlike the $C_5H_5^+$ system, the gas-phase carbon atom scrambling between **162** and the benzyl cation (**172**) [238] proceeds not via the pyramidal structure but rather via intermediate structures **170** and **171** (activation energies in kcal/mol calculated by the MINDO/3 method are as follows: 57.5 for **162** → **170**, 2.8 for **170** → **171**, and 2.1 for **171** → **172** [239].

MNDO calculations [247] show that whereas the tropylium cation has D_{7h} structure in the ground state, the structure of the lowest excited state is characterized by the bond length alternation. Conversely, the cycloheptatrienide anion has a structure of D_{7h} symmetry in the lowest singlet state without bond alternation, similar to that of the ground state of the tropylium cation. Based on structural criteria of aromaticity (see Section 2.3), these results may be viewed as more evidence that there is a possibility of aromaticity ⇌ antiaromaticity inversion in the ground and lowest excited states [55].

The cycloheptatrienide trianion $C_7H_7^{3-}$ (**173**) satisfies the $(4n + 2)$ rule. However, judging from MNDO results [127], the filling of the antibonding π-MOs in this trianion leads to a bond length of $R(CC) = 1.475$ Å, which is too great for an aromatic system, as well as to a very high value of $\Delta H_f = 386.3$ kcal/mol for **173** (cf. $\Delta H_f = 30.1$ kcal/mol for **166**. As for the stabilization of the trianion by lithium counterions, it results in nonplanar structures of C_s symmetry, **174** and **175** [127]:

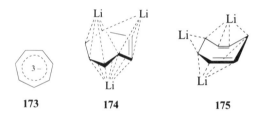

173   174   175

4.2.4   Cyclononatetraenyl Cation and Cyclononatetraenide Anion

The 10 π-electron anion **176** should be classified as aromatic. As was pointed out in Section 4.1.2, the manifestations of aromaticity in [10]annulene that is isoelectronic with the cyclononatetraenide anion $C_9H_9^-$ (**176**) are suppressed by steric effects. Now the question arises as to whether the same occurs in the $C_9H_9^-$ anion. Similar to [10]annulene, for the anion in question the following structures are conceivable: all-*cis* (**176**), single-*trans* (**177**), and double-*trans* (**178**) (cf. **111**–**113**, respectively).

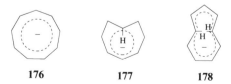

176   177   178

The all-*cis* structure (**176**) is more stable than the single-*trans* isomer (**177**), which is formed upon treatment of chlorobicyclo[6.1.0] nonatriene (**179**) with potassium to transform them into **176** [244–246]:

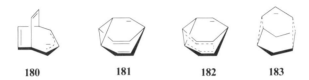

179   177   176

The planar D_{9h} structure of $C_9H_9^-$ has been confirmed by the low-temperature ^{1}H NMR spectrum [244, 245]. According to MNDO and AM1 calculations [247], structure **176** is more stable by 19.2 and 10.8 kcal/mol, respectively, than structure **177**. The latter isomer has a nonplanar structure (the difference in energy between the planar and nonplanar structures amounts to 19.1 (MNDO) and 10.8 (AM1) kcal/mol [247]). Like the $(CH)_{10}$ structure **113**, the double-*trans* structure **178** is highly nonplanar and its energy is 32.6 kcal/mol too high relative to **176**. The fact that the cyclononatetraenide anion has, unlike [10]annulene, a planar structure without bond alternation confirms the finding (see Chapter 2) that charged species have a more strongly pronounced aromatic character compared to the corresponding isoelectronic neutral molecules (see also [1, 248]).

180   181   182   183

The aromatic D_{9h} structure (**176**) is more stable than the isomeric $C_9H_9^-$ structures (**180**–**183**) (Table 4.15).

For structure **177**, the topomerization (4.23) is characteristic [246]—cf. Eq. (4.12)—which develops faster ($\Delta G^{\neq} = 22.1 \pm 0.1$ kcal/mol, 300 K [246]) relative to the isomerization into **176**.

$$\cdots \rightleftharpoons \quad \rightleftharpoons \quad \rightleftharpoons \quad \rightleftharpoons \cdots \qquad (4.23)$$

TABLE 4.15   Calculated Relative Energies (in kcal/mol) of Some $C_9H_9^-$ Isomers [247, 250, 251]

Method	Isomer					
	176	177	180	181	182	183
MINDO/3	0	—	51.5	62.9	83.9	79.3
MNDO	0	19.2	10.5	27.2	63.4	65.3
STO-3G//MINDO/3	0	—	8.4	25.2	72.7	60.6

As for the benzannelated cyclononatetraenide anion, its all-*cis* isomer is not as stable as the nonplanar single-*trans* isomer (by 3.9 (MNDO), 7.6 (AM1) kcal/mol [247]) into which it in fact transforms upon heating [249].

The cyclononatetraenyl cation has been studied less extensively. The attention of researchers has been concentrated primarily on the polycyclic species of the $C_9H_9^+$ cations. Noteworthy is the lesser stability of the antibicycloaromatic longicyclic cation **184** (bicyclo- and antibicycloaromaticity is treated in [252], see also Chapter 3) as compared to the isomeric structures **185–186a** (by, respectively, 3.67, 9.95, and 9.97 kcal/mol; STO-3G energies for the MNDO optimized geometries) [253, 254]. However, C_{2v} structure **184** corresponds to a first order saddle point (transition structure) at MP4SDQ/6-31G* and has 4.6 kcal/mol higher energy than 9-barbaryl cation (**185**) [254]. Nonclassical 1,4-bishomotropylium cation (**186a**) is a minimum which is 8.3 kcal/mol lower in energy than **185** at MP4SDQ/6-31G*. Cation **186** has 13.2 kcal/mol higher energy than that of **186a** at MP2/6-31G* [254].

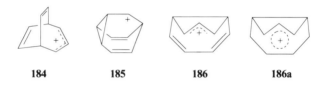

184	**185**	**186**	**186a**

4.3   ANNULENOANNULENES

As has been noted in Chapter 3, the Hückel rule was in fact derived in reference to the monocyclic systems only. It would, however, be of interest to examine its applicability to the simplest polycyclic systems, namely, the bicyclic systems formed by the fusion of two rings, the so-called [*M*]annuleno [*N*]annulenes. We confine our investigation to the neutral systems, annulenoannulenes with one shared bond. If both fused rings are even-membered, the annulenoannulene is an alternant conjugated hydrocarbon, but if they have an odd number of carbon atoms, it represents a nonalternant system. For the former, three Kekule resonance structures are conceivable, **187a–187c**, while for the latter only two are possible, **188a** and **188b** [255, 256]:

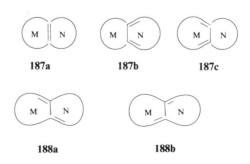

If in an alternant annulenoannulene either ring is a $(4n + 2)$ π-electron system, this species is aromatic. The fusion of two $(4n)$ π-electronic rings gives rise to either a nonaromatic system or a weakly antiaromatic system or else a weakly aromatic one despite the fact that the perimeter of such a molecule is $(4n + 2)$ π-electronic. The first system of this type, butalene (**189**), turns out to be antiaromatic. The data of Table 4.16 show that the degree of aromaticity of [$4n + 2$] annuleno [$4n + 2$]annulenes is close to that of the individual rings, rather than to that of the peripheral system. For example, for naphthalene the HSRE per π-electron, equal to 0.055β, is closer to that of benzene (0.065β) than to the value of HSRE for [10] annulene (0.026β) [255].

189

The specific features of the electronic structure and geometry associated with aromaticity can be distinctly revealed through comparison of the properties of isomeric annulenoannulenes instead of the customary separate examination of a series of alternant and nonalternant species. For example, the following isomers may be compared: C_6H_4, **189** and **190**; C_8H_6, **191**, **192**, and **193**; $C_{10}H_8$, **194**, **195**, and **196**.

C_6H_4

189   **190**

C_8H_6

191   **192**

193

$C_{10}H_8$

194   **195**

196

The structures of the corresponding isomeric dehydroannulenes $C_{2n}H_{2n-2}$ will also be discussed. We start with the simplest annulenoannulene, propalene, which has no isomeric structures of the annulenoannulene type.

TABLE 4.16   **Hess–Schaad Resonance Enrgies (HSREs) per π-Electron (in eV) of [n]Annulenes and Alternant and Nonalternant Annulenoannulenes [255, 256]**

		Alternant					Nonalternant			
		[N]Annuleno		[N]Annuleno			[N]Annuleno		[N]Annuleno	
[N] Annulene		[N]Annulene		[M]Annulene			[N]Annulene		[M]Annulene	
N	HSRE	N	HSRE	N	M	HSRE	N	HSRE	N	M	HSRE
4	−0.268	4	−0.067	4	6	−0.027	3	−0.100	3	5	0.055
6	0.065	6	0.055	4	8	−0.029	5	−0.018	3	7	−0.016
8	−0.060	8	−0.007	4	10	−0.020	7	−0.004	5	7	0.023
10	0.016	10	0.021	4	12	−0.019	9	0.000	5	9	−0.004
12	−0.011	12	0.001	6	8	0.005					
14	0.012	14	0.013	6	10	0.033					
16	−0.006	16	0.003	6	12	0.009					
18	0.010										

4.3.1   [3] Annuleno[3]annulene (Propalene)

According to RE values (see Tables 2.1 and 4.16), propalene is to a considerable degree antiaromatic, though less so than cyclobutadiene. The PPP [257], MINDO/3 [258], MNDO [259], and *ab initio* [258, 259] calculations indicate for propalene a C_{2h} ground-state structure with bond alternation (**197**). The structure of D_{2h} symmetry (**198**) has a higher energy (Table 4.17) and corresponds to a transition state of the bond-shift isomerization [258–260].

$$\tag{4.24}$$

197   **198**

The ISE of propalene is –187.5 kcal/mol [259]. Note that the principal contribution to the value of ISE comes from the strain. The antiaromatic destabilization estimated as the difference between the heats of the hydrogenation reactions, ΔH (Eq. (4.25)) − 2 ΔH (Eq. (4.26)), amounts to 26 kcal/mol (MNDO) [259]:

$+ 2H_2 \longrightarrow$ $\Delta H = -167.8$ kcal/mol

$+ H_2 \longrightarrow$ $\Delta H = -70.9$ kcal/mol (4.25)

At MP2/6-31G*//MP2/6-31G* the triplet $^3B_{1g}$ state (D_{2h} structure **198**) is 20.9 kcal/mol higher in energy than the $^1A_{1g}$ singlet **198** (Table 4.17). A nonplanar C_{2v} structure (**199**) is a minimum, which has 18.2 kcal/mol higher energy than that of **197**.

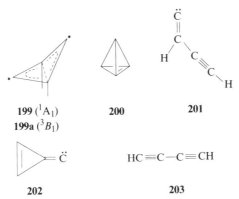

199 ($^1A_1$)
199a ($^3B_1$)   **200**   **201**

202   **203**

The structure of the valence isomer of propalene, dehydrotetrahedrane (**200**), is not a minimum [258]. Interestingly, the perdehydrotetrahedrane C_4 (T_d) corresponds to a minimum [261, 262], in contrast to **200**.

Antiaromatic propalene is destabilized relative to the open-chain isomers (Table 4.17). Moreover, it has a low kinetic stability. For example, the barrier of the isomerization of **198** into carbene **201** is only 0.9 kcal/mol (MNDO [259]). For the triplet $^3B_{1g}$ state of propalene (the D_{2h} structure), the carbon–carbon bond lengths are close in value, thus pointing to an aromatic character in this state.

The resonance energy for the $^3B_{1g}$ state of the D_{2h} propalene structure is positive (8.0 kcal/mol, MNDO method; the reference system is the three-membered ring plus a methyl radical (**198a**) [62]).

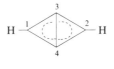

198a

Owing to the dynamic spin-polarization, the excited singlet $^1B_{1g}$ state of propalene has a lower energy than the triplet $^3B_{1g}$ state (by 1.3 kcal/mol, 6-31G + CI calculation [263]). The 2 π-electron propalene dication that satisfies the Hückel rule has a planar rhombic D_{2h} structure (**204**) without bond alternation (R(C1C3) = 1.386, R (C3C4) = 1.593 Å, 6-31G*) [264, 265].

H — ⟨ ⟩ — H

204

TABLE 4.17 Relative Energies (in kcal/mol) of the C_4H_2 Structures Calculated by Semiempirical and *Ab Initio* Methods

Structure	MINDO/3 [258]	MNDO [259]	AM1[b]	4-31G// STO-3G [258]	4-31G// 4-31G [259]	MP2/4-31G// 4-31G [259]	MP2/6-31G*//MP2/ 6-31G*[b]
197	0	0	0	0	0	0	(0)[d]
198 ($^1A_{1g}$)	7.7	13.5	10.3	10.2	15.5	—	9.3 (3)
198 ($^3B_{1g}$)	12.7[a]	−4.7	−5.6		2.3		20.9
199	9.3	11.8[b]	10.0	−6.9			18.2 (0)[e]
200	50.6	—	—	30.7			—[c]
201	−15.6	−59.1	−74.7	—	−56.5	−29.4	—
202	45.1	−34.8	−8.3	—	−20.8	8.4	—
203	−56.3	−128.2	−136.2	—	−100.7	−91.9	−61.4 (0)

[a]According to the MINDO/3 [258] and MP2/6-31G* computations, the lowest triplet state of propalene has a nonplanar C_2 structure. The D_{2h} ($^3B_{1g}$) structure is a transition state at HF/6-31G*.

[b]M. N. Glukhovtsev and P. v. R. Schleyer, unpublished results.

[c]The geometry optimization leads to transformation of the dehydrotetrahedrane structure **200** into **199**.[b]

[d]Number of imaginary frequencies calculated at MP2/6-31G* is given in parentheses. While at HF/3-21G singlet **198** is a transition state, CASSCF(4,4)/3-21G calculations (all π-orbitals are active) show singlet **198** to be a second order saddle point.[b]

[e]At MP2/6-31G*//MP2/6-31G*, C_{2v} triplet **199a** is 64.7 kcal/mol higher in energy than **197**.[b]

4.3.2   [4]Annuleno[4]annulene (Butalene) and [3]Annuleno[5]annulene

The first title compound (**189**) and its isomer, the second compound (**190**), have been generated and trapped as intermediates in a number of reactions [266, 267] (for a review see [268]), such as the following:

Butalene is an antiaromatic system (Table 4.16); even though it corresponds to a minimum on the PES, its 6 π-electron isomer, p-benzyne (**205**), has an energy lower by 35.9 (MINDO/3) [269], 26.0 (MNDO 3 $\times$ 3 CI) [270], 23.8 (MNDOC) [194], or 77.2 kcal/mol (4-31G) [271]. According to HSRE values (Table 4.16), bicyclo[3.1.0]hexa-1,3,5-triene (**190**) has an aromatic character. Indeed, notwithstanding the strain effects due to the presence of the three-membered ring, structure **190** is more stable than **189** by 21.0 kcal/mol (MNDO, 3 $\times$ 3 CI) [270].

205

Whereas for the butalene molecule alternation of the peripheral bond lengths is characteristic, in the case of **190**, it is insignificant. The bond lengths in angstrom (Å) units, MNDO, 3 $\times$ 3 CI, are as follows [270]:

189a   **189b**

As is apparent from these data, the structure of butalene is close to D_{2h} symmetry, and its molecule may be represented as **189a**. The D_{2h} structure **189b**, corresponding to a Kekule structure of type **187a**, has an energy much higher than that of **189a** [260]. The destabilization of **189b** relative to **189a** is consistent with the greater antiaromaticity of the former (DRE(**189a**) = – 4.3 while DRE(**189b**) = – 39.1 kcal/mol, found from the data of the PPP-type SCF calculations [260]).

The **189** → **205** isomerization is allowed by orbital symmetry [272]; its activation barrier is as low as 4.6 (MINDO/3) [269], 3.0 kcal/mol (MNDO, 3 × 3 CI) [270]. The barrier of the **190** → **206** interconversion is somewhat higher, amounting to 9.1 kcal/mol:

ΔH = – 26.0 kcal/mol (MNDO, 3 × 3 CI [270])

205

ΔH = – 5.7 kcal/mol (MNDO, 3 × 3 CI [270])

206

The lowest triplet state of butalene (**189**) has, judging from RE calculations, an aromatic character (7.9 kcal/mol, MNDO) [62].

4.3.3 [5]Annuleno[5]annulene (Pentalene), [4]Annuleno[6]annulene, and [3]Annuleno[7]annulene.

The first of the title compounds (**192**) should, as judged from RE values (Tables 2.1 and 4.16), be classified as antiaromatic. Unsubstituted pentalene has not been isolated so far; however a number of its derivatives are known (for reviews see [273, 274]). In particular, under conditions of matrix isolation, at 78 K, 2-methylpentalene has been obtained [275]; it is highly unstable and is very susceptible to dimerization. According to structural criteria of aromaticity (see Section 2.3.3), the C_{2h} pentalene structure with bond alternation should be more stable than the D_{2h} structure (**192**). Indeed, X-ray data on 1,3,5-tri-*tert*-butyl pentalene and dimethyl-4,6-di-*tert*-butylpentalene-1,2-dicarboxylate indicate planar bicyclic structures with bond alternation [276]. The same conclusion is reached after comparison of the experimental absorption spectra (UV–vis) of 1,3,5-tri-*tert*-butylpentalene [277] and 1,3-dimethylpentalene [278] with calculation results [240, 260, 277, 279, 280]. As has been shown by MINDO/3 [279] and *ab initio* [281] calculations, D_{2h} structure (**207**) is a transition state in the automerization of the C_{2h} structure (**192**).

(4.27)

192, C_{2h} **207**, D_{2h} **192a**

Computational data show the activation barrier of isomerization—Eq. (4.27)—to be greater than 10 kcal/mol (13.4, MINDO/3 [279]; 13.6 kcal/mol, *ab initio* 4-31G [281]). These values exceed that for the automerization of 1,3,5-tri-*tert*-butylpentalene, equaling 4 kcal/mol (^{13}C NMR spectroscopy) [277]).

Apart from the automerization passing through the D_{2h} structure, another mechanism has been proposed for polypentalenyls, for example, for 2,2'-bipen-talenyl (**207a**) [282]. With the use of second-order perturbation theory within the framework of the Hückel MO formalism, the automerization of **207a** was shown to pass via a particular localized structure (**207b**), instead of passing through the high-energy, fully delocalized D_{2h} structure (**207c**).

207a   207b   207c

The D_{2h} structure of the lowest triplet $^3B_{1g}$ state corresponds to a minimum on the PES and has an energy lower by 1.7 kcal/mol than the D_{2h} structure (**207**) of the ^1A_g state (4-31G) [282]. As opposed to the latter, the former structure should be assigned on the basis of RCI calculations to the aromatic class (RCI (^1A_g) = 1.25, RCI($^3B_{1g}$) = 1.60, SINDO1 method) [55]. The energy of the aromatic stabilization calculated for the lowest triplet state of pentalene by the MNDO method equals 27.1 kcal/mol [62]. Similar to the case of pro-palene (Section 4.3.1), the lowest excited singlet state $^1B_{1g}$ of pentalene (D_{2h} structure) has, because of the dynamic spin-polarization (see Section 4.1.1), a lower energy than the lowest triplet state (by 12.3 kcal/mol, STO-3G + CI) [263, 283]. The open-shell state ($^1B_{1g}$) turns out to be lower in energy by 13.1 kcal/mol compared to the closed-shell $^1A_{2g}$ state of the D_{2h} structure of pentalene (STO-3G + CI) [283]. Hence the ground state of the D_{2h} structure of pentalene is in effect an open-shell state ($^1B_{1g}$).

Unlike the D_{2h} ground-state structure (**207**), the D_{2h} structure of the lowest excited singlet state is stable against bond distortion into a structure of lower symmetry (whereas $E_1 - E_0$ is calculated by the PPP method to be 0.35 eV, the difference $E_2 - E_1$ equals, by the same method, 3.23 eV, exceeding the critical value of 1.2 eV; see Section 2.3.3 [52]).

The stabilization of the D_{2h} pentalene structure against bond distortion is facilitated by π-electron-donating substituents [273, 284]. In predicting the existence of bond alternation in substituted conjugated hydrocarbons [285], the index I was suggested (which we changed to IBA, seeing that the symbol I was used by us earlier in a different sense):

$$\text{IBA} = \frac{1.2A^2}{(E_1 - E_0)} \tag{4.28}$$

where $(E_1 - E_0)$ is the energy gap (in eV) and the quantity A^2 is given by

$$A^2 = 1 - \left(\frac{C_{lr}\beta_{CX}}{e_l - e_X}\right)^2 - \left(\frac{C_{hr}\beta_{CX}}{e_h - e_X}\right)^2. \tag{4.29}$$

where e_h and e_l are the energies of HOMO and LUMO, C_{lr} and C_{hr} are the atomic orbital coefficients of the rth atom in LUMO and HOMO, respectively, and β_{CX} is the resonance integral between the parent molecule and the substituent. At IBA < 1 the original totally symmetrical structure is stable with respect to distortions into a lower-symmetry structure with bond alternation, while for IBA > 1 it is unstable.

As may be seen from the schemes below that the HOMO and LUMO of pentalene are localized at different carbon atoms. From Eqs. (4.28) and (4.29) the conclusion can be drawn that the introduction of donor substituents at the 1, 3, 4, 6 positions of the pentalene structure will reduce the IBA, that is, stabilize the delocalized structure, as is indeed observed in 1,3-bis(dimethylamino) pentalene [100, 284]. In contrast, the π-electron-donating substitution at 2,5 positions will increase the instability of the D_{2h} structure with respect to bond distortion.

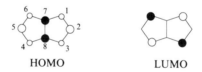

HOMO   LUMO

The RCI value of 1.56 [55] implies that the 10 π-electron pentalene dianion (**208**) should be classified as aromatic. Unlike unsubstituted pentalene, its dianion was prepared (in the early 1960s) [286]; as judged from the ^{1}H NMR spectrum, it is a diatropic system with planar bicyclic structure that is stable to distortion into a C_{2h} structure [52]. The planar D_{2h} structure of the pentalene dianion with the peripheral CC bond lengths close in value has also been confirmed by MNDO calculations of its dilithium salt [287].

Benzocyclobutadiene has an antiaromatic character slightly greater than that of pentalene (Table 4.16). Like pentalene, unsubstituted benzocyclobutadiene has been generated only under conditions of matrix isolation at 20K [288]; it easily undergoes dimerization (for a review see [99]).

208   **191**   **209**   **210**

For benzocyclobutadiene the following structures are conceivable: **209** and **191** of the same type as **187b** and **187c**, respectively as well as **210** of the **187a** type. Structures **209** and **210**, both containing a cyclobutadiene fragment, have higher energies than structure **191**, which contains a fragment of 1,2-dimethylenecyclobutadiene [260, 281]. The CC bond length alternation in **191**, characteristic of antiaromatic structures for both the four-membered and six-membered ring, may be explained by a greater absolute energy of the antiaromatic destabilization relative to the energy of the aromatic stabilization of benzene [281].

The idea of benzocyclobutadiene as a π-system comprising a 6 π-electron six-membered ring and an isolated π-bond [240] is in accord with X-ray data on substituted benzocyclobutadiene (**211**) [289]. In **211** the C1C6 bond is the shortest of all the CC bonds in this molecule in spite of the influence of two *t*-Bu groups, which might have been expected to result in the lengthening of this particular bond.

211   **193**

The energy of the aromatic stabilization of the lowest triplet state of benzocyclobutadiene equals, according to the NNDO calculations [62], 16.2 kcal/mol.

Bicyclo[5.1.0]octatetraene (**193**) has an energy higher by 37.6 kcal/mol compared to pentalene (MNDO) [281], which may be attributed to the strain energy of the three-membered ring in **193**. Dehydrocyclooctatetraene (**212**) is less stable than **193** (Table 4.18); the barrier to its conversion into **193** lies below 10 kcal/mol [281]. The activation energy of the isomerization of pentalene into the less stable structure **214** is 55.1 kcal/mol (UMNDO) [281]. The reverse isomerization (cyclization) has to overcome quite a low barrier of only 8 kcal/mol and, in view of the overestimation of the biradical stability for which the UMNDO scheme is known, it is probable that **214**, similarly to **212**, does not in fact correspond to a minimum on the PES.

Structures **215** and **217** of allenic type possess lower energies relative to **212** and **214**, respectively (Table 4.18).

212   **213**   **214**   **215**

TABLE 4.18   Relative Energies (in kcal/mol) of Some C_8H_6 Valence Isomers Calculated by the MNDO Method

Structure	MNDO [240, 280]	MNDO 3×3 CI [280]	UMNDO [280]
191	3.6	3.6	8.5 (12.4)
192	0	0	0 $(0)^b$
193	37.6	39.6	39.2 (37.7)
209	30.0	27.4	a
212	a	a	54.7 (53.5)
213	a	a	38.1 (32.1)
214	a	a	47.0 (44.8)
215	84.6	—	—
216a	48.0	43.3	—
216b	39.9	37.4	—
217	75.3	—	—
218	33.4	34.8	33.8 (38.3)
219	43.6	44.1	70.7

^{a}According to calculation (MNDO, MNDO 3×3 CI, UMNDO), the corresponding minimum is absent on the PES.
^{b}The relative energies of the lowest triplet state structures are given in parenthesis; for the D_{2h} structure of the lowest triplet state of pentalene, $\Delta H_f = 88.3$ kcal/mol and for the C_{2h} structure of the ground state of pentalene, $\Delta H_f = 70.7$ kcal/mol(UMNDO) [280].

        216   **217**

Cyclooctatrienyne (**218**) has been prepared as a reactive intermediate, and a number of its derivatives are known (for a review see [160]). Structure **216** is not planar; two configurations of it, **216a** and **216b**, are conceivable of C_s and C_{2v} symmetry, respectively [281]. Structures **215**–**217** as well as structures **219** of cumulene type are less stable than structure **218** (Table 4.18).

    216a   **216b**   **218**   **219**

The activation barrier of the isomerization **216b** → **191** amounts to 19.6 kcal/mol (MNDO) [281].

4.3.4 |6|Annuleno|6|annuleno (Naphthalene), |5|Annuleno|7|annulene (Azulene), and |4|Annuleno|8|annulene

Based on RE values (Tables 2.1 and 2.14), one may conclude that naphthalene (**194**) belongs to a typically aromatic system. Such an assignment is corroborated by experimental data on its physical and chemical properties: naphthalene is diatropic, it can undergo electrophilic substitution, and, conversely, it is not susceptible to reactions with dienophiles [1, 2, 177]. Unlike the bicyclic structure of butalene formed by fusion of two four-membered rings, the naphthalene structure is of **187a** type. The central bond in **194** is nearly as long as the C2C3 bond and it is shorter than the C1C9 bond (data on the following scheme are taken from an X-ray study [290]; in parentheses 6-31G-calculated values are given [291]):

220, C_{2v}

Geometry optimizations enabled, in the case of benzocyclobutadiene, structures **191, 209,** and **210** to be found [260, 281]; by contrast, in calculations on naphthalene making use of both the semiempirical [292] and nonempirical [291] methods, all attempts to arrive through geometry optimization at the C_{2v} structure (**220**) failed, with calculation results pointing to the D_{2h} structure (**194**). An *ab initio* calculation using the minimal basis set has shown [293] that the C_{2v} structure (**220**) with a model geometry (with CC double bonds of 1.34 Å and CC single bonds of 1.52 Å) has an energy higher by 19.4 kcal/mol than the D_{2h} structure in which all CC bonds are equal to 1.40 Å.

The aromaticity of azulene, which is isomeric with naphthalene, is less pronounced (Tables 2.1 and 4.16). Its resonance energy calculated from Eq. (2.2) is 9.1 kcal/mol, while for naphthalene it amounts to 43 kcal/mol (minimal basis set) [293]. Other evidence indicating lesser aromaticity of azulene is as follows: the results of the simplest PMO calculations [294]; the HOMO–LUMO energy gap (see Section 2.5.1.2) equals 10.55 eV for naphthalene and 8.43 eV for azulene (6-31G) [291]; and the DRE of naphthalene is 33.7 kcal/mol while for azulene it is 6.7 kcal/mol (MMP2) [295].

221, C_{2v} **196a**, C_s

The structure of naphthalene has, according to experimental and theoretical data, D_{2h} symmetry. As for the structure of azulene, X-ray data are inconclusive while analysis of numerous calculations [240, 291, 293, 296, 297] has shown that more or less reliable results as to relative energies of the C_{2v} and C_s structures (**221** and **196a**), may be obtained only when extended basis sets are used and electron correlation is included [169]. For **221**, $\lambda_{max} > \lambda_{crit}$ (see Section 2.3.3); that is, this structure should be stable against distortion into structure **196a** [298]. Note that while 6-31G*// MINDO/3 calculations indicate greater stability of C_s structure (**196a**) (by 2.8 kcal/mol) [297], the inclusion of electron correlation reverses this relation so that ultimately **221** is indeed more stable by 7.57 kcal/mol (MP2/6-31G//STO-3G) [169] (for **221** bond lengths given on the above scheme were calculated with the 6-31G basis set [169]).

The crystal structure of azulene is disordered and one should not disregard the possibility of crystal forces inverting the insignificant difference between the energies of structures **196a** and **221** [240]. Azulene has a smaller value of RE than naphthalene, which is consistent with the greater stability of the latter. The experimental difference between the energies of **194** and **196** is 37.4 kcal/mol [299] and *ab initio* calculations yield the values of 47.3 (6-31G) [291], 45.4 (6-31G*) [300], and 37.7 kcal/mol (MP2/6-31G) [291]. Similarly to the valence isomers of benzene (Section 4.1.1), the valence isomers **222–228** of aromatic naphthalene and azulene have higher energies (under the numbers of structures in parenthesis and brackets[4] relative energies in kcal/mol are given compared to naphthalene calculated by the MNDO method) [301–303]; the difference between the energies of naphthalene and azulene calculated by this method equals 39.0 kcal/mol [302].

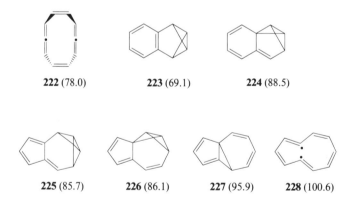

222 (78.0)   223 (69.1)   224 (88.5)

225 (85.7)   226 (86.1)   227 (95.9)   228 (100.6)

Whereas the structure of *p*-benzyne (**205**), according to the MNDO calculation, is more stable than the bicyclic structure of antiaromatic butalene [270], the aromatic bicyclic structure of azulene is, on the contrary, much more stable than the biradical structure (**228**), which is formed as a result of breaking of the

---

[4] MNDO results are given in parenthesis and UMNDO results are in square brackets.

transannular bond in azulene. The bisallenic structure (**222**) emerging as a result of the allowed ring opening of naphthalene has a higher energy than **194**. The barrier to the ring opening reaction is 98 kcal/mol (MNDO) [303]. The barrier of the homolytic cleavage of the transannular bond in azulene is also fairly high (75.2 kcal/mol, according to MNDO calculations [301, 303]). The activation energy of the rearrangement of naphthalene to **224** amounts to 112.0 kcal/mol (MNDO); the overall activation energy of the **194** → **223** isomerization equals 88.7 kcal/mol (UMNDO) [300]. For the conversion **196** → **225,** the activation energy is close to 75 kcal/mol (MNDO) [301, 303]. Thus aromatic naphthalene and azulene are, similar to benzene, kinetically and thermodynamically more stable relative to other valence isomers of $C_{10}H_8$. The isomerization of naphthalene and azulene into the valence isomers **222–228** is associated with overcoming considerable activation barriers.

Unlike pentalene, see Eq. (4.27), thermal rearrangement of aromatic naphthalene (**194**) and azulene (**196**) can proceed at high temperatures only (1035°C for reaction (4.30) and 440°C for reaction (4.31)) [34].

$$(4.30)$$

$$(4.31)$$

Along with the azulene-to-azulene isomerization, there is a competitive reaction that is, as a rule, preferable, namely, the rearrangement to naphthalene [34]. The rearrangement of azulene to naphthalene proceeds at temperature above 35°C [34]; it may be vibrationally activated by IR radiation from a continuous-wave CO_2 laser [304].

$$(4.32)$$

Several mechanisms have been suggested for such reactions [34, 300, 305]. The rearrangements may continue by two or more parallel pathways. Activation barriers of these rearrangements are quite considerable. For example, experimental data show that the activation energy of the **196** → **194** isomerization amounts to about 49 [306] or 60–65 kcal/mol [302]. In the mechanism of the naphthalene automerization based on MNDO calculations, the activation energy for the first step equals 102 kcal/mol [303] (values under the structural formulas represent ΔH_f in kcal/mol calculated by the MNDO method, those under the arrows are ΔH_f of transition-state structures).

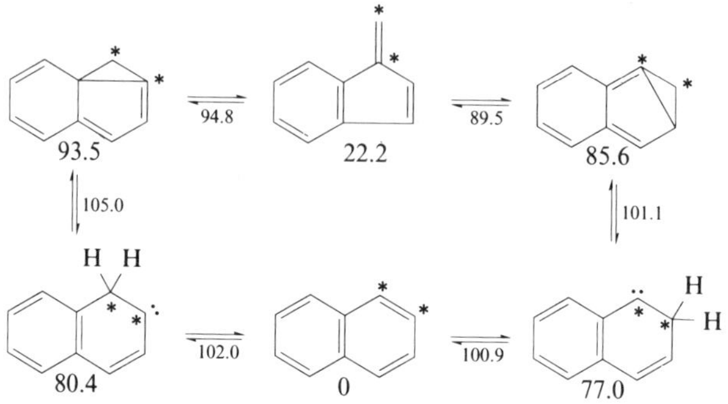

For bicyclo[6.2.0]decapentaene, an isomer of naphthalene and azulene, three structures, **195**, **229**, and **230**, are conceivable corresponding typologically to **187c**, **187b**, and **187a**:

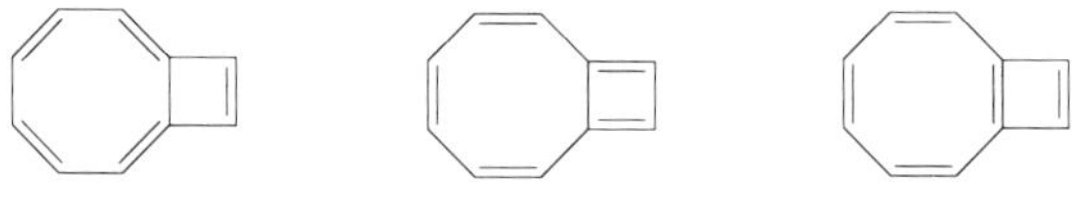

Bicyclo[6.2.0]decapentaene was obtained as a result of the thermal electrocyclic ring opening of the tricyclic valence isomer **231** (for a review see [307]); it was also obtained through oxidation of the η^4-tricarbonyl-iron complex (**233**) [308]. It is thermally a good deal more stable than benzocyclobutadiene. At 100°C in benzene this molecule is slowly dimerized to give cyclooct[c]octalene [307].

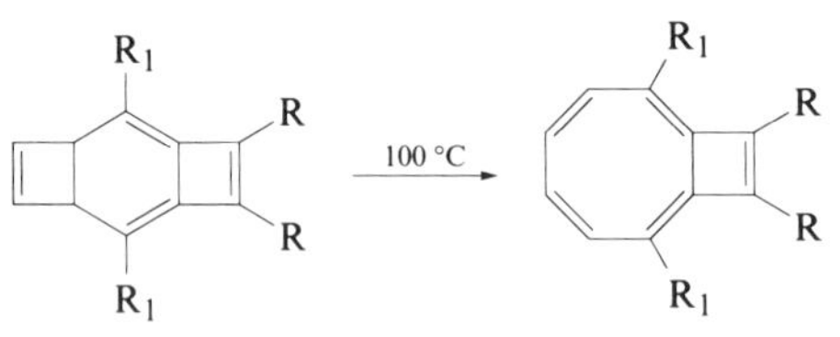

(**195**) R = R$_1$= H
(**232**) R = Ph, R$_1$ = CH$_3$
(**232a**) R = CH$_3$, R$_1$ = H

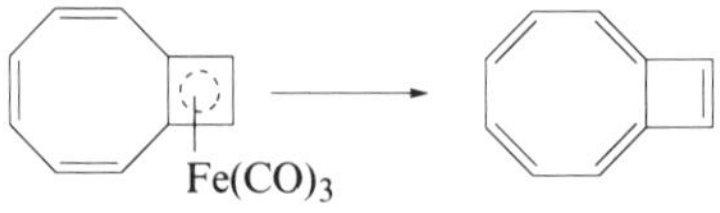

An X-ray study of **232** indicates a type **195** structure having a transannular bond of quite considerable length [309] (bond lengths according to STO-3G calculations on **195** are given in parentheses [310]):

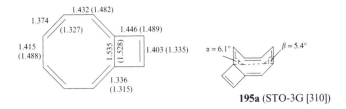

195a (STO-3G [310])

Thus **195** possesses a 10 π-electron peripheral system [240, 307, 310]. According to MNDOC calculations [310], structures **229** and **230** have an energy higher by 11.3 and 22.5 kcal/mol, respectively, than that of structure **195**.

MINDO/3 calculations [154] have shown structure **195** to be more stable than **230** by 25 kcal/mol. Judging from SINDO1 calculations [55], structure **195** is more stable than **229** and **230** by 22.8 and 26.3 kcal/mol, respectively.

Structure **195** is slightly puckered (STO-3G, MNDOC [310]; for STO-3G folding angles see **195a**, but the puckering potential is flat and the barrier of isomerization through the flat structure does not exceed 0.6 kcal/mol.

Analysis of the electron delocalization degree shows that for the four-membered ring in **195a** the index D is 18.5%—see Eq. 2.115; that value is slightly greater than in the case of cyclobutadiene; for an eight-membered ring $D = 42.7\%$, while for a 10 π-electron peripheral 10-membered ring it is 52.8%, which is greater than in the case of butadiene. Whereas the planar structure (**195**) should, according to RE calculations (Table 4.16), be assigned to the aromatic class, the value of D calculated for the equilibrium nonplanar structure (**195a**) of C_1 symmetry indicates loss of antiaromaticity and a weakly aromatic character [310, 311]. Indeed the RE value of **195a** calculated from the value of MNDOC enthalpy of the homodesmotic reaction (4.43), with the RE values of 1,2-methylenecyclobutadiene and cyclooctatetraene taken into account, equals 4.4 ± 1.5 kcal/mol.

$$\text{(structures)} \qquad (4.33)$$

Thus **195a** is stabilized by partial 10 π-electron delocalization along the periphery of the bicyclic carbon framework [240, 310]; it can be regarded as weakly aromatic. A model planar structure with a totally delocalized 10π-electron system (all $R(CC) = 1.394$ Å, the transannular bond equals 1.549 Å) is 3.8 kcal/mol higher in energy than **195a** (MNDOC) [310].

As has been pointed out by Cremer et al. [310], the RE value for **195** of – 4 kcal/mol obtained in MMP2 calculations [295] may apparently be attributed to an underestimation of the length of the transannular bond of the four-membered ring, which results in a negative value of the RE.

Structures **229** and **230** have negative RE values (– 6.9 and – 18.1 kcal/mol, respectively). These values are calculated, for example, as follows: RE(**229**) = RE(**195a**) – (ΔH_f (**229**) – ΔH_f(**195a**)) [310].

The RCI values equally evidence antiaromaticity of structures **229** and **230** (RCI(**229**) = 1.03 and RCI(**230**) = 0.96, while RCI(**195**) = 1.24) [55]. The barriers of isomerization of these structures into **195** are estimated (MNDOC) to lie in the 2–5 kcal/mol range [310]; that is, their experimental detection under oxidation conditions of **233** is hardly possible.

The lowest triplet state structure of bicyclo[6.2.0]decapentaene and its dianion does not possess aromatic character (the respective RCI values are 1.17 and 1.14) [55]. As judged from MINDO/3 calculations [154], structure **195** is less stable by 48 kcal/mol than that of naphthalene, which possesses greater aromaticity. Thermal rearrangement of **232a** at 680°C leads to the formation of a mixture of 1,2- and 2,3-dimethylnaphthalenes as well as 1,2-dimethylazulene [1]:

Thus bicyclo[6.2.0]decapentaene, similar to the preceding [4*n*]annuleno [4*n*]annulene, butalene, has a structure with peripheral (4*n* + 2) π-electron delocalization and a considerable length of the transannular bond. Note that thanks to such peripheral delocalization, bicyclo[6.2.0]decapentaene turns out to be slightly aromatic.

The data presented in Section 4.3 show that [4*n*]annuleno[4*n*]annulenes have a structure characterized by peripheral π-electron delocalization and a considerable length of the central bond. A central bond with a length close to that of the single bond is also typical of the structures of nonalternant annulenoannulenes. Note that these may be characterized not only by peripheral delocalization, as in 10 π-electron azulene, but also by bond fixation, as in 4*n* π-electron propalene and pentalene.

By contrast, alternant [4*n* + 2]annulenoannulenes possess structures of type **187a**. As for benzo[4*n*]annulenes, they are characterized by the delocalization within the six-membered ring. The exocyclic bonds linking that ring to the 4*n* π-electron fragment are of much greater length than that of the CC bond in benzene.

All these regularities, which may be expressed in a sort of "arithmetic of aromaticity" [312], are dictated by the "effort" to produce (4*n* + 2) π-electron delocalized systems.

The results on calculations on annulenes and annulenoannulenes show that the role of the electron correlation effects in determining relative stability of high-symmetry structures and of structures of lower symmetry with bond alternation may be different. According to that role, these molecules may formally be divided into two groups [283, 313]:

1. The molecules for which the relative stability of both a fully symmetrical structure and a bond-distortion one, predicted by the calculations in the

framework of HF approximation, is enhanced by taking into account the effect of electron correlation (pentalene, heptalene) [283].

2. The molecules whose relative stability, calculated by using the HF approximation, is reversed at the computational level that takes the electron correlation into account, as in the case of calculations on [18]annulene and [10]annulene [186] or on octalene [313]. For these molecules the stability of the symmetrical structure relative to the distortion into the structure with bond alternation may be revealed by the calculations beyond HF approximation.

REFERENCES

1. P. J. Garrat, *Aromaticity*, Wiley, New York, 1986.

2. D. Lloyd, *Non-benzonoid Conjugated Carbocyclic Compounds,* Elsevier, Amsterdam, 1984.

3. T. Bally and S. Masamune, *Tetrahedron*, **36**, 343 (1980).

4. G. Maier, *Angew. Chem. Int. Ed. Engl.*, **27**, 309 (1986).

5. E. Vogel, in H. Nozaki (Ed.), *Current Trends in Organic Synthesis*, Pergamon Press, Oxford, 1983.

6. M. V. Gorelik, *Usp. Khim.*, **58**, 197 (1990).

7. O. L. Chapman, C. L. McIntosh, and J. Pacansky, *J. Am. Chem. Soc.*, **95**, 614 (1973).

8. A. Krantz, C. Y. Lin, and M. D. Newton, *J. Am. Chem. Soc.,* **95**, 2774 (1973).

9. B. Ya. Simkin and M. N. Glukhovtsev, *Khim. Zhyzn,* 22 (1987).

10. D. W. Whitman and B. K. Carpenter, *J. Am. Chem. Soc.,***102**, 4272 (1980).

11. L. J. Schaad, B. A. Hess, and C. S. Ewig, *J. Org. Chem.*, **47**, 2904 (1982).

12. B. A. Hess, L. J. Schaad, and P. Carsky, *Pure Appl. Chem.*, **55**, 253 (1983).

13. B. A. Hess, L. J. Schaad, P. Carsky, and R. Zahradnik, *Chem. Rev.*, **88**, 709 (1986).

14. O. L. Chapman, D. L. Cruz, R. Roth, and J. Pacansky, *J. Am. Chem. Soc.*, **95**, 1337 (1973).

15. G. Maier, H.O. Kalinowski, and K. Euler, *Angew. Chem. Int. Ed. Engl.*, **21**, 693 (1982).

16. M. Saunders, M. H. Jaffe, and P. Vogel, *J. Am. Chem. Soc.*, **93**, 2558 (1971).

17. A. M. Orendt, B. R. Arnold, J. G. Radziszewski, J. C. Facelli, K. D. Malsch, H. Strub, D. M. Grant, and J. Michl, *J. Am. Chem. Soc.*, **110**, 2648 (1988).

18. W. T. Borden, E. R. Davidson, and P. Hart, *J. Am. Chem. Soc.*, **100**, 338 (1978).

19. H. Kollmar and V. Staemmler, *J. Am. Chem. Soc.*, **99**, 3583 (1977).

20. J. A. Jafri and M. D. Newton, *J. Am. Chem. Soc.*, **100**, 5012 (1978).

21. H. Algren, N. Correia, A. Flores-Riveros, and H. J. A. A. Jensen, *Int. J. Quant. Chem. Symp.*, **19**, 237 (1986).

22. P. Čarsky, V. Špirko, B. A. Hess, and L. J. Schaad, *J. Chem. Phys.*, **92**, 6069 (1990).

23. K. Nakamura, Y. Osamura, and S. Iwata, *Chem. Phys.*, **136**, 67 (1989).

24. R. Janoschek and J. Kalcher, *Int. J. Quant. Chem. Symp.*, **38**, 653 (1990).

25. B. A. Hess, P. Čársky, and L. J. Schaad, *J. Am. Chem. Soc.*, **105**, 695 (1983).

26. D. W. Whitman and B. K. Carpenter, *J. Am. Chem. Soc.*, **104**, 6473 (1982).

27. B. K. Carpenter, *J. Am. Chem. Soc.*, **105**, 1700 (1983).

28. M. J. S. Dewar, K. M. Merz, and J. J. P. Stewart, *J. Am. Chem. Soc.*, **106**, 4040 (1984).

29. M. J. Huang and M. Wolfsberg, *J. Am. Chem. Soc.,* **106**, 4039 (1984).

30. M. N. Glukhotsev and B. Ya. Simkin, *Zh. Strukt. Khim.*, **26**, 168 (1985).

31. G. Maier, R. Wolf, and H.-O. Kalinowski, *Angew. Chem.,* **104**, 764 (1992).

32. M. N. Glukhovtsev, B. Ya. Simkin, and V. I. Minkin, *Zh. Org. Khim.*, **20**, 886 (1984).

33. E. A. Halevi, *Orbital Symmetry and Reaction Mechanism. The OCAMS View*, Springer, Berlin, 1992.

34. L. T. Scott, *Acc. Chem. Res.*, **15**, 52 (1982).

35. J. J. Gajewski, *Hydrocarbon Thermal Isomerizations*, Academic, New York, 1981.

36. L. T. Scott, N. H. Roefols, and T. H. Tsang, *J. Am. Chem Soc.*, **109**, 5456 (1987).

37. V. I. Minkin, I. A. Yudilevich, R. M. Minyaev, and G. V. Orlova, *Dokl. Akad. Nauk SSSR*, **305**, 358 (1989).

38. M. J. S. Dewar and S. Kirschner, *J. Am. Chem. Soc.*, **97**, 2932 (1975).

39. S. Oikawa, M. Tsuda, Y. Okamura, and T. Urabe, *J. Am. Chem. Soc.*, **106**, 6751 (1984).

40. E. Lindholm, *Trans. Faraday Soc.,* **54**, 200 (1972).

41. M. Tronc and L. Malegat, in F. Lahmani, (Ed.), *Photophysics and Photochemistry above 6 eV*, Elsevier, Amsterdam, 1985, p. 203.

42. A. Stranger and K. P. C. Vollhardt, *J. Org. Chem.*, **53**, 4890 (1988).

43. C. R. Brundle, M. B. Robin, and N. A. Kuebler, *J. Am. Chem. Soc.*, **94**, 1446 (1972).

44. J. Almlof and K. Faegri, *J. Chem. Phys.*, **79**, 2284 (1983).

45. J. M. O. Matos, B. O. Roos, and P. A. Malmquist, *J. Chem. Phys.*, **86**, 1458 (1987).

46. M. H. Palmer and I. C. Walker, *Chem. Phys.*, **133**, 113 (1989).

47. P. J. Hay and I. Shavitt., *J. Chem. Phys.*, **60**, 2865 (1974).

48. L. D. Ziegler and B. S. Hudson, The Vibronic Spectroscopy of Benzene in E. C. Lim (Ed.), *Excited States*, Vol. 5, Academic, New York, 1982, p. 42.

49. S. Koto, *J. Chem. Phys.*, **88**, 3045 (1988).

50. M. Meisi and R. Janoschek, *J. Chem. Soc. Chem.Commun.*, 1066 (1986).

51. E. J. P. Malar and K. Jug, *J. Chem. Phys.*, **88**, 3008 (1988).

52. T. Nakajima, A. Toyota, and M. Kataoka, *J. Am. Chem. Soc.*, **104**, 5610 (1982).

53. W. J. Buma, J. H. van der Waals, and M. C. van der Hemert, *J. Am. Chem. Soc.*, **111**, 86 (1989).

54. Y. Osamura, *Chem. Phys. Lett.*, **145**, 541 (1988).

55. K. Jug and E. J. P. Malar, *J. Mol. Struct. (THEOCHEM)*, **153**, 221 (1987).

56. J. H. van der Waals, *Mol. Cryst. Lif. Cryst.*, **50**, 301 (1979).

57. L. Salem., *Electronics in Chemical Reactions: First Principles*, Wiley, New York, 1982.

58. A. F. Voter and W. A. Goddard, *J. Am. Chem. Soc.*, **108**, 2830 (1986).

59. W. T. Borden, *Effects of Electron Repulsion in Diradicals*, W. T. Borden (Ed.), *Diradicals,* Wiley, New York, 1982, pp. 1–72.

60. F. Fratev, V. Monev, and R. Janoschek, *Tetrahedron*, **38**, 2929 (1982).

61. J. I. Aihara, *Bull. Chem. Soc. Jpn.*, **51**, 1778 (1978).

62. N. C. Baird, *J. Am. Chem. Soc.*, **94**, 4941 (1972).

63. J. Wirz, A. Kerbs, H. Schalstieg, and H. Angliker, *Angew. Chem. Int Ed. Engl.*, **20**, 192 (1981).

64. F. Bickelhaupt and W. H. de Wolf. *Recl. Trav. Chim. Pays-Bas*, **107**, 459 (1988).

65. E. Fahrenhost, *Tetrahedron Lett.*, 6465 (1973).

66. A. Balaban, *Rev. Roum. Chim.*, **18**, 635 (1973).

67. W. E. Billups and M. M. Haley, *Angew. Chem. Int. Ed. Engl.*, **28**, 1711 (1989).

68. A. Greenberg and J. F. Leibman, *Strained Organic Compounds*, Academic, New York, 1978

69. M. Y. Zhang, C. Wesdemotic, M. Marchetti, P.O. Danis., J.C. Ray, B.K. Catpenter, and F. W. McLafferty, *J. Am. Chem. Soc.*, **111**, 8341 (1989).

70. J. M. Schulman and R. L. Disch, *J. Am. Chem. Soc.*, **107**, 5059 (1985).

71. D. S. Warren and B. M. Gimarc, *J. Am. Chem. Soc.*, **114**, 5378 (1992).

72. S. W. Staley and T. D. Norden, *J. Am. Chem. Soc.*, **111**, 445 (1989).

73. B. A. Hess, W. D. Allen, D. Michalska, L. J. Schaad, and H. F. Schaefer, *J. Am Chem. Soc.*,**109**, 1615 (1987).

74. M. N. Glukhovtsev and P. v. R. Schleyer, *Int. J. Quant.Chem.*, **43**, 119 (1993).

75. G. Fitzgerald, P. Saxe, and H. F. Schaefer, *J. Am. Chem. Soc.*, **105**, 690 (1983).

76. D. Bryce-Smith and A. Gilbert, *Tetrahedron*, **32**, 1309 (1976).

77. J. A. Barltrop and J. D. Coyle, *Excited States in Organic Chemistry*, Wiley, New York, 1975.

78. E. E. van Temelen, *Acc. Chem. Res.,* **5**, 186 (1972).

79. R. G. Pearson, *J. Am. Chem. Soc.*, **110**, 2092 (1988).

80. J. J. Gajewski and A. M. Gortla, *J. Am. Chem. Soc.*, **104**, 335 (1982).

81. J. E. Bartmess, *J. Am. Chem. soc.*, **104**, 335 (1982).

82. J. E. Bartmess and S. S. Griffith, *J. Am. Chem. Soc.*, **112**, 2931 (1990).

83. A. R. Katritzky and C. M. Marson, *Tetrahedron Lett.*, **32**, 1309 (1976).

84. A. R. Katritzky, C. M. Marson, G. Palenic, A. E. Koziol, H. Luce, M. Karelson, B. C. Chen, and W. Brey, *Tetrahedron*, **44**, 3209 (1988).

85. D. R. Howton and E. R. Buchman, *J. Am Chem. Soc.*, **78**, 4011 (1956).

86. M. N. Glukhovtsev, to be published.

87. J. D. Dill, A. Greenberg, and N. Liebman, *J. Am. Chem. Soc.*, **101**, 6814 (1979).

88. V. I. Minkin, R. M. Minyaev, and Yu. A. Zhdanov, *Nonclassical Structures of Organic Compounds*, Mir, Moscow, 1987.

89. G. Maier, S. Pfriem, U. Schäfer, K. D. Malsch, and R. Matusch, *Chem. Ber.*, **114**, 3965 (1981).

90. G. Maier, S. Pfriem, U. Schäfer, and K. D. Malsch, *Angew. Chem.*, **90**, 552 (1978).

91. A. Schweig and W. Thiel, *J. Am. Chem. Soc.*, **101**, 4742 (1979).

92. H. Block, B. Roth, and G. Manier, *Chem. Ber.*, **117**, 172 (1984).

93. D. T. Clark and A. Herrison, *Chem. Phys.*, **62**, 353 (1981).

94. B. A. Hess and L. J. Schaad, *J. Am. Chem. Soc.*, **107**, 865 (1985).

95. H. Kollmar, E. Carrion, M. J. S. Dewar, and R. C. Bingham, *J. Am. Chem. Soc.*, **103**, 5292 (1981)

96. M. C. Bohm and R. Gleiter, *Tetrahedron Lett.*, 1179 (1978).

97. R. Hoffmann, *J. Chem. Soc. Chem. Commun.*, 241 (1969).

98. R. Gompper and G. Seybold, *Angew. Chem. Int. Ed. Engl.*, **7**, 824 (1968).

99. K. P. C. Vollhardt, *Topics Curr. Chem.*, **59**, 114 (1975).

100. R. Gompper and H. V. Wagner, *Angew. Chem. Int. Ed. Engl.*, **27**, 1437 (1988).

101. R. Wiess and J. N. Murrell, *Tetrahedron*, **27**, 2877 (1971).

102. O. M. Herrera and I. M. Brinn, *J. Chem. Soc.Perkin Trans. II*, 1683 (1986).

103. G. R. Stevenson, S.S. Zigler, and R.C. Reiter, *J. Am. Chem. Soc.*, **103**, 6057 (1981).

104. G. K. S. Prakash, T. N. Rawdah, and G. A. Olah, *Angew. Chem. Int Ed. Engl.*, **22**, 401 (1983).

105. R. M. Pagni, *Tetrahedron*, **40**, 4161 (1984).

106. J. R. Appling, G. W. Burdick, M. J. Hayward, L. E. Abbey, and T. F. Moran, *J. Phys. Chem.*, **89**, 13 (1985).

107. E. Wasserman, R. E. Hutton, V. J. Kuck, and E. A. Chandross, *J. Am. Chem. Soc.*, **96**, 1965 (1974).

108. M. J. S. Dewar and M. K. Holloway, *J. Am. Chem. Soc.*, **106**, 6619 (1984).

109. K. Lammertsma and P. v. R. Schleyer, *J. Am. Chem. Soc.*, **105**, 1049 (1983).

110. K. Krogh-Jespersen, *J. Am. Chem. Soc.*, **113**, 417 (1991).

111. M. N. Glukhovtsev, B. Ya. Simkin, and V. I. Minkin, *Zh. Org. Khim.*, **26**, 2249 (1990).

112. J. S. Miller, D. A. Dixon, J. C.Calabrese, C. Vazquez, P. J. Krusic, M. D. Ward, E. Wasserman, and R. L. Harlow, *J. Am. Chem. Soc.*, **112**, 381 (1990).

113. A. L. Buchachenko, *Usp. Khim.*, **59**, 529 (1990).

114. J. M. Chance, B. Kahr, A. B. Buda, J. P. Toscano, and K. Mislow, *J. Org. Chem.*, **53**, 3226 (1988).

115. J. Thomaides, P. Maslak, and R. Breslow, *J. Am. Chem. Soc.*, **110**, 3970 (1988).

116. J. S. Miller, D.A. Dixon, and J. C. Calabrese, *Science*, **240**, 1185 (1988).

117. D. A. Dixon, J.C. Calabrese, R. L. Harlow, and J. S. Miller, *Agnew. Chem. Int. Ed. Engl.*, **28**, 92 (1989).

118. H. Hogeveen and P. W. Kwant, *Tetrahedron Lett.*, 1655 (1973).

119. H. Hogeveen and E. M. G. A. Kruchten, *J. Org. Chem.*, **46**, 4301 (1978).

120. K. Krogh-Jespersen and P. v. R. Schleyer, *J. Am. Chem. Soc.*, **100**, 4301 (1978).

121. J. Chandresekhar, P. v. R. Schleyer, and K. Krogh-Jespersen, *J. Comput. Chem.*, **2**, 356 (1981).

122. M. Bremer and P. v. R. Schleyer, *J. Am. Chem. Soc.*, **111**, 1147 (1989).

123. L. Radom and H. F. Schaefer, *J. Am. Chem Soc.*, **99**, 7522 (1977).

124. V. I. Minkin, R. M. Minyaev, M. E. Kletzky, and M. N. Glukhovtsev, *Zh. Org. Khim.*, (*Engl. Transl.*), **27**, 1583 (1991).

125. M. Nakayama, H. Ishikawa, T. Nakano, and O. Kikuchi, *J. Mol. Struct.*, (THEOCHEM), **184**, 369 (1989).

126. P. v. R. Schleyer and N. J. R.v.E. Hommes, to be published.

127. P. v. R. Scheleyer, D. Wilhelm, and T. Clark. *Organomet. Chem.*, **281**, C17 (1985).

128. A. Sekiguchi, K. Ebata, C. Kabuto, and H. Sakura, *J. Am. Chem. Soc.*, **113**, 1464 (1991).

129. J. S. McKennis, L. Brener, J. R. Schweiger, and R. Pettit, *J. Chem. Soc. Chem. Commun.*, 365 (1972).

130. G. von Boche, H. Etzordt, M. Marsch, and W. Thiel, *Angew, Chem.*, **94**, 141 (1982).

131. P. J. Gerrat and R. Zahler, *J. Am. Chem. Soc.*, **100**, 7753 (1978).

132. T. Clark, D. Wilhelm, and P. v. R. Schleyer, *Tetrahedron Lett.*, **23**, 3547 (1982).

133. A. Skancke and I. Agranat, *Nouv. J. Chim.*, **9**, 577 (1985).

134. M. N. Glukhovtsev, B. Ya. Simkin, and V. I. Minkin, *Zh. Org. Khim.*, **23**, 1317 (1987).

135. B. A. Hess, C.S. Ewig, and L. J. Schaad, *J. Org. Chem.*, **50**, 5869 (1985).

136. G. van Zandwijk, R. A. J. Janssen, and H. M. Buck, *J. Am. Chem. Soc.*, **112**, 4155 (1990).

137. R. C. Haddon, *Pure Appl. Chem.*, **58**, 129 (1986).

138. L. A. Paquette, *Tetrahedron*, **31**, 2855 (1975).

139. G. I. Fray and R. G. Saxton, *The Chemistry of Cyclooctatetraene and its Derivatives*, Cambridge University Press, New York, 1978.

140. M. Traetteberg, *Acta Chem.Scand.*, **20**, 1724 (1966).

141. T. Wolfskill, *Int. J. Quant. Chem. Symp.*, **22**, 739 (1988).

142. M. J. S. Dewar and K. M. Merz, *J. Chem. Soc. Chem Commun.*, 343 (1985).

143. L. A. Paquette, Y. Hanzavwa, K. J. McCullough, B. Tagle., W. Swenson, and J. Clardy, *J. Am. Chem Soc.*, **103**, 2262 (1981).

144. L. A. Paquette, *Pure Appl. Chem*, **54**, 987 (1982).

145. M. N. Glukhovtsev, B. Ya. Simkin., and V. I. Minkin, *Zh. Org. Khim.*, **22**, 212 (1986).

146. M. J. S. Dewar and K. M. Merz. *J. Phys. Chem.*, **89**, 4739 (1985).

147. K. Hassenrück, H. D. Martin, and R. Walsh, *Chem. Rev.*, **89**, 1125 (1989).

148. R. C. Bingham, M. J. S. Dewar, and D. H. Lo, *J. Am. Chem. Soc.*, **97**, 1285 (1975).

149. D. A. Hrovat and W. T. Borden, *J. Am. Chem. Soc.*, **114**, 5879 (1992).

150 A. T. Balaban and M. Banciu, *J. Chem. Educ.*, **61**, 766 (1984).

151. O. Ermer, F. G. Kläner, and M. Wette, *J. Am. Chem. Soc.*, **108**, 4908 (1986).

152. L. A. Paquette. T. Z. Wang, and C. E. Cottrell, *J. Am. Chem. Soc.*, **109**, 3730 (1987).

153. T. W. C. Mak and W. K. Li, *J. Mol. Struct.*, (THEOCHEM), **89**, 281 (1982).

154. J. Spanget-Larsen, *J. Mol. Struct.* (THEOCHEM), **89**, 281 (1982).

155. H. N. C. Wong and W. K. Li, *J. Chem. Res.* (Suppl), 302 (1984).

156. H. Durr, G. Klauck., K. Petetrs, and H. v. G. Schnering, *Angew. Chem. Int. Ed. Engl.*, **22**, 332 (1983).

157. R. E. Cobbledick and F. W. B. Einstein, *Acta Crystalloger. Sect. B* **33**, 2339 (1977)

158. F. W. B. Einstein, A. C. Willis, W. R. Cullen, and R. L. Soulen, *J. Chem. Soc. Chem Commun.*, 526 (1981).

159. N. Z. Haung and F. Sondheimer, *Acc. Chem Res.*, **15**, 96, (1982).

160. J. F. M. Oth, *Pure Appl. Chem.*, **25**, 573 (1971).

161. L. A. Paquette and J. M. Gardlikl, *J. Am. Chem Soc.*, **102**, 5033 (1980).

162. L. A. Paquette and T. Z. Wang, *J. Am. Chem. Soc.*, **110**, 3663 (1988).

163. L. A. Paquette, T. Z. Wang, J. Luo, C. E. Cottrell, A. E. Clough, and L. B. Anderson, *J. Am. Chem. Soc.*, **112**, 239 (1990).

164. R. Naor and Z. Luz, *J. Chem Phys.*, **76**, 5662 (1982).

165. M. E. Squillacote and A. Bergman, *J. Org. Chem.*, **51**, 3910 (1986).

166. E. Vogel, H. Kiefer, and W. R. Roth, *Angew. Chem. Int. Ed. Engl.*, **3**, 83 (1964).

167. D. Loos and J. Leška, *Col. Czech. Chem. Commun.*, **45**, 187 (1980).

168. L. Farnell, J. Kao, L. Radom, and H. F. Schaefer, *J. Am. Chem,. Soc.*, **103**, 2147 (1981).

169. R. C. Haddon and K. Raghavachari, *J. Am. Chem. Soc.*, **104**, 3516 (1982).

170. R. C. Haddon and K. Raghavachari, *J. Am. Chem. Soc.*, **107**, 289 (1985).

171. S. Masamune, K. Hojo, G. Bigam, and D. L. Raberstein, *J. Am. Chem. Soc.*, **93**, 4966 (1971).

172. S. Masamune and N. Darby, *Acc. Chem. Res.*, **5**, 272 (1972).

173. L. Farnell and L. Radom, *J. Am. Chem. Soc.*, **104**, 7650 (1982).

174. D. Cremer and B. Dick, *Angew. Chem.*, **94**, 877 (1982).

175. A. Sabljić and N. Trinajstić, *J. Org. Chem.*, **46**, 3457 (1981).

176. W. C. Herndon and C. Parkanyi, *Tetrahedron*, **38**, 2551 (1982).

177. G. R. Stevenson, *Stabilization and Destabilization of Aromatic and Antiaromatic Compounds* in J. L. Liebman and A. Greenberg (Eds.), *Molecular Structure and Energetics*, Vol. 3, VCH, New York, 1986, pp. 57–83.

178. J. H. Noordlik, T. E. M. van der Hark, J. J. Mooij, and A. K. K. Klaassen, *Acta Crystallogr. Sect. B*, **30**, 833 (1974).

179. G. A. Olah, J. S. Staral, and L. A. Paquette, *J. Am. Chem. Soc.*, **98**, 1267 (1976).

180. H. Spiesecke and W. G. Schneider, *Tetrahedron Lett.*, 468 (1961).

181. D. Wilhelm, T. Clark, P. v. R. Schleyer, and A. G. Davies, *J. Chem. Soc. Chem. Commun.*, 558 (1984).

182. G. W. Klumpp, J. Fleischnauer, and W. Schlener, *Rec. Trav. Chim. Pays-Bas*, **101**, 208 (1982).

183. C. C. Chaing and I. Paul, *J. Am. Chem. Soc.*, **94**, 4741 (1972).

184. D. Loos and J. Leška, *Col. Czech. Chem. Commun.*, **49**, 920 (1984).

185. K. Jug and E. Fasold, *J. Am. Chem. Soc.*, **109**, 2263 (1987).

186. H. Baumann, *J. Am. Chem. Soc.*, **100**, 7196 (1978).

187. J. M. Hernando, J. J. Quirante, and F. Enriquez, *J. Mol. Struct*, (THEOCHEM), **204**, 2081 (1990).

188. J. Bergman, F. L. Hirshfeld, D. Rabinovich, and G. M. J. Schmidt, *Acta Crystalloger.*, **19**, 227 (1965).

189. H. Baumann and J. F. M. Oth, *Helv. Chim. Acta*, **65**, 1885 (1992).

190. F. Sondheimer and R. Wolovsky, *Tetrahedron Lett.*, 3 (1959).

191. M. J. S. Dewar, R. C. Haddon, and P. J. Student, *J. Chem. Soc. Chem. Commun.*, 569 (1974).

192. M. J. S. Dewar and M. L. McKee, *Pure Appl. Chem.*, **52**, 1431 (1980).

193. R. C. Haddon, *Chem. Phys. Lett.*, **70**, 210 (1980).

194. W. Thiel, *J. Am. Chem. Soc.*, **103**, 1420 (1981).

195. D. Cremer and W. Thiel, *J. Comput. Chem.*, **8**, 48 (1987).

196. A. Korth, M. L. Marcon, D. A. Mendis, F. R. Krueger, A. K. Richter, R. P. Lin, D. L. Mithell, K. A. Anderson, C. W. Carlson, H. Reme, J. A. Sauvand, and C. d'Uston, *Nature*, **337**, 53 (1989).

197. M. R. Wasielewski and R. Breslow, *J. Am. Chem. Soc.*, **98**, 4222 (1976).

198. H. G. Köser, G. E. Renzoni and W. T. Borden, *J. Am. Chem. Soc.*, **105**, 6359 (1983).

199. J. E. Bartmess, J. Kestler, W. T. Borden, and H. G. Köser, *Tetrahedron Lett.*, **27**, 5931 (1986).

200. F. H. Allen, *Tetrahedron*, **38**, 645 (1982).

201. P. C. Burgers, J. L. Holmes, A. A. Mommers, and J. E. Sznlejko, *J. Am. Chem. Soc.*, **106**, 521 (1984).

202. G. Winkelhofer, R. Janoschek, F. Fratev, G. W. Spitzagel, J. Chadrasekhar, and P. v. R. Schleyer, *J. Am. Chem. Soc.*, **107**, 332 (1985).

203. L. Radom, P. C. Hariharan, J. A. Pople, and P. v. R. Schleyer, *J. Am. Chem. Soc.*, **98**, 10 (1976).

204. J. Leszczynski, B. Weiner, and M. C. Zerner, *J. Phys. Chem.*, **93**, 139 (1989).

205. A. C. Hopkinson and M. H. Lien, *J. Am. Chem. Soc.*, **108**, 2843 (1986).

206. K. Raghavachari, R. A. Whiteside, J. A. Pople, and P. v. R. Schleyer, *J. Am. Chem. Soc.*, **103**, 5649 (1981).

207. M. W. Wong and L. Radom, *J. Am. Chem. Soc.*, **111**, 6976 (1989).

208. W. -K. Lee and N. V. Riggs, *J. Mol. Struct.* (THEOCHEM), **257**, 189 (1992).

209. H. M. Rosenstock, K. Draxl, B. W. Steiner, and J. T. Herron, *J. Phys. Chem. Ref. Data* (*Suppl.*), **1**, 6 (1977).

210. T. D. Norden, S. W. Staley, W. H. Tayler, and M. D. Harmony, *J. Am. Chem. Soc.*, **108**, 7912 (1986).

211. T. J. Lee, A. Willettes, J. F. Gaw, and N. C. Handy, *J. Chem. Phys.*, **90**, 4330 (1989).

212. W. -K. Li. *J. Chem. Res.* (*S*), 220 (1988).

213. N. C. Craig, J. Pranata, S. J. Reinganum, and J. R. Stevens, *J. Am. Chem. Soc.*, **108**, 4378 (1986).

214. T. Takada and K. Ohno, *Bull. Chem. Soc. Jpn.*, **52**, 334 (1979).

215. I. W. Levin and R. A. R. Pearce, *J. Chem. Phys.*, **69**, 2196 (1978).

216. M. N. Glukhovtsev, B. Ya. Simkin., and V. I. Minkin, *Metalorg. Khim.* (*USSR*), **3**, 1289 (1990); E. D. Jemmis, P. Buzek, M. N. Glukhovtsev, P. S. Gregory, A. Kos, and P. v. R. Schleyer, to be published.

217. W. J. Hehre, L. Radom, P. v. R. Schleyer, and J. A. Pople, *Ab Initio Molecular Orbital Theory*, Wiley, New York, 1986.

218. M. S. Gordon, P. Boudjouk, and F. Anwari, *J. Am. Chem. Soc.*, **105**, 4972 (1983).

219. M. Barzaghi and C. Gatti, *J. Chem. Phys.*, **84**, 783 (1987).

220. D. H. Aue, M. T. Bowers, and M. T. Bowers (Ed.), *Gas Phase Ion Chemistry*, Vol. 1. Academic, New York, 1970, p.1. in Ch. 1

221. J. B. Pedley and J. Rylance, *CATCH Tables*, University of Sussex, 1977.

222. B. A. Hess, L. J. Schaad, and P. Čársky, *Tetrahedron Lett.*, **25**, 4721 (1984).

223. E. R. Davidson and W. T. Borden, *J. Chem. Phys.*, **67**, 2191 (1977).

224. W. T. Borden and E. R. Davidson, *Acc. Chem. Res.*, **14**, 69 (1981).

225. P. v. R. Schleyer, E. Kaufmann, G. W. Spitznagel, R. Janoschek, and G. Winkelhofer, *Organometallics*, **5**, 79 (1986).

226. M. Saunder, R. Berger, A. Jaffe, J. M. McBride, J. O'Neil, R. Breslow, J. M. Hoffman, C. Perchonek, E. Wasserman, R. S. Hutton, and V. J. Kuck, *J. Am. Chem. Soc.*, **95**, 3017 (1973).

227. M. J. S. Dewar and R. C. Haddon, *J. Am. Chem. Soc.*, **95**, 5836 (1973).

228. M. J. S. Dewar and R. C. Haddon, *J. Am. Chem. Soc.*, **96**, 255 (1974).

229. W. T. Borden and E. R. Davidson, *J. Am. Chem. Soc.*, **101**, 3771 (1979).

230. J. Feng, J. Leszczynski, B. Weiner, and M. C. Zerner, *J. Am. Chem. Soc.*, **111**, 4648 (1989).

231. J. N. Murrell and K. J. Laidler, *Trans. Faraday Soc.*, **64**, 371 (1968).

232. V. G. Enchev and I. N. Kanev, *Bulg. Acad. Sci.*, **21**, 378 (1988).

233. H. J. Köhler and H. Lischka, *J. Am. Chem. Soc.*, **101**, 3479 (1979).

234. M. N. Glukhovtsev, P. v. R. Schleyer and B. Reindl, *Mendeleev Comm.*, 100 (1993).

235. R. H. Mitchell., N. A. Khailifa, and T. W. Dingle, *J. Am. Chem Soc.*, **113**, 6696 (1991).

236. W. D. Stohrer and R. Hoffmann, *J. Am. Chem Soc.*, **94**, 1661 (1972).

237. M. N. Glukhovtsev, B. Ya. Simkin, and V. I. Minkin, *unpublished results.*

238. H. Schwarz, H. Thies, and W. Franke, in M. A. A. Ferreira (Ed.)., *Ionic Processes in the Gas Phase*, Reidel, New York, 1984, p. 267.

239. C. Cone, M. J. S. Dewar, and D. Landman, *J. Am. Chem. Soc.,* **99**, 372 (1977).

240. C. Glidwell and D. Lloyd, *Tetrahedron*, **40**, 4455 (1984).

241. G. R. Clarc and P. J. Palenik, *J. Organomet. Chem.*, **50**, 185 (1973).

242. A. W. Zwaard and H. Kloosterziel, *Recl.Trav. Chim. Pays-Bas*, **100**, 126 (1981).

243. A. W. Zwaard, A. M. Brouwer, and J. J.C. Mulder, *Recl. Trav.Chim. Pays-Bas*, **101**, 137 (1982).

244. T. J. Katz and P. J. Garrat, *J. Am. Chem. Soc.*, **85**, 2852 (1963).

245. G. Boche, D. Martens, and W. Danzer, *Angew. Chem. Int. Ed. Engl.*, **8**, 984 (1969).

246. G. Boche, H. Weber, and A. Bieberhach, *Chem. Ber.*, **111**, 2833 (1978).

247. A. Sygula, U. Edlung, and P. W. Rabideau, *J. Chem. Res. (Suppl)*, 312 (1987).

248. S. Kuwajima and Z. G. Soos, *J. Am. Chem. Soc.*, **109**, 107 (1987).

249. A. G. Anastassiou and R. C. Griffith, *J. Am. Chem. Soc*., **96**, 611 (1974).

250. M. B. Huang, O. Goscinski, G. Jonsall, and P. Ahlborg, *J. Chem. Soc. Perkin Trans II*, 1327 (1984).

251. M. B. Huang, *Tetrahedron*, **41**, 5209 (1985).

252. M. J. Goldstein and R. Hoffmann, *J. Am.Chem. Soc.*, **93**, 6193 (1971).

253. M. B. Huang and G. Jonsall, *Tetrahedron*, **41**, 6055 (1985).

254. D. Cremer, P. Svensson, E. Kraka, and P. Ahlberg, *J. Am. Chem. Soc.*, **115**, 7445 (1993).

255. B. A. Hess , L.J. Schaad, and I. Agranat, *J. Am. Chem. Soc.*, **100**, 5268 (1978).

256. I. Agranat, B. A. Hess, and L. J. Schaad, *Pure Appl. Chem.*, **52**, 1339 (1980).

257. A. Toyota and T. Nakajima, *Theor. Chim. Acta*, **53**, 297 (1979).

258. B. Ya. Simkin, M. N. Glukhovtsev, and V. I. Minkin, *Zh. Org. Khim.*, **28**, 1337 (1982).

259. J. G. Andrade, J. Chandrasekhar, and P. v. R. Schleyer, *J. Comput. Chem.*, **2**, 207 (1981).

260. B. Ya. Simkin and M. N. Glukhovtsev, *Zh. Strukt,. Khim.*, **24**(3), 23 (1983).

261. M. N. Glukhovtsev and B. Ya. Simkin, *Zh. Strukt. Khim.*, **28**(6), 27 (1987).

262. D. Michalska, H. Chojnacki, B. A. Hess, and L. J. Schaad, *Chem. Phys. Lett.*, **141**, 376 (1987).

263. S. Koseki, T. Nakajima, and A. Toyota, *Canad. J. Chem.*, **63**, 1572 (1985).

264. K. Lammertsma, J. A. Pople, and P. v. R. Schleyer, *J. Am. Chem. Soc.*, **108**, 7 (1986).

265. K. Lammertsma, *J. Am. Chem. Soc.*, **108**, 5127 (1986).

266. R. Breslow, J. Napieski, and T. C. Clarke, *J. Am. Chem. Soc.*, **97**, 6275 (1975).

267. W. N. Washburn, R. Zahler, and I. Chen, *J. Chem. Soc.*, **100**, 5863 (1978).

268. P. B. Dervan and D. A. Dougherty, Nonconjugated Diradicals as Reactive Intermediates in W. T. Borden (Ed.), *Diradicals*, Wiley, New York, 1982, p. 107.

269. M. J. S. Dewar and W. -K. Li, *J. Am. Chem. Soc.*, **96**, 5569 (1974).

270. M. J. S. Dewar, G. P. Ford, and C. H. Reynolds, *J. Am. Chem. Soc.*, **105**, 3162 (1983).

271. J. O. Noell and M. D. Newton, *J. Am. Chem. Soc.*, **101**, 51 (1979).

272. R. Hoffmann, A. Imamura, and W. J. Hehre, *J. Am. Chem. Soc.*, **90**, 1499 (1969).

273. K. Hafner, *Pure Appl. Chem.*, **28**, 153 (1971).

274. A. R. Knox and F. G. Stone, *Acc. Chem. Res.*, **7**, 321 (1974).

275. K. Hafner, R. Donges, E. Goedecke, and R. Kaiser, *Angew. Chem.*, **95**, 362 (1983).

276. B. Kitschke and H. J. Lindler, *Tetrahedron Lett.*, **29**, 2511 (1977).

277. B. Bischoff, R. Gleiter, K. Hafner, K. H. Knauer, J. Spanget-Larsen, and H. U. Süss, *Chem. Ber.*, **111**, 932 (1978).

278. K. Hafner and H. U. Süss, *Angew. Chem.*, **85**, 626 (1973).

279. B. Ya. Simkin, M. N. Glukhovtsev, and V. I. Minkin, *Zh. Org. Khim.*, **28**, 1345 (1982).

280. M. J. S. Dewar and K. M. Merz, *J. Am. Chem. Soc.*, **107**, 6175 (1985).

281. Y. Jean, *Nouv. J. Chim.*, **4**, 11 (1980).

282. E. Heilbronner and S. Shaik, *Helv. Chim. Acta*, **75**, 539 (1992).

283. M. Kataoka, S. Koseki, T. Nakajima, and K. Iida, *Nouv. J. Chim.*, **9**, 135 (1985).

284. K. Hafner, *Pure Appl. Chem.*, **54** , 939 (1982).

285. M. Kataoka, T. Ohmae, and T. Nakajima, *J. Org. Chem.*, **51**, 358 (1986).

286. T. J. Katz and M. Rosenberg, *J. Am. Chem. Soc.*, **84**, 867 (1992).

287. D. Wilhelm, J. L. Courtneidge, T. Clark, and A. G. Davies, *J. Chem. Soc. Chem. Commun.*, 810 (1984).

288. M. K. Shepherd, *Cyclobutarenes. The Chemistry of Benzocyclobutadiene, Biphenylene, and Related Compounds*, Elsevier, Amstredam, 1991.

289. W. Winter and H. Straub, *Angew. Chem. Int. Ed. Engl.*, **17**, 127 (1978).

290. D. W. S. Cruickshank and R. A. Sparks, *Proc. Soc., London Ser. A*, **258**, 270 (1960).

291. R. C. Haddon and K. Raghavachari, *J. Chem. Phys.*, **79**, 1093 (1983).

292. J. Kruszewski, *Pure Appl. Chem.*, **52**, 1525 (1980).

293. H. Kollmar, *J. Am. Chem. Soc.*, **101**, 4832 (1979).

294. M. J. S. Dewar and R. C. Dougherty, *The Perturbation Theory in Organic Chemistry*, Plenum Press, New York, 1975.

295. N. L. Allinger and Y. H. Yuh, *Pure Appl. Chem.*, **55**, 191 (1983).

296. M. J. S. Dewar, *Pure Appl.Chem.*, **44**, 767 (1975).

297. R. C. Haddon and Z. Wasserman, *Nouv. J. Chim.*, **5**, 357 (1981).

298. G. Binsch, E. Heilbronner, and J. N. Murrell. *Mol. Phys.*, **11**, 305 (1966).

299. J. D. Cox and G. Pilcher, *Thermochemistry of Organic and Organomettalic Compounds*, Academic, New York, 1970.

300. J. M. Schulman, R. C. Peck., and R. L. Disch, *J. Am. Chem Soc.*, **111**, 5675 (1989).

301. M. J. S. Dewar and K. M. Merz, *J. Am. Chem. Soc.*, **107**, 6111 (1985).

302. M. J. S. Dewar and K. M. Merz, *J. Am. Chem. Soc.*, **108**, 5142 (1986).

303. M. J. S. Dewar and K. M. Merz, *J. Am. Chem. Soc.*, **108**, 5146 (1986).

304. L. T. Scott, M. A. Kirms, and B. C. Earl, *J. Chem. Soc. Chem. Commun.*, 1373 (1983)

305. L. T. Scott, *J. Org. Chem.*, **49**, 3021 (1984).

306. E. Heilbronner, in D. Ginsburg (Ed.), *Nonbenzenoid Aromatic Compounds*, Interscience, New York, 1959.

307. M. Oda, *Pure Appl.Chem.*, **58**, 7 (1976).

308. D. Kawka, P. Mues, and E. Vogel, *Angew. Chem. Int. Ed. Engl.*, **22**, 1003 (1983).

309. C. Kabuto and M. Oda, *Tetrahedron Lett.*, **21**, 103 (1980).

310. D. Cremer, T. Schmidt, and C. W. Bock, *J. Org. Chem.*, **50**, 2684 (1985).

311. E. Kraka and D. Cremer, Chemical Implication of Local Features of the Electron Density Distribution in Z. B. Maksić (Ed.). *Theoretical Models of Chemical Bonding*, Part 2, Springer, Berlin, 1990, pp. 453–542.

312. C. Glidewell and D. Lloyd, *J. Chem. Educ.*, **63**, 306 (1986).

313. S. Koseki, M. Kataoka, M. Nanamura, T. Nakajima, and T. Toyota, *J. Org. Chem.*, **49**, 2988 (1984).

5

HETEROAROMATICITY

5.1 GENERAL TRENDS OBSERVED IN THE CHANGE OF AROMATIC CHARACTER DUE TO HETEROSUBSTITUTION

The cyclic electron (bond) delocalization associated with certain specificities in the physical and chemical properties of compounds described by the term "aromaticity" is not confined to the conjugated hydrocarbons only. It also characterizes systems containing other atoms from the periodical table. Next we consider some of the questions that arise when examining the aromaticity of heterocyclic compounds.

How does the aromatic (antiaromatic) character of such cyclic compounds change depending on the type of the heteroatom? Are the conjugated heterocyclic compounds characterized by a specificity, described by the term "heteroaromaticity," which is in principle different from that of the corresponding hydrocarbon analogs [1] ? Can there be heterocyclic molecules that possess a greater aromaticity than that of the parent conjugated hydrocarbon ?

We wish to start searching for answers to these and other questions by examining key representatives of the aromatic and antiaromatic classes, that is benzene and cyclobutadiene.

The heteroatoms may be divided into three types—X, Y, and Z—depending on the number of electrons (2, 1 or 0) present in the p_z orbital of the sp^2-hybridized atom contained in the ring. [2–4].

Depending on the type of heteroatom (X, Y, or Z) that replaces the —CH= group in the original hydrocarbon, heterocyclic structures may be designed with either the same number of π-electrons as in the parent molecule, or with a lesser or greater number of these. In the latter case, a change in the number of π-

electrons must reverse the aromatic (antiaromatic) character relative to the hydrocarbon analog:

Such a substitution may involve substantial changes in the structure and may disturb the cyclic electron (bond) delocalization.

Clearly, in view of the diversity in the types of heterocyclic compounds, one may hardly expect that all the manifestations of their aromaticity (antiaromaticity) could be rationalized in terms of some simple regularities. We shall therefore attempt to trace certain characteristic trends in the dependence of the aromaticity on the type of heteroatoms, their number and position in the molecular structure. Our reasoning will be based on the nature of the aromaticity criteria (Chapter 2) and on the electron-count rules (Chapter 3).

The replacement of the —CH= group in an alternant aromatic hydrocarbon with a heteroatom gives rise to the alternation of the π-electron density at the ring atoms. As a result, the aromaticity should, according to the Julg criterion, Eq. (2.55), be somewhat reduced.

Indeed, the values of A found from the data of MNDO calculations on benzene (**1**), borabenzene (**5**), and pyridine (**6**) are 1.0, 0.939, and 0.971, respectively [5]. HSE calculations indicate (Table 2.2) that pyridine is, in aromatic character, only slightly inferior to benzene. The values of the resonance energy (HSRE, TRE, CCMRE; see Table 2.1) point to a lessening of the aromaticity in the order benzene > pyridine (**10**) > pyrazine (**11**); note that the change in the RE value is in this case insignificant. Nearly the same trend is observed for the structural index ΔN (Table 2.7), according to which the aromaticity relative to benzene is for **6** 82%, for **7** 67%, for **8** 75%, and for pyridazine (**12**) 65% [1].

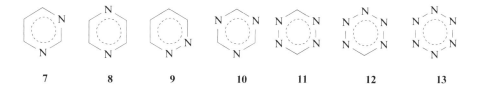

7   8   9   10   11   12   13

The aromaticity of azines is reduced relative to benzene, as is evidenced by the RCI values [6] (Table 2.7) as well as by the RE values calculated from the energies of hydrogen-transfer reactions [7] (Table 5.1). For example, in the case of pyridine the MP3/6-31G** calculated energy of the homodesmotic reaction (5.1) equals –1.8 kcal/mol. Since the RE of benzene determined from the hydrogenation enthalpies is 36 kcal/mol, pyridine's RE will, accordingly, be 34.2 kcal/mol.

14

$$ \text{(5.1)} $$

TABLE 5.1   Aromaticity Indices of Azines

Compound	RCI[6]	RE $[10]^a$ MP3/6-31G** //6-31G*	ΔH_{diss} [7] Calculated^b	Observed
Benzene	1.751	36	153	143
Pyridine	1.731	34.2	117	105
Pyrazine	1.739	32.0	74	70
Pyrimidine	1.727	32.6	78	70
Pyridazine	1.716	26.1	50	50
s-Triazine	1.724	24.8	40	40
s-Tetrazine	1.735	15.3	– 54	– 56
Hexazine	1.792	—	– 213.2 ^{c}	

^{a}Calculated from the energies of hydrogen-transfer reactions, for example, Eq. (5.1).
^{b}Based on the MP3/6-31G**//6-31G* energy changes.
^{c}For hexazine the MP3/6-31G**//6-31G* dissociation energies are given without ZPVE correction. Respective values for benzene and pyridine are 165.8 and 124.8 kcal/mol [7]. At MP4SDTQ/6-31G**//MP2/6-31G** + ZPE(MP2/6-31G*), the energies of dissociations of hexazine (**14**) (into $3N_2$) and of benzene (into $3HC \equiv CH$) are – 214.5 and 146.7 kcal/mol, respectively [17, 18].

Thus azine, with the exception of possibly *sym*-tetrazine (**11**), have fairly similar resonance energies (Table 5.1). Even though the π-density is concentrated near the nitrogens, the π-density distribution per unit volume of the elements (HF/6-31G** calculations [7]) is not changed in any significant manner when CH is replaced with N.

The relatively low (compared to benzene) RE values of pyridazine, *s*-triazine, and *s*-tetrazine (see Table 5.1) are explained primarily by changes in the σ-system that occur in passing from the conjugated system to the reference system, that is by the factors such as the compression energy (see discussion on the so-called empirical resonance energies, Chapter 2, Section 2.5.1.1).

Hexazine (**13**), which is the final product of the successive azasubstitution of benzene, possesses an even greater aromaticity than benzene: this is indicated by the values of RCI [6] in Table 5.1, DRE (28.2 kcal/mol for **13**, 20.0 kcal/mol for benzene [8]), and QMRE (102.5 kcal/mol for **13**, 85.2 kcal/mol for benzene [9]). However, several reports of the possible generation of **13** [10–12] have been challenged [13]. Computations show that the C_2 open-chain N_6 structure is much more stable than **13** [13–16]. Hexazine like hexaphosphabenzene P_6 [17], favors the nonplanar D_2 twist-boat structure (**14**) [18]. Total stabilization energy (HSE) for hexazine is negative ($- 17.6$ kcal/mol at MP4SDTQ/6-31G**//MP2/6-1G**) [18]. Instability of hexazine, in particular, with respect to the dissociation into $3N_2$ (Table 5.1) is associated with the specificity of its σ-electronic system [19–21].

Like hexazine, *s*-tetrazine (**11**) and pentazine (**12**) are thermodynamically unstable to dissociation into HCN and N_2 (see Table 5.1 and MNDO results [20, 22]). At MP2/6-31G**//MP2/6-31G**, pentazine (**12**) is 132.2 kcal/mol unstable towards the dissociation into HCN and molecular nitrogen [17]. In contrast to hexazine, **12** has a planar C_{2v} structure (a minimum at MP2/6-31G* [17]).

The aromaticity of azoles rises, according to the values of the structural index ΔN [1] and of HSRE [23], with the increase in the number of nitrogen atoms in the ring. The aromaticity relative to benzene estimated from the value of the ΔN index is as follows: for pyrrole (**15**) 37%, for imidazole (**16**) 43%, for pyrazole (**17**) 61%, for 1,2,4-triazole (**18**) 71%, and for tetrazole (**19**) 81% [1]. The values of HSRE for pyrrole, imidazole, and pyrazole are 0.234, 0.251, and 0.330, respectively (in β units) [23].

| 15 | 16 | 17 | 18 | 19 |

When atoms are involved whose electronegativity differs considerably from that of carbon, it may appreciably reduce aromaticity (e.g., judging from the structural index ΔN, the aromaticity of the pyrylium cation (**20**) is 43% that of benzene). In the π-electron-excessive five-membered ring heterocycles (**15**, **21**, and **22**) the aromaticity decreases with the increase in the electronegativity of the heteroatom [1, 24] (Tables 2.1 and 5.2).

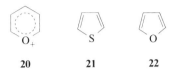

As has been noted in Chapter 3, when the difference between the electronegativities of the heteroatom and the carbon is large enough, the $(4n + 2)$ rule may lose its validity [25].

Calculated values of the HSE (see Section 2.2.9) [26] indicate that the aromatic character is decreased with the increase in the difference between electronegativities of the neighboring atoms. For example, for the series benzene (**1**), *sym*-triazine (**10**), borazine (**23**), and boroxine (**24**) the ratio between their 4-31G calculated HSE values is roughly 3:2:1:0 [27].

      H   H

      B   B

HN   NH   O   O

HB   BH   HB   BH

     N   O

     H

     23   **24**

As one goes to main-group heteroatoms of the second and further rows, the aromatic character of the heterosubstituted benzenes is diminished compared to the parent hydrocarbons. This is apparent from the ISE and HHSE values (see Section 2.2.9) presented in Table 5.2. The same regularities characterize the series of pyrrole isologs C_4H_4XH (X = N—Sb).

The heterosubstitution in antiaromatic molecules, with the number of π-electrons remaining unchanged, may remove the degeneracy of the incompletely filled π-levels, for example, in the D_{nh} structures of antiaromatic annulenes and monocyclic conjugated ions, and may lead to stabilization of a molecule as a whole [28–30]. This approach represents one of the routes for obtaining stable derivatives of cyclobutadiene. Besides the introduction of donor and acceptor substituents [28, 31] (see Chapter 4), the stabilization of a 4 π-electronic four-membered ring may be achieved by the heterosubstitution to give a **25**-type system, where X and Y are the atoms of different electronegativity. For a half-filled π-system, structure **25** is more stable than structure **26** (see Fig. 5.1) [29, 30]. As distinct from **26**, for **25** an effective stabilization of the e_g π-MO is possible. This MO possesses large amplitudes with more electronegative atoms.

TABLE 5.2   The ISE and HHSE Values (in kcal/mol) Calculated with 3-21G* Basis Set [26] for Some Monoheteroatomic Cycles

Compound	ISE[a]	HHSE
Benzene	61.27[b]	25.99 (23.5)[c, d]
Silabenzene	46.84	17.97
Germabenzene	46.32	16.75
Stannabenzene	42.82	12.20
Phosphabenzene	56.63	23.30 (22.9)[d]
Pyridine	64.19	25.78
Arsabenzene	53.87	21.58 (21.8)[d]
Stibabenzene	50.52	18.56
Pyrrole	43.80	5.16
Phosphole	16.49	−1.53
Arsole	16.81	−0.18
Stibole	14.33	−1.92
Furan	35.19	5.01
Thiophene	32.54	10.07
Selenophene	27.78	7.89
Tellurophene	22.19	4.31

[a]Corrected for ZPVE; the 3-21G basis set was used for the first row atoms.
[b]The experimental values of ISE for benzene and pyridine are 64.1 ± 1.7 kcal/mol and 65.5 ± 1.6 kcal/mol, respectively [36]. At HF/6-31G*, HSE (benzene) = 24.8 kcal/mol (21.6 ± 1.5, exptl.) and HSE (pyridine) = 25.4 kcal/mol [36].
[c]Experimental 22.2 kcal/mol; at MP4SDTQ/6-31G**//MP2/6-31G**, HHSE (benzene) = 20.3 kcal/mol [17].
[d]HHSE values calculated at HF/6-31G* are given in parentheses [35].

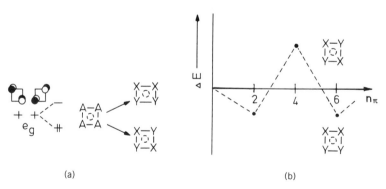

(a)   (b)

Figure 5.1   (a) The lifting of degeneration of the e_g π-MOs in cyclobutadiene and its heterocyclic analogs. One of the e_g π-MOs that is localized at the atoms with higher electronegativity is stabilized. This is possible for the **30**-type structures, in contrast to the **31**-type structures [35, 36]. (b) The dependence of relative stability of structures **30** and **31** on the filling of π-levels [35], obtained by the method of moments.

Thus the larger the difference between the electronegativities of the atoms X and Y, the more sizeable is the energy splitting.

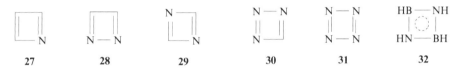

Such criteria of aromaticity as the value of ΔE (HOMO–LUMO) and the difference between the energies of the ground and the lowest singlet excited states (Section 2.5.1.2) imply that the heterosubstituted cyclobutadienes will have lesser antiaromaticity as compared to the parent hydrocarbon. Indeed, according to calculations of the ISE for azetes and of HHSE (Table 5.3, **27–31**) [32–34],1,3-diazete (**29**) is less antiaromatic than cyclobutadiene and it is more stable than 1,2-diazete (**28**) [33, 37] (cf. Fig.5.1). Also, the TRE values point to a smaller antiaromaticity of azetes relative to cyclobutadiene [38]. *Ab initio* calculations of the HSE for **28** and **32** [27] show that as the difference between the electronegativities of the ring atoms grows, the degree of antiaromaticity is indeed lowered: HSE(**28**) = – 95.0 while HSE(**32**) = – 51.9 kcal/mol (4-31G [27]).

Antiaromatic monocyclic ions and antiaromatic annulenoannulenes can equally be stabilized by heterosubstitution. For example, unlike the cyclopentadienyl cation (**33**), the 1,3-diazacyclopentadienyl cation (**34**) can be isolated in the form of a stable crystalline salt, such as (**35**) [28] (see Fig. 5.2).

TABLE 5.3   HHSE Values Calculated at MP4SDTQ/6-311 + G**//MP2/6-31G** for Azasubstituted Cyclobutadienes [34]

Compound	HHSE, kcal/mol
Cyclobutadiene	– 77.7[a]
Azete, **27**	– 72.7
1,3-Diazete, **29**	– 64.7
Triazete, **30**	– 57.8
Tetrazete, **31**	– 60.8

[a]These HHSE values do not include strain energy contributions. For cyclobutadiene, the strain energy is estimated to be 32 kcal/mol [39].

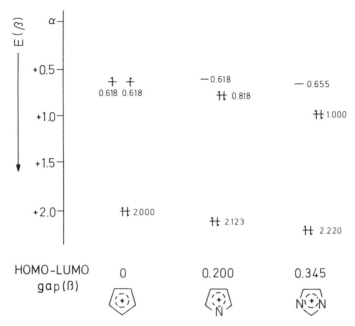

Figure 5.2   HOMO–LUMO energy gaps of (aza) cyclopentadienyl cation calculated by HMO. (Adapted from [34]).

The stabilization through the heterosubstitution is not confined to the monocyclic systems, it can be achieved for bicyclic and polycyclic antiaromatic species as well. For example, HSRE [23] and TRE [40] calculations indicate that 2,5-diazapentalene (**37**) is not antiaromatic, like pentalene (**36**), but rather non aromatic: HSRE(**36**) = – 0.14, HSRE(**37**) = – 0.007, and TRE(**36**) = – 0.064 (in β units) [23, 40]. The finding is supported by the structural criterion of λ_{max} (see Eq. (2.72)): λ_{max}(**36**) = 2.357 and λ_{max}(**37**) = 1.677 [41]. These results, together with experimental data such as those showing thermal stability of derivatives of **37** [28], indicate the absence of the antiaromatic destabilization in 2,5-diazapental-ene.

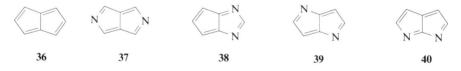

The stabilization of **37** provides an example of the so-called topological charge stabilization [42]. The maximal stabilization energy is attained in this case through insertion of heteroatoms in such sites of conjugated hydrocarbon with nonuniform charge densities (nonalternant hydrocarbons, dications and dianions of alterant hydrocarbons) as to enable the   electronegativities of heteroatoms to match the pattern of charge densities in the isoelectronic hydrocarbon species. In the case of heteroatoms with large electronegativities compared to carbon, the maximal stabilization is achieved when heteroatoms are introduced at sites characterized by the greatest charge densities in the iso-electronic conjugated hydrocarbons. Thus in pentalene (**36**), the π-charge densities are 0.82 (CI) and 1.17(C2) [42]. Whereas 2,5-diazapentalene (**37**) may be regarded as nonaromatic, the isomeric 1,3-diaza-3,6-diaza-, and 3,4-diazapentalenes (**38–40**) have TRE values (– 0.376, – 0.328, and – 0.336, respectively) that are even more negative than that of **36** [23].

Viewing the nitrogen substitution as a perturbation of the corresponding hydrocarbon, one may derive a quantitative relationship between the RE value and those of the Coulomb integral (which, upon substitution, changes to $\Delta\alpha_i$) and of the charge at the ith site. For the HSRE per π-electron [31]

$$\Delta HSRE(PE) = (q_i - q_i \,(ref))\Delta\alpha_i \qquad (5.2)$$

where $q_i(ref)$ is the value of the ith charge in the reference structure. When $q_i > 1$, the nitrogen substitution must result in an increased HSRE (in β units), as is indeed exemplified by structure **37**. Conversely, for $q_i < 1$, the value of HSRE falls as a result of the azasubstitution, as is the case in **40**.

5.2. AROMATICITY AND THE TREND TOWARD PYRAMIDALIZATION OF THE HETEROATOM

As was pointed out in Section 2.3.4, the intention to link the antiaromatic character of a molecule with a greater stability of its nonplanar structure compared to the planar one is not fully justified.

It was noted in Chapter 4 that, according to *ab initio* calculations, the cyclo-propenide anion $(CH)_3^-$ has a nonplanar structure (**41**) with pyramidalized carbon atoms whose preferability over the 4 π-electron planar structure (**42**) is the result of the antiaromaticity of the latter. Thus one expect that 2-azirine, which is isoelectronic with $(CH)_3^-$, will have a nonplanar structure (**43**) with pyramidalized nitrogen atoms. Indeed, semiempirical [46] and nonempirical [47–50] calculations support this assumption; note that the barrier of the nitrogen inversion turns out to be considerably greater (37.7 kcal/mol, MP3/6-31 + G//6-31 + G [49]) than in NH_3 (5.8 kcal/mol, experiment [51]).

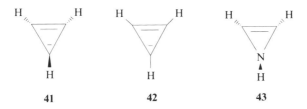

41   42   43

N	P	P	As
H	H	H	H
15	**44**	**44a**	**45**

The planar bond configuration of the nitrogen atom in pyrrole is usually explained in terms of aromaticity. The pyramidalization of the phosphorus and arsenic atoms in phosphole (**44**) and arsole (**45**) was taken to be the consequence of their much lower aromaticity relative to pyrrole [52] (see Table 5.4).

The question as to the correctness of such an explanation of the nonplanarity of 2-azirine and the planarity of pyrrole was analyzed by Mo et al. [53]. The evolution of the MOs as a function of the pyramidalization angle in nitrogen was traced. It has turned out that for both 2-azirine and pyrrole the nitrogen pyramidalization leads to the stabilization of all π-MOs of the planar structure and, conversely, to the destabilization of several σ-type MOs. In the case of pyrrole, the stabilization of the π-orbitals proves insufficient to counterbalance the concomitant destabilization of the σ-orbitals. For 2-azirine, the situation is reversed. Consequently, the planar structure of pyrrole and the nonplanar structure of 2-azirine are primarily dictated by the σ-frame [53]. As for the nonplanar geometries of phosphole and arsole, they also are determined not so much by the loss of the π-aromaticity as by the σ-system [54, 55]. Moreover, as shown by Jug and Köster [56] for five- and six-membered heterocycles (furan, pyrrole, pyrazole, azines, etc.), the σ-system dictates the structure with bond equalization, thereby enforcing delocalization of π-electrons. Note, however, that the stabilization of the π-MOs considered by Mo et al. [53] may, in the case

TABLE 5.4 Topological Resonance Energies (TREs), Hess–Schaad Resonance Energies (HSREs), and Ring-Current Indices (RCIs) for Some Odd-Membered Heterocycles (TREs and HSREs are given in β units)

Compound	TRE [40, 43]	HSRE [43, 44]	RCI [6, 45]
Borirene	0.165	—	1.614
Borole	−0.080	—	1.140
Borepin	0.026	—	—
2-Azirine	−0.129	—	1.031
Pyrrole	0.040	0.039	1.463
1H-Azepine	−0.029	−0.036	—
Oxirene	−0.109	—	1.030
Furan	0.007	0.007	1.430
Oxepine	−0.004	−0.006	—
Thiirene	−0.107	−0.114	—
Thiophene	0.033	0.032	—
Thiepine	−0.023	−0.029	—

of nonplanar distortions of geometry, be accompanied by a reduction of the aromaticity, as is the case in benzene distortions [57] (see Section 2.3.1). The aromatic cyclic π-electron (bond) delocalization stabilizes the planar structure with bond equalization [57]: the problem is that, in addition to that effect, other effects may exist, which may eventually overshadow it.

Thus we conclude that the preferability of the planar or nonplanar geometry of a heterocyclic depends on a number of factors, including the aromaticity (antiaromaticity), which may not be the most important factor. In any case, this factor should not be disregarded if one wishes to obtain a correct overall energy balance. Note, for example, that the aromaticity is reflected in the values of inversion barriers. Thus for antiaromatic 2-azirine, the nitrogen inversion barrier is 37.7 kcal/mol, while in the case of its saturated analog, aziridine (**46**), this barrier, calculated at the same level, is 14.9 kcal/mol [49].

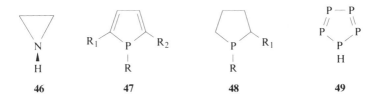

The phosphorous inversion barrier in phosphole (**44**) is 19.8 kcal/mol at MP4SDTQ/6-31G*//MP2/6-31G* [59]. Planar C_{2v} structure **44a** is the transition state for this inversion. The phosphorous inversion barrier, according to data on P-substituted phospholes (**47**) [58], is about 15.5 kcal/mol (the value of such a low barrier does not depend essentially on the type of substituent), but in its saturated analog (**48**) the barrier is 36 kcal/mol. At the same time, pentaphosphole (**49**) favors a planar C_{2v} structure, which is a minimum at MP2/6-31G* [59], in contrast to the planar C_{2v} structure of phosphole (**44a**).

Unfortunately, the restricted space of this book prevents us from discussing particular compounds. Some information may be found in [60–75].

REFERENCES

1. A. F. Pozharskii, *Khim. Geterotsikl. Soed (USSR)*, 867 (1985).

2. A. T. Balaban, *Pure Appl. Chem.*, **52**, 1409 (1980).

3. A. T. Balaban, *J. Chem. Inf. Comput. Sci.*, **25**, 334 (1985).

4. A. T. Balaban, M. Banciu, and V. Ciorba, *Annulenes, Benzo-, Homo-, Heteroannulenes and Their Valence Isomers*, CRC Press, Boca Raton, FL, 1985.

5. M. N. Glukhovtsev, *unpublished results.*

6. K. Jug, *J. Org. Chem.*, **48**, 1344 (1983).

7. K. B. Wiberg, D. Nakaji, and C. M. Breneman, *J. Am. Chem. Soc.*, **111**, 4178 (1989).

8. B. Ya. Simkin, M. N. Glukhovtsev, and V. I. Minkin, *Zh. Org. Khim.*, **24**, 24 (1988).

9. G. Ohanessian, P. C. Hiberty, J. M. Lefour, J. P. Flament, and S. S. Shaik, *Inorg. Chem.*, **27**, 2219 (1988).

10. E. Hagon and M.Simic, *J. Am. Chem. Soc.*, **92**, 7486 (1970).

11. V. Pizak and H. Wendt. *Ber. Bunsenges. Phys. Chem.*, **83**, 481 (1979).

12. A. Vogler, R.E. Wright, and H. Kunkley, *Angew. Chem.*, **92**, 745 (1980).

13. T.-K. Ha and M. T. Nguyen, *Chem. Phys. Lett.*, **195**, 179 (1992).

14. H. Huber, T.-K. Ha, and M. T. Nguyen, *J. Mol. Struct.* (THEOCHEM), **105**, 351 (1983).

15. M. T. Nguyen, *J. Phys. Chem.*, **94**, 6923 (1990).

16. R. Engelke, *J. Phys. Chem.*, **94**, 6924 (1990).

17. M. N. Glukhovtsev and P. v. R. Schleyer, *J. Am. Chem. Soc.*, to be published.

18. M. N. Glukhovtsev and P. v. R. Schleyer, *Chem. Phys. Lett.*, **198**, 547 (1992).

19. S. Inagaki and N. Goto, *J. Am. Chem. Soc.*, **109**, 3234 (1987).

20. S. D. Peyerimhoff and R. J. Buenker, *J. Chem. Phys.*, **48**, 354 (1968).

21. J. S. Wright, *J. Am. Chem. Soc.*, **96**, 4753 (1974).

22. M. J. S. Dewar, *Pure Appl. Chem.*, **44**, 767(1975).

23. B. A. Hess, L. J. Schaad, and C. W. Holyoke, *Tetrahedron*, **31**, 295 (1975).

24. E. Uggerud, *J. Chem. Soc. Perkin Trans. II*, 1857 (1986).

25. I. Gutman and A. K. Mukherjee, *Indian J. Chem.*, **27a**, 487 (1988).

26. K. K. Baldridge and M. S. Gordon, *J. Am. Chem. Soc.*, **110**, 4204 (1988).

27. R. C. Haddon, *Pure Appl. Chem.*, **52**,1129 (1982).

28. R. Comper and H. U. Wagner, *Angew. Chem. Int. Ed. Engl.*, **24**, 1437 (1988).

29. J. K. Burdett, S. Lee, and T. J. McLarnan, *J. Am. Chem. Soc.*, **107**, 3083 (1985).

30. J. K. Burdett, E. Canadell, and T. Hugnbanks, *J. Am. Chem. Soc.*, **108**, 3971 (1986).

31. B. A. Hess and L. J. Schaad, *J. Org. Chem.*, **41**, 3058 (1976).

32. M. N. Glukhovtsev, B. Ya. Simkin, and V. I. Minkin, *Zh. Org. Khim.*, **19**, 1353 (1983).

33. M. N. Glukhovtsev, B. Ya. Simkin, and V. I. Minkin, *Zh. Strukt. Khim.*, **28** (N4), 28 (1987)

34. M. N. Glukhovtsev and P. v. R. Schleyer, *unpublished results.*

35. C. W. Bock, M. Trachtman, and P. George, *Struct. Chem.*, **1**, 345 (1990).

36. P. George, C. W. Bock, and M. Trachtman, *Tetrahedron Lett.*, **26**, 5667 (1985).

37. I. Alkorta, J. Elguero, I. Rozas, and A. T. Balaban, *J. Mol. Struct.*, (THEOCHEM), **208**, 63 (1990).

38. S. Singh, R. K. Mishra, and B. K. Mishra, *Indian J. Chem.*, **27a**, 653 (1988).

39. B. A. Hess and L. J. Schaad, *J. Am. Chem. Soc.*, **105**, 7500 (1983).

40. I. Gutman, M. Milun, and N. Trinajstić, *J. Am. Chem. Soc.*, **99**, 1692 (1977).

41. G. Binsch and I. Tamir, *J. Am. Chem. Soc.*, **91**, 2450 (1969).

42. B. M.Gimarc, *J. Am. Chem. Soc.*, **105**, 1979 (1983).

43. B. A. Hess, L. J. Schaad, and C. W. Holyoke, *Tetrahedron*, **28**, 3657 (1972).

44. B. A. Hess and L. J. Schaad, *J. Am. Chem. Soc.*, **95**, 3907 (1973).

45. K. Jug, *J. Org. Chem.*, **49**, 4475 (1984).

46. M. J. S. Dewar and C. A. Ramsden, *J. Chem. Soc. Chem. Commun.*, 688 (1973).

47. P. Čársky, B. A. Hess, and L. J.Schaad, *J. Am. Chem. Soc.*, **105**, 396 (1983).

48. S. M. Zavoryev and R.-I. I. Rakauskas, *Litov. Fiz.* **27**, 241 (1987).

49. M. Alcami, J. L. G. dePaz, and M. Yanez, *J. Comput. Chem.*, **10**, 468 (1989).

50. Y.-G. Byun, S. Saebo, and C. U. Pittman, *J. Am. Chem. Soc.*, **113**, 3689 (1991).

51. J. D. Smalen and J. A. Ibers, *J.Chem. Phys.*, **36**, 1914 (1962).

52. M.H. Palmer, *J. Chem. Soc. Perkin Trans. II*, 974 (1975).

53. O. Mo, M. Yanez, and J. Elguero, *J. Mol. Struct.* (THEOCHEM), **201**, 17 (1989).

54. N. D. Epiotis and W. Cherry, *J. Am. Chem. Soc.*, **98**, 4365 (1976).

55. W. Schäfer, A. Schweig, and F. Mathey, *J. Am. Chem. Soc.*, **98**, 407 (1976).

56. K. Jug and A. M. Köster, *J. Am. Chem. Soc.*, **112**, 6772 (1990).

57. M. N. Glukhovtsev, B. Ya. Simkin, and V. I. Minkin, *Zh. Org. Khim.*, **25**, 673 (1989).

58. W. Egan, R. Tang, G. Zon, and K. Mislow, *J. Am. Chem. Soc.*, **93**, 6205 (1971).

59. M. N. Glukhovtsev, and P. v. R. Schleyer, *unpublished results*.

60. M. J. Cook, A. R. Katritsky, and P. Linda, *Adv. Heterocycl. Chem.*, **17**, 255 (1974).

61. C. A. Ramsden, *Tetrahedron*, **33**, 3203 (1977).

62. A. J. Ashe, *Acc. Chem. Res.*, **11**, 153 (1978).

63. A. G. Anastassiuo and H. S. Kasmai, *Adv. Heterocycl.Chem.*, **23**, 55(1978).

64. M. Torres, E. M. Lown, H. E. Gunning, and O.P. Strausz, *Pure Appl. Chem.*, **52**, 1623 (1980).

65. K. Dimroth, *Acc. Chem. Res.*, **15**, 58 (1982).

66. A. F. Pozharskii, *Theoretical Principles of Heterocyclic Chemistry*, Khimiya, Moscow, 1985.

67. G. E. Herberich and H. Ohst, *Adv. Organomet. Chem.*, **25**, 199 (1986).

68. B. Ya. Simkin and M. N. Glukhovtsev, *Khim.Geterotsikl. Soed.*, 1587 (1989).

69. M. N. Glukhovtsev and B. Ya. Simkin., *Metallorg. Khim.* (*USSR*), **5**, 1063 (1990).

70. V. I. Minkin, M. N. Glukhovtsev, and B. Ya. Simkin, *Adv. Heterocycl. Chem.*, **56**, 303 (1993).

71. G. Leroy, M. Sana, and C. Wilante, *Theor. Chim. Acta.*, **85**, 155 (1993).

72. V. Jonas and G. Frenking, *Chem. Phys. Lett.*, **210**, 211 (1993).

73. W. W. Schoeller and T. Busch, *Angew. Chem. Int. Ed. Engl.*, **32**, 617 (1993).

74. K. B. Wiberg, D. Y. Nakaji, and K. M. Morgan, *J. Am. Chem. Soc.*, **115**, 3527 (1993).

75. M. N. Glukhovtsev, P. v. R. Schleyer, and C. Maerker, *J. Phys. Chem.*, **97**, 8200 (1993).

6

HOMOAROMATICITY

6.1 GENERAL OUTLINE

In 1956 Applequist and Roberts [1] assumed that the rupture of a cyclic conjugation due to the insertion of a saturated fragment (such as CH_2) partly preserves the aromatic stabilization of the original compound. They drew attention to the unusual stability of the cyclobutenylium ion (**1**). Winstein, who earlier introduced the concepts of "homoallylic" bonding and "homoconjugation" [2], generalized the idea proposed in [1] as follows: "the aromatic stabilization of conjugated systems with $(4n + 2)$ π-electrons may not be destroyed by the insertion of one or more intervening groups" [3]. He suggested the term "homoaromaticity."

One of the formal ways of designing a homoaromatic system consists in the replacement of the double bond of an aromatic compound with a cyclopropane fragment. Such a replacement allows one to pass from benzene to homobenzene (**2**) or from the cycloheptatrienyl cation to the homotropylium cation (**3**).

Another way is to insert a CH_2 group into the corresponding aromatic ring. The third formal way of obtaining a homoaromatic system consists of the protonation of antiaromatic structures. For example, structures **1–3** may be conceived of as protonated cyclobutadiene, cycloheptatrienyl anion, and cyclooctatetraene. The concept of homoaromaticity has quickly spread in organic chemistry (for reviews see [4–8]). A good deal of experimental data have been interpreted with its aid, and yet, up to the present day, no conclusive answer has been given to the following question. Is the stabilization of the initial aromatic system really preserved or is the term in question of semantic value, representing no actual physical effect? Even so, the concept of the homoaromaticity has been extended, giving rise to derivatives, such as bishomo-aromaticity, trishomoaromaticity, and homoantiaromaticity. Some examples of the compounds representing, respectively, these types of homoaromatic stabilization or destabilization are given below.

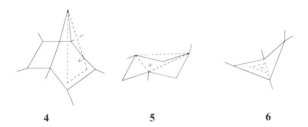

4   5   6

According to the Goldstein–Hoffmann formulation ·[9], the π-electron topology of homoaromatic and homoantiaromatic structures is a particular case of the pericyclic topology (see Chapter 3).

3a   3b

The key structure in the problem of homoaromaticity is the homotropylium cation (**3**). The synthesis of its derivatives and the study of their properties have been the subject of more than 100 publications (for reviews see [4, 5, 7]). For the first time, **3** was prepared by Pettit and co-workers [10] by treatment of cyclo-octatetraene with concentrated sulfuric acid or $SbCl_5/HCl$ in nitromethane. Perhaps the most surprising fact revealed by experimental observations is the downfield and upfield shifts of, respectively, *exo* (5.86 ppm) and *endo* (− 0.73 ppm) protons of the methylene group in the ^1H NMR spectrum. This led Pettit to reject the classical structure (**3a**), suggesting instead the structure of bicyclo [5.1.0] octadienyl cation (**3b**), which afterward was portrayed as **3** in order that the cyclic six-electron delocalization might be retained. Thus it was held that the ring currents induced by magnetic field determined the striking difference between the chemical shifts of the *exo* and *endo* protons of the methylene group. This view was backed by the results derived from the study of NMR spectra of the metal carbonyl complexes **7** and **8**:

δ −0.15 H   H δ 3.37 δ 1.35 H   H δ 1.53

Mo(CO)$_3$ Fe(CO)$_3$

7   **8**

In the molybdenum complex (**7**) six electrons are retained in the π-system, while iron complex (**8**) has only four π-electrons, which are not completely delocalized over the seven-membered ring, so that no ring currents are present. However, it should be emphasized once again (see Chapter 2) that the presence of an induced ring current in a molecule, whether it be detected in its ^{1}H NMR spectrum or determined directly with diamagnetic susceptibility measurements, is but one and, indeed, a relatively poor criterion of aromaticity [11], which has also been shown in the case of homoaromatic structures. The most effective tool in studying the actual effects of homoaromaticity has been the quantum chemical approach. Our analysis starts with the examination of the results gained by employing qualitative approaches.

6.2 QUALITATIVE APPROACHES

The formal approach based on the HMO method and applied by Winstein leads to the conclusion that with the insertion of the CH$_2$ group the delocalization is largely retained. Indeed, taking the resonance integral β_1 of the 1,3-bond in **3** to be nonzero and equal to $0.5\beta_0$, it is easy to calculate the energy of delocalization of **3**, namely, $2.423\beta_0$, which is comparable with that of the cycloheptatrienyl cation amounting to $2.988\beta_0$.

The qualitative theory of the homoaromatic and homoantiaromatic structures was formulated independently and at almost the same time by several researchers [12–20] proceeding from the analysis of orbital interactions. In most cases, the interactions between Walsh orbitals of the cyclopropane ring and π-orbitals of the polyene fragment were considered. The most important orbital interactions arise between the degenerate pair of cyclopropane HOMOs (e_s and e_a) and the HOMO (π) and LUMO (π^*) of the polyene.

For the simplest homoantiaromatic system, bicyclo[2.1.0]pentene, the frontier orbital interactions are shown in Fig. 6.1. The interaction between the π- and e_s-MOs leads to a four-electron destabilization. The two-electron

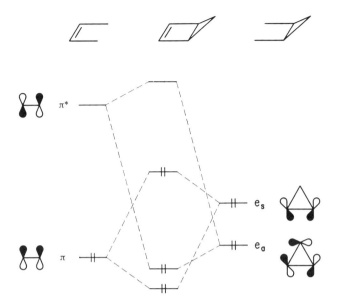

Figure 6.1 Frontier orbital interactions in bicyclo[2.1.0]pentene. Reprinted with permission from W. L. Jorgensen, *J. Am. Chem. Soc.*, **98**, 6784 (1976).

stabilization brought about by the interaction of the e_a with π^*-MOs is not sufficient for offsetting the π-e_s repulsion because of the large energy gap between the e_a-π^*-MOs and their small (relative to the e_s-MO) wavefunction amplitudes at the interacting atoms. Thus the total interaction between the cyclopropyl orbitals and the π-orbitals of ethylene turns out to be destabilizing.

In the case of the homoaromatic structures **2** and **3**, the situation is different, Whereas for homoaromatic structures the overlap of the e_s-orbital of cyclopropane with the LUMO of the polyene is due to their different symmetry, zero, in the case of homoaromatic structures this overlap becomes appreciable, determining the ability for stabilization. Indeed, the LUMOs given below of the butadiene and of the pentadienyl cation are symmetrical. The strongest interaction with the e_s-orbital of cyclopropane arises in the **1** and **3** structures because the LUMOs of polyene are nonbonding.

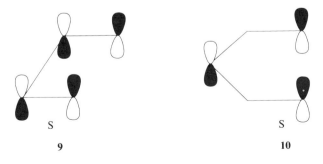

The two-electron stabilizing interaction is associated with the transfer of the electron density from the e_s- to π-orbitals. Since the e_s-MO is highly bonding for the fused bonds, the loss of the electron density is predicted to result in the lengthening of these. Hence the lengths of the fused bonds must be larger than that of the C—C bond in the cyclopropane ring.

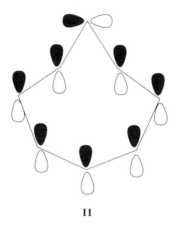

11

A somewhat different qualitative treatment of homoaromatic systems was suggested by Hehre [12, 16]. He proceeded from another pattern of fragmentation of the parent molecule, namely, into methylene and polyene C_7H_7. The orbital interaction of the open cyclopropane ring with the polyene fragment is considered. By analogy with [21, 22], both the methylene orbital and the orbitals of the neighboring CH groups are taken into account. In this case, the following picture of bonding arises, which indicates a transformation of the Hückel aromatic system to the Möbius aromatic eight-electron structure [23], since in this interpretation the number of electrons populating the basic orbital system increases by two. Both approaches are by no means contradictory; on the contrary, they complement each other. Similar to the case of cyclopropane, the qualitative schemes allow one to investigate the interaction of the cyclobutane ring with the polyene fragment with the aim of determining the possibility of manifestations of the homoaromaticity in compounds **12–15**.

| 12 | 13 | 14 | 15 |

The examination of the interaction between the orbital of cyclobutane and polyene has shown that in systems **12**, **14** and **15**, which formally are also homoaromatic, the difference between the length of the fused bond and that of the C—C bond in the cyclobutadiene ring must be substantially less than the difference between the length of the C—C bond in cyclopropane and that of the fused bond in corresponding homoaromatic structures [14, 19, 20]. Direct quantum mechanical calculations [19] have confirmed this conclusion.[1] Thus the cyclopropane ring holds a unique position among the saturated cycles as to the magnitude of its interaction with polyene fragments.

Based on the theory of orbital interactions, one may predict [13] that the homoaromaticity must manifest itself most strongly in cations and anions and to a much lesser degree in neutral compounds. This conclusion rests on the estimates of the difference between the HOMO and LUMO energy levels. We shall see later whether the conclusions of qualitative theories are compatible with those derived from quantitative semiempirical and nonempirical quantum mechanical calculations.

6.3   HOMOAROMATIC CATIONS

6.3.1   Homotropylium Cation

6.3.1.1   Structure of the Homotropylium Cation   The first *ab initio* calculation on the homotropylium cation made by Hehre using the STO-3G basis set [16, 24] led to an unexpected result: the length of the C(1)–C(3) homoaromatic bond turned out to be a mere 1.512 Å, which practically coincided with the length of the C—C bond in the cyclopropane ring calculated with the same basis set (1.502 Å). The data obtained favored structure **3b** rather than **3**. Haddon carried out a MINDO/3 calculation with complete geometry optimization [15] (in early calculations [16, 24] (see also review [25]) seven carbon atoms were located in the plane), which yielded the value of 1.621 Å for the fused (homoaromatic) bond, whereas the cyclopropane bond in the same scheme was 1.504 Å. This result made it possible to retain Winstein's pattern, that is, structure **3**.

As a rule, the STO-3G and MINDO/3 methods give bond lengths of carbocations that are in agreement with each other, which suggested the idea of an additional *ab initio* calculation with complete geometry optimization [26]. As a result, geometry parameters have been obtained in qualitative agreement with the MINDO/3 data. Thus both the semiempirical and *ab initio* schemes have consistently supported the conclusions of the qualitative theories as to the lengthening of the homoaromatic bond over the free cyclopropane. The conjugated ring is a coplanar, which accords well with the requirement [13, 15] of maximal overlap of Walsh orbitals of cyclopropane with the π-orbital array

---

[1]According to the MINDO/3 calculations [9], the lengths of the fused bonds in **13**, **14** and **15** are 1.582, 1.580, and 1.575, respectively. Thus in going from formally homoantiaromatic **13** to homoaromatic **14** and **15**, its length does not change appreciably (MINDO/3 calculation for cyclobutane yields 1.526 Å).

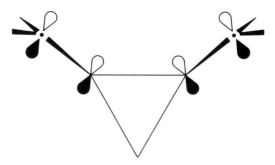

Figure 6.2 Homoaromatic orbital overlap between the frontier MOs of cyclopropane and the termini of the polyenyl fragment. Reprinted with permission from R. C. Haddon, *J. Org. Chem.* **44**, 3608 (1979), American Chemical Society.

of the ring (Fig. 6.2). Note that, according to the computational data [26], in the nonhomoaromatic structure **15**, the seven-membered cycle stays nearly planar, which once again shows lack of interaction with the cyclobutane ring.

The structure of the cyclooctatrienyl cation (**3a**) with a planar eight-membered cycle (C_{2v} symmetry) does not correspond to a minimum on the PES and it is relevant only in the examination of the reaction pathway of the bridge-flipping process shown to occur [27, 28] in the solution of **3** with the barrier of ~ 23 kcal/mol and as a reference state in the determination of the degree of aromaticity [29].

$$\text{3} \quad \rightleftharpoons \quad [\text{3a}]^{\neq} \quad \rightleftharpoons \quad \text{3}' \qquad\qquad (6.1)$$

At the same time, an unexpected minimum (**3c**) was found on the PES of $C_7H_8^+$ [26] whose structure differs in principle from **3** in the C1–C7 distance (STO-3G gives for 2.303 Å for **3c**).

3c

No X-ray structure studies of salts of the unsubstituted homotropylium cation have been reported. However, geometry parameters of the derivatives have been determined [7, 8, 30–32]. The length of the C1–C3 bond in the 2-hydroxyhomotropylium cation (**16**) is 1.626 Å, however, a sharp lengthening of this bond (to 2.284 Å) occurs in the 1-ethoxyhomotropylium cation (**17**).

16   17

A qualitative explanation of these findings is suggested by Haddon [26], who argued that the electron-donating substituents in the 2, 4, and 6 positions must stabilize the structural form **3b**, whereas in the 1, 5, 7, and 3 positions their effect must be opposite and structure **3c** will be favored.

In [26, 33, 34] two more approaches were taken for the analysis of the geometry cháracteristics of the homotropylium cation, one of which involves the construction of LMOs while the other is based on the topological theory of molecular structure. The pattern of the LMO [26] clearly favors an electronic structure corresponding to bicyclo[5.1.0.]octadienyl cation (**3b**). The analysis of charge density distribution $\rho(r)$ provides a basis for a rigorous definition of homoaromaticity [33, 34]. A system with $(4n + 2)$ π-electrons may be considered as homoaromatic if it fulfills the following criteria:

1. The system is closed by a 1,3-bond path with a bond critical point $r_b(\text{Cl, C3})$ and $H_b < 0$ $(H(\mathbf{r}_b) \equiv H_b)$ is the energy density in the interatomic region; sign of $H(\mathbf{r})$ determines whether accumulation of charge at a point $\mathbf{r}$ is stabilizing $(H(\mathbf{r}) < 0)$ or destabilizing $(H(\mathbf{r}) > 0$ [34]).
2. The bond order n (see Eq. (2.113)) of the 1,3-bond is $0 < n < 1$.
3. The π-character of the 1,3-bond as measured by the bond ellipticity ε (see Eq. (2.112)) is larger than that of cyclopropane.
4. The major axis of $\varepsilon(1,3)$ overlaps effectively with those of the neighboring bonds.

Cremer and co-workers [33] found that the critical point characteristics of the C1–C3 bond of the homotropylium cation and C—C bond of the cyclo-propane ring are similar. Hence a stretching of the homoaromatic bond in **36** must result in its easy rupture contrary to the second energy minimum found by Haddon [35]. The above-described inconsistencies led Haddon to a detailed study of the PES of the homotropylium cation making use of *ab initio* calculations with an extended (6-31 G + 5D) basis set and correlation effects included in terms of the MP theory. The results indicate that the double minimum potential obtained at the HF level is an artifact of this HF approach [29, 35]. The PES of the homotropylium cation is extremely flat—to the point that experimental determination of its structure may be difficult [35].

These results necessitate a reappraisal [26] of the two-well potential along the C1–C3 bond and, accordingly, a different interpretation of experimental results [31, 32]. According to Haddon [35], the low force constant for stretching the homoconjugate bond warrants the assumption that the molecular structure should be highly polarizable along this coordinate. If that is the case, the short and long homoconjugated bonds in **16** and **17** reflect the extreme sensitivity of the molecular geometry to electronic effects.

The flatness of the potential energy curve at the 1,3-equilibrium distance has been supported by the MP2/6-31G* and MP4SDQ/6-31G**//MP2/6-31G* calculations [29]. These calculations of the PES of the homotropylium cation **3** along the coordinate $R(1,3)$ show a single minimum potential curve with a minimum at 2.03 Å. The stretching constant $k(C1C3)$ is only 0.2 mdyn/Å [29]. Maximum equalizations of positive charge and of bond lengths in the seven-membered ring at $R(1,3) = 2.03$ Å manifest efficient six π-electron delocalization involving through-space 1,3 interactions.

As has already been noted, the $C_8H_9^+$ structures are acoplanar; that is their π-system is disturbed. For the analysis of particularly distorted π-systems the POAV (π-orbital axis vector) scheme was devised [35–37] (see Section 3.1). It is particularly convenient for analyzing homoaromatic molecules, for example, the homotropylium cation. The most important conclusions based on the POAV scheme are as follows:

1. The strength of the homoconjugated bond of the homotropylium ion stays fairly large for the studied range of its variation.
2. The homoconjugated bond is the strongest π-bond in the molecule when its length is less than 1.8 Å.
3. The resonance integrals of the π-orbitals are largely equalized in the region of $1.8 \leq R_{1-3} \leq 2.0$ Å, indicating the validity of the structural criterion of aromaticity.

Conclusion 2 refutes the widespread opinion according to which the homoconjugated bond is a weakened π-bond [2–4]. Winstein's assertion that the cyclic delocalization is interrupted in only one region of the molecule also needs verification. It appears that the conjugated distortion effects are uniformly distributed over all the bonds of the homotropylium cation.

The POAV analysis enabled the dependence of the ring-current magnitudes on the length of the homoaromatic bond to be studied. The calculated ring current turned out to be quite strong for the whole range of the R_{1-3} values. Even for $R_{1-3} = 2.6$ Å, it made up 39% of the value for the tropylium cation, and in the region of $1.6 \leq R \leq 1.95$ Å, this proportion rose to 90%. These facts account for the similarity of the chemical shifts for the 2-hydroxyhomotropylium cation (**16**) and the 1-ethoxyhomotropylium cation (**17**) [31, 32].

Based on MNDO and *ab initio* calculations with correlation effects taken into account, Barzaghi and Gatti [38–40] supported the above conclusions. Considerable lengthening of the homoaromatic bond, as compared to the C—C bond of cyclopropene, led them to regard the homotropylium cation as a homoaromatic Möbius system (**11**). Having performed calculations on a large number of substituted homotropylium cations, the authors concluded that practically both structures (of Möbius and Winstein types) coexist on the PES and their relative stability depends essentially on substituents. The stability of the Möbius structure may be explained on the basis of the topological analysis of the electron density distribution [38]. The experimental calculated magnetic properties of **3**, that is, the ^{13}C chemical shifts, the magnetic susceptibility χ, and the difference $\sigma H_a - \sigma H_b$ between the ^1H chemical shifts of the *endo* proton H_a and the *exo* proton H_b at C8, give evidence that **3** is a cyclic system with substantial electron delocalization [29].

6.3.1.2 Energetic Criteria The energy of the isodesmic bond separation reaction [41] (Eq. (6.2)) is for the homotropylium cation the measure of its stabilization:

$$\mathbf{3} + 7CH_4 + CH_3^+ \rightarrow 3C_2H_6 + 2C_2H_5^+ + 3C_2H_4 \qquad (6.2)$$

Which is relative to the tropylium cation (**18**) [38]:

$$\mathbf{18} + 6CH_4 + CH_3^+ \rightarrow 2C_2H_6 + 2C_2H_5^+ + 3C_2H_4 \qquad (6.3)$$

Both these reactions are endothermic: the respective heats calculated with the 3-21G basis set amount to $- 80$ kcal/mol. Thus the homoaromatic tropylium cation retains (to 69%) the aromatic character of the tropylium cation.

There are other approaches to evaluating the stability of **3**; for example, one may estimate the heat of reaction (6.4). According to MINDO/3 calculations, it equals 11.4 kcal/mol:

$$\mathbf{3} + C_2H_6 \longrightarrow \quad + \qquad (6.4)$$

The earlier approach [16] delineated by reaction (6.5) appears to be less correct; the right-hand and left-hand parts of this equation contain structures with different strain energies:

$$\mathbf{3} + 6CH_4 + CH_3^+ \longrightarrow \quad + 2C_2H_6 + 2C_2H_5^+ + 2C_2H_4 \qquad (6.5)$$

As noted earlier, from the qualitative point of view, structural form **15** cannot be stabilized because of homoaromaticity. From this fact, two more schemes follow by which the stabilization of the homotropylium cation can be estimated; namely, the determination of the energy of reaction described by (6.6),

$$\mathbf{3} + \quad \longrightarrow \quad \mathbf{15} + \qquad (6.6)$$

and a comparison between the heats of reactions (6.7) and (6.8),

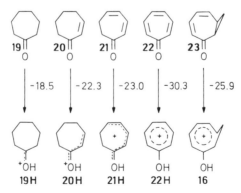

$$3 + \bigcirc \longrightarrow \bigcirc\!\!\!\!\diagup + \mathbf{18} \tag{6.7}$$

$$\mathbf{15} + \bigcirc \longrightarrow \bigcirc\!\!\!\square + \mathbf{18} \tag{6.8}$$

MINDO/3 calculations estimated the endothermicity of reaction (6.6) to be 14.4 kcal/mol [20]. The estimation of homoaromatic stabilization of **3** using **3a** as a reference state gives 4 kcal/mol (MP4SDQ/6-31G*) [29].

An elegant experimental approach to the determination of energy effects caused by the homoaromaticity was used by Childs' group [42]. Heats of protonation were measured for a series of seven-membered unsaturated ketones. These are represented by the heats of transfer (ΔH_{tr}) from CCl$_4$ to FSO$_3$H [43], see Fig. 6.3. Consider ΔH_{tr} for cations **19–22**.

The introduction of the first double bond considerably increases ΔH_{tr} (by 3.8 kcal/mol), while the second double bond essentially does not affect it (an increase of 0.7 kcal/mol). One would have expected the latter effect upon addition of yet another double bond (structure **22**). However, a sharp increase in ΔH_{tr} is actually observed in this case (7.3 kcal/mol), which is accounted for by the increased stability of the hydroxysubstituted aromatic tropylium cation (**22H**) formed as a result of the protonation of compound **22**. In a similar manner, by comparing ΔH_{tr} for the transformations **21 → 21H** and **23 → 16**, one may estimate the degree of the aromatic stabilization of **16**. It equals 2.9 kcal/mol, that is, about 40% of the stabilization of the corresponding tropy-

Figure 6.3 Heats of transfer of various ketones from CCl$_4$ to FSO$_3$H. All values shown are in kcal/mol. Reprinted with permission from R. F. Childs, *Acc. Chem. Res.*, **17**, 347 (1984), American Chemical Society.

lium cation. Note, however, that the potential surface of $C_8H_9^+$ and probably that of $C_8H_9OH^+$ have many minima, which may affect the above conclusion.

Thus all criteria discussed above clearly indicate a homoaromatic character of the homotropylium cation.

6.3.2   Other Homoaromatic Cations: Does Homoantiaromaticity Exist?

The homotropylium cation is viewed as the prototype of a homoaromatic system [29,35]. In other words, it is the reference species (like benzene for the concept of aromaticity), relative to which the properties of all homoconjugated cyclic compounds may be gauged. The best studied examples of these are the homoaromatic cation (**1**) and its cyclobutyl analog (**19**) as well as the homoantiaromatic bicyclo [3.1.0] hexenyl cation (**20**) and its cyclobutyl analog (**21**).

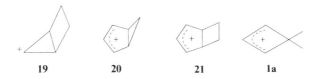

| 19 | 20 | 21 | 1a |

The length of the homoaromatic bond in the homocyclopropenylium cation is, according to MP2/6-31G* calculations [44], 1.741 Å, which is close to the experimental value of 1.775 Å [45] for the tetramethyl-substituted derivative. This bond does not have a critical point of (3,-1) type [33]. The homoaromatic stabilization in **1** is smaller than in **3**. According to the experimental findings by Olah and co-workers [46], the free energy of activation for bridge flipping in **1** is 8.4 kcal/mol, while in **3** this process develops with a much greater energy of activation, namely, 22.3 kcal/mol. If the difference between the energies of **3** and **3a** and, analogously, of **1** and **1a** is taken as the energy of the homoaromatic stabilization, the homocyclopropenyl cation will make up ~ 40% that of **3**. Calculation data are in good agreement with these conclusions. For example, MP4/6-31G* computations gave the difference between **1** and **1a** as 9.73 kcal/mol [44].

Unlike **1**, its cyclobutyl analog (**19**) does not correspond to a minimum on the PES [20] and relaxes without an energy barrier into a nearly planar structure of the cyclopentenyl cation. The bond length in this case equals 2.253 Å (MINDO/3) and 1, 3-π overlap is negligible. Thus the interaction of the cyclopropane ring orbitals with the vacant p-orbitals is an important factor in the stabilization of **1**.

The homoconjugative interactions might be expected to bring about cyclic delocalization of $4n$ π-electrons as well. Is it possible to speak of homoantiaromaticity in this case ? Indeed, since $4n$ π-electron delocalization should result in destabilization, homoconjugative interactions should be unfavorable. Thus such a system will try to avoid homoantiaromatic destabilization. Topological analysis of the electron density distribution in bicyclo[3.1.0]hexenyl cation (**20**) shows that the ring of the six outer bonds forms a conjugative system [33, 34]. The bond orders Eq. (2.113) sum to 6.8, equivalent to four single bonds and a π-system of approximately six electrons.

$$\mathbf{20} + \square \longrightarrow \mathbf{21} + \triangle \qquad (6.9)$$

The comparison of **20** and **21** reveals no considerable destabilization of **20**, while such a destabilization should be expected if **20** had homoantiaromatic character. The bond lengths common to the two cycles in **20** and **21**, according to MINDO/3 results [20], are practically the same, 1.563 and 1.580 Å. The energetic estimate of the stability of **20** by means of Eq. (6.9) confirms the conclusion drawn from the structural analysis. MINDO/3 results show that the endothermicity of Eq. (6.9) amounts to 1.4 kcal/mol (cf. the heat of reaction (6.6)). Calculated by the same scheme, the heat of the isodesmic reaction (6.10) equals – 2.1 kcal/mol:

$$\mathbf{20} + C_2H_6 \longrightarrow \qquad + \triangle \qquad (6.10)$$

The absence of substantial destabilization, which might be expected for structure **20**, if it were homoantiaromatic, led to the assignment of this compound to the nonhomoaromatic class [20]. This conclusion was supported by Olah and co-workers on the basis of the NMR study [47].

Thus, in the case of the cations, the antiaromatic destabilization appears to be practically absent, for example, as illustrated by the fact that bishomoantiaromatic bicyclo[3.3.0]octadienediyl dication (**22**) is the most stable isomer on the PES of $C_8H_8^{2+}$ [48]. In this case, some other factors should be decisive in achieving stability.

22   **22a**   **22b**

In summary, the conclusion can be drawn that potentially homoconjugative systems with $4n$ π-electrons avoid homoantiaromaticity by adopting an electronic structure with $(4n + 2)$ delocalized π-electrons [33, 34]. This conclusion is also valid for neutral homoconjugated species. This is exemplified by bicyclo[2.1.0]pent-2-ene (**22a**). Analysis of the electron density distribution in **22a** indicates that approximately six π-electrons delocalized on the perimeters of the six-membered ring [34].

Radical cations that have $(4n + 1)$ π-electrons may benefit from homoaromatic stabilization [49]. For example, bishomoaromatic character has been assigned to bishomoheptafulvene radical cation **22b** [49].

6.4   DOES HOMOAROMATICITY TAKE PLACE FOR ANIONS?

In spite of the qualitatively predicted [9, 13] equality of the homoaromatic stabilization of cations and anions, it is in the latter either quite insignificant or nonexistent altogether. The simplest potentially homoaromatic anions are the homocyclopentadienide anion (**23**), the homocyclobutadiene dianion (**24**) and the homocyclooctatetraene dianion (**25**). The global minimum on the PES of $C_6H_7^-$ is presented not by structure **23** but rather by that of the cyclohexadienide anion (**23a**) [26, 38, 50], where the C(1)–C(5) bond length equals 2.507 Å (6-31G*). The local minimum corresponding to **23** is destabilized by 23.2 kcal/mol (MP2/6-31G*//6-31G) [38].

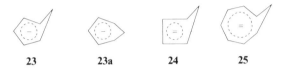

23	**23a**	**24**	**25**

The heats of isodesmic the reactions (6.11)–(6.13) also provide evidence for the low stability of **23** [38]:

$$\text{(structure)} + 4CH_4 + CH_3^- \xrightarrow{\Delta H = 87 \text{ kcal/mol}} C_2H_6 + 2C_2H_5^- + 2C_2H_4 \quad (6.11)$$

$$\textbf{23a} + 5CH_4 + CH_3^- \xrightarrow{63 \text{ kcal/mol}} 2C_2H_6 + 2C_2H_5^- + 2C_2H_4 \quad (6.12)$$

$$\textbf{23} + 7CH_4 + CH_3^- \xrightarrow{10 \text{ kcal/mol}} 4C_2H_6 + 2C_2H_5^- + C_2H_4 \quad (6.13)$$

Experimental data [50] indicate a planar nonaromatic structure **23a**. Structure **24** is destabilized to an even greater degree (by 71.0 kcal/mol) relative to the cyclopentadiene dianion. In contrast to the COT dianion, which can be stored at room temperature in THE for years (see Chapter 4), the homo[8]annulene dianion will not endure in this solvent except at low temperatures. The energy of reaction (6.14) determined through the experimental electron affinities of homocyclooctatetraene (homoCOT), *cis*-bicyclo[6.1.9], and the homoCOT anion radical shows **25** to be much less thermodynamically stable than the COT dianion relative to the corresponding neutral species [51]:

$$\text{(structure)}^{2-} + \text{(structure)} \longrightarrow \text{(structure)} + \text{(structure)}^{2-} \quad (6.14)$$

$\Delta H° = 50$ kcal/mol (gas phase); 79 kcal/mol (solid with cation = K^+).

Estimates for the PES of $C_9H_{10}^{2-}$ are not known. However, indirect evidence (the C(1)–C(8) bond length in **25** is practically equal to that in cyclopropane [26], with a small difference between chemical shifts of the *exo* and *endo* protons [52]) points to the nonaromatic character of structure **25**.

A lengthy discussion has been going on for some time on the problem of the stabilization of bishomoaromatic anions [53–60]. Experimental research indicates increased stability of the bicyclo [3.2.1] octa-3,6-dien-2-yl anion (**26**). Thus the H/D exchange rate in **27** is $10^{4.5}$ times faster than in **28** [53]. The ^{1}H [54] and ^{13}C [55] NMR spectra show a $\delta = 2.3$ downfield shift of H6 and H7 of anion **26** relative to the corresponding protons of **27** and a $\delta = 43.8$ upfield shift of C6 and C7 of **26** as compared to **27**. Moreover, Washburn [56], by measuring the pK_a values of compounds **27–29**, has found that anion **26** is more stable than anion **30** by >12.2 kcal/mol, while ion **31** is more stable than **32** by only 8.7 kcal/mol. This finding was, incidentally, claimed to be the first quantitative estimate of the anionic homoaromatic stabilization.

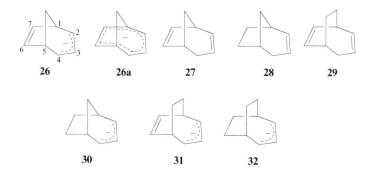

All relevant computational studies [57–59], with the exception of [60], concur in the view that the bishomoaromaticity in anionic systems is an elusive phenomenon. Based on a comparison of geometrical, electronic and energetic properties of the six π-electron anion **26** and the four π-electron cation **33** with those of anion **30** and cation **34**, calculated by semiempirical and nonempirical methods, Schleyer and co-workers did not find any additional cyclic delocalization in **26** [58].

For example, the length of the allylic bond C(2)–C(3) is the same in **26** and **30** (1.395 Å), which is also the case with compounds **33** and **34** (1.406 Å). Moreover, the "homoaromatic" C(2)–C(7) bond is shorter in **26** than in **33** by a mere 0.003 Å. A comparison between the shapes of the HOMOs for **26** and **33** as well as between the electron density distributions has confirmed the absence of the π-electron delocalization.

At the same time, reaction (6.15), according to *ab initio* (STO-3G) calculations, is exothermic (the heat of reaction equals 4.2 kcal/mol), while process (6.16) is, on the contrary, endothermic:

$$26 + 28 \rightarrow 27 + 30 \qquad \Delta E = 4.2 \text{ kcal/mol} \qquad (6.15)$$

$$33 + 28 \rightarrow 27 + 34 \qquad \Delta E = 2.9 \text{ kcal/mol} \qquad (6.16)$$

The gas-phase experiments by Lee and Squires [61] in which the heat of reaction (6.15) was determined, gave the value of 9.5 ± 2.0 kcal/mol, which is in satisfactory agreement with calculations.

The energy contribution from the bishomoaromaticity and the inductive effect of double bonds may be separated by calculating the enthalpies of reactions (6.17)–(6.20) [58].

In anions **35** and **37** additional double bonds are oriented in such a way as to switch off the HOMO and LUMO interaction with the allyl anion (zero HOMO–LUMO overlap because of different symmetry). Despite this, the stabilization of **37** is practically the same as **26**, and that of **35** is twice as high. This shows that in the stabilization of **26** the inductive effects are operative rather than the bishomoaromaticity.

After studying the interaction of the allyl anion with ethene by the *ab initio* STO-3G method, Brown and co-workers [60] disagreed with the claims in [57–59]. It should be noted, however, that their results were based on a study of the allyl anion–ethene complex within the repulsive region.

The conclusion drawn in [57, 58] were confirmed by the Roos group [59] through MCSCF calculations on anion **26** and its complex with the cation Li^+.

The geometrical structure of **26** is the same whether determined by an X-ray diffraction study [62] or by means of quantum chemical calculations [59]. The length of the olefinic bond is identical in ethene and the anion. The geometry of the allylic anion is hardly altered upon its inclusion in **26**. The interaction of **26** with the cation Li$^+$ is not in effect distinguishable from the interaction of Li$^+$ with the allyl anion. Thus the enhanced stability of **26** cannot be accounted for by the effect of the cation.

There is one more surprising result, namely, the finding that there is a substantial rehybridization of atoms C6 and C7, which have a lesser(!) negative charge than the carbon atoms in ethene. Because of the rehybridization, atoms H6 and H7 come out of the plane of the double bond and the overlap between the π-systems of the olefinic and carbanionic bridges is diminished(!).

The enhanced stability of **26** over **30** can be dealt with in terms of a simple electrostatic model that explains it as stemming from the interaction of the quadrupole moments of ethane and ethene with the charges located in carbanion bridges of **26** and **30**. The quadruple–charge interaction showed a relative stabilization of the charge in the carbanionic bridge by 5.8 kcal/mol upon going from ethane to ethene. Furthermore, in view of a difference between polarizabilities of ethane and ethene the additional inductive and dispersion stabilization of **26**, equal to ~1 kcal/mol, should also be taken into account. The total stabilization of ~ 7 kcal/mol is comparable to the experimental value.

Thus, experimentally observed, potentially bishomoaromatic anions, such as **39**–**42**, do indeed possess increased stability, which apparently is not explained by an interaction between π-systems but rather by electronic interactions that ought to be taken into account in a accurate manner.

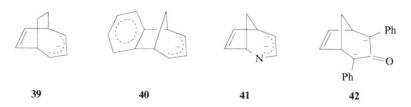

39   **40**   **41**   **42**

6.5   CAN NEUTRAL MOLECULES MANIFEST HOMOAROMATICITY?

Direct quantum chemical calculation at various levels of approximation [20, 38, 63] bear witness, in complete agreement with qualitative predictions [9, 13], to a merely insignificant homoaromatic stabilization of noncharged molecules. The elusiveness of the neutral homoaromaticity concept was elegantly demonstrated by calculating the energy profile of the model reaction of the trimerization of acetylene into benzene [64]. Thus the interaction between π-systems of the molecules in equilibrium geometries is destabilizing, since filled–filled orbital interactions are much stronger than the filled–vacant ones. This finding suggested a quite unexpected conclusion: the homoconjugative interactions among three neutral proximal closed-shell π-systems are in fact destabilizing! This conclusion was later confirmed by Bach et al. [65].

A possibility of neutral homoaromaticity is not ruled out at all. However, it is suggested that it takes place in the case when two interacting π-moieties are substituted by strong donors (the first moiety) and by strong acceptors (the second one). This brings about stabilizing HOMO–LUMO interactions.

A lively discussion is now taking place on the homoaromaticity of triquinacene **43** [66–70]. Studying the heats of hydrogenation of **43**, **44**, and **45** to hexahydrotriquinacene **46**, Liebman et al. [66] have found an extra stabilization of ~ 4.5 kcal/mol for **43** attributed to the homoaromaticity (the heats of hydrogenation, ΔH_h in kcal/mol, are shown above the arrows):

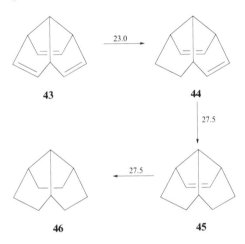

This result is unexpected since the spectroscopic (IR, UV, CD, photoelectron) as well as structural data offer no evidence whatever for the homoaromatic character of triquinacene (for a detailed review see [67]).

Ab initio (MP2/6-31G*//HF/6-31G*) [67] and MM2 force field calculations [68] could not reproduce the experimental trends in ΔH_h. Hence one or more of the experimental heats of hydrogenation must have been in error by amounts well outside the expected limits, or the calculations are wrong.

Dewar and Holder [69], making use of different computational schemes (MM2, AM1, 6-31G*//3-21G), have arrived at a more plausible interpretation of the differences between theory and experiment. They suggested that the anomaly in the experimental heats of formation results from the differing tendency of **44 – 46** to relieve strain by twisting. The most stable conformation of **44– 46** are twisted; that is the saturated five-membered rings have no planes of symmetry. The heat of hydrogenation for each step in series **43** → **44** t → **45** t → **46** t (with "t" standing for twisted) can be divided into two parts, one corresponding to the heat of hydrogenation under conditions where the rings are kept untwisted throughout and the other being the relief of strain energy due to twist-ing. If **43** is not aromatic, the "untwisted" heats of hydrogenation should be equal, because any change in strain energy should be the same in each step. If **43** is homoaromatic, the value for **43** → **44** should be correspondingly smaller than that for **44** → **45** or **45** → **46**. But drawback of this analysis [69] is that none of the methods employed reproduces the experimental results.

The MM3 force field calculations [70] are able to reproduce the experimen-tal trends in ΔH_h. The Dewar–Holder approach [69] was adapted and relaxations due to overall twisting have been considered in the framework of the MM3 scheme [70]. A detailed analysis of the MM3 energy components indicates that the increase in the steric energy between **43** and the optimized **44** is mainly caused by nonbonded 1,4-interactions such as those involving *endo*-hydrogens of the saturated cyclopentane and carbons attached to the cyclopentane ring. The cyclopentane moieties in **45** and **46** can have envelope conformations due to increased backbone flexibility not found in **44**. Since the 1,4-interactions are reduced by the more even distribution of saturated ring dihedrals, these envelope conformations are favorable. The hydrogenation energies of **44** and **45** are larger than that of **43** primarily for this reason [70]. Thus it is unnecessary to invoke explanations based on homoaromaticity. Here, once again, we are confronted with the situation when the "apparent" and even "experimentally detected" homoaromatic stabilization is in actual fact caused by effects quite different from those previously assumed.

There is no evidence that 1,3,5-cycloheptatriene **47** is a homoaromatic system [63].

47   **48**

But in contrast to **47**, the analysis of the charge density distribution $\rho(r)$ (see Section 2.5.2) in norcaradiene **48**, the valence tautomer of **47**, indicates the bond C1–C6 to be labile ($n = 0.85$) with a rather large ellipticity, indicating substantial π-character [71]. Thus the homoaromaticity criteria mentioned in Section 6.3.1 are fulfilled. The six π-electron cyclic delocalization system may be formed due to the interaction of the electrons forming this bond with other π-electrons of the six-membered ring. Norcaradiene is 5.4 kcal/mol higher in energy **47**. If the strain energy of cyclopropane (27.5 kcal/mol) is taken into account, the energy difference between **47** and **48** containing a strained three-membered ring seems to be very small [64]. This has been assigned to homoaro-matic stabilization of norcaradiene [34, 71].

Noteworthy is the work by Scott et al. [72] (see also [73]), in which possibil-ity of homoaromaticity stabilization in cyclic polyacetylene-decamethyl[5]peri-cyclyne **49** was discussed [72]. Experimental estimation of the homoaromatic stabilization of **49** with the use of the heats of hydrogenation of **49** and of the series of acyclic alkynes indicates **49** to have the homoaromatic stabilization of 6 kcal/mol [72, 74]. However, RHF/3-21G calculations of DRE show that **49** is antihomoaromatic by 1 kcal/mol [75].

49

If the homoconjugation (orbital interactions) for such systems is well documented and rests on solid theoretical groups [70, 73], further research will be needed to corroborate or refute the attribution of the thermodynamic stabilization to homoaromaticity in neutral molecules.

REFERENCES

1. D. E. Applequist and J. D. Roberts, *J. Am. Chem. Soc.*, **78**, 4012 (1956).

2. S. Winstein and R. Adams, *J. Am. Chem. Soc.*, **70**, 838 (1948).

3. S. Winstein, *J. Am. Chem. Soc.*, **81**, 6524 (1959).

4. S. Winstein, *Rev. Chem. Soc.*, **23**, 141 (1969).

5. L. A. Paquette, *Angew. Chem. Int. Ed. Engl.*, **17**, 106 (1978).

6. P. M. Warner, *Topics Nonbenzenoid Aromatic Chem.*, 2 (1976).

7. R. F. Childs, *Acc, Chem. Res.*, **17**, 347 (1984).

8. R. F. Childs, M. Mahendran, S. D. Zweep, G. S. Shaw, S. K. Chadda, N.A.D.Burke, B. E. George, R. Faggiani, and C. J. L. Lock, *Pure Appl. Chem.*, **58**, 111 (1986).

9. M. J. Goldstein and R. Hoffmann, *J. Am. Chem. Soc.*, **93**, 6193 (1971)

10. J. L. Rosenburg, J. E. Mahler, and R. Pettit, *J. Am. Chem. Soc.*, **84**, 2842 (1962).

11. M. Barfield, D. M. Grand, and D. Ikkenberry, *J. Am. Chem. Soc.*, **97**, 6956 (1975).

12. W. J. Hehre, *J. Am. Chem. Soc.*, **95**, 5807 (1973).

13. R. C. Haddon, *Tetrahedron Lett.*, 2797 (1974).

14. R. C. Haddon, *Tetrahedron Lett.*, 4303 (1974).

15. R. C. Haddon, *Tetrahedron Lett.*, 863 (1975).

16. W. J. Hehre, *J. Am. Chem. Soc.*, **96**, 5207 (1974).

17. A. J. P. Devaquet and W. J. Hehre, *J. Am. Chem. Soc.*, **96**, 3644 (1974).

18. W. L. Jorgensen and W. T. Borden, *J. Am. Chem. Soc.*, **95**, 6649 (1973).

19. W. L. Jorgensen, *J. Am. Chem. Soc.*, **97**, 3082 (1975).

20. W. L. Jorgensen, *J. Am. Chem. Soc.*, **98**, 6784 (1976).

21. R. Hoffmann, *Tetrahedron Lett.*, 2907 (1970).

22. H. Günter, *Tetrahedron Lett.*, 5173 (1970).

23. E. Heilbronner, *Tetrahedron Lett.*, 1923 (1964).

24. W. J. Hehre, *J. Am. Chem. Soc.*, **94**, 8908 (1972).

25. L. Radom, D. Poppinger, and R. C. Haddon, *Carbonium Ions,* **5**, 2303 (1976).

26. R. C. Haddon, *J. Org. Chem.*, **44**, 3608 (1979).

27. S. Winstein, C. G. Kreiter, and J. I. Brauman, *J. Am. Chem. Soc.*, **88**, 2047 (1966).

28. S. Winstein, *Rev. Chem. Soc.*, **23**, 141 (1969).

29. D. Cremer, F. Reichel, and E. Kraka, *J. Am. Chem. Soc.*, **113**, 9459 (1991).

30. R. Destro, T. Pilati, and M. Simonetta, *J. Am. Chem. Soc.*, **98**, 1999 (1976).

31. R. F. Childs, A. Varadarajan, C. J. L. Lock, R. Faggiani, C. A. Fyve, and R. E. Wasylishen, *J. Am. Chem. Soc.*, **104**, 2452 (1982).

32. R. F. Childs, R. Faggiani, C. J. L. Lock, and M. Mahendran, *J. Am. Chem. Soc.*, **108**, 3613 (1986).

33. D. Cremer, E. Kraka, T. S. Slee, R. F. W. Bader, C. D. H. Lau, T. T. Nhuyen-Dang, and P. J. MacDougall, *J. Am. Chem. Soc.*, **105**, 5069 (1983).

34. E. Kraka and D. Cremer, Chemical Implementation of local Features of the Electron Density Distribution in Z. B. Maksic (Ed.), *The Concept of the Chemical Bond*, Part 2, Springer, Berlin, 1990, pp. 453–542.

35. R. C. Haddon, *J. Am. Chem. Soc.*, **110**, 1108 (1988).

36. R. C. Haddon, *J. Am. Chem. Soc.*, **109**, 1676 (1987).

37. R. C. Haddon, *Acc. Chem. Res.*, **21**, 243 (1988).

38. M. Barzaghi and C. Gatti, *J. Chim. Phys.*, **84**, 783 (1987).

39. M. Barzaghi and C. Gatti, *J. Mol. Struct.*, (THEOCHEM), **167**, 275 (1988).

40. M. Barzaghi and C. Gatti, *J. Mol. Struct.*, (THEOCHEM), **166**, 431 (1988).

41. W. J. Hehre, R. Ditchfield, L. Radom, and J. A. Pople, *J. Am. Chem. Soc.*, **92**, 4796 (1970).

42. R. F. Childs, D. L. Mulholland, A. Varadarajan, and S. Yeroushalmi, *J. Org. Chem.*, **48**, 1431 (1983).

43. E. M. Arnett and J. W. Larsen, *J. Am. Chem. Soc.*, **91**, 1438 (1969).

44. R. C. Haddon and K. Raghavachari, *J. Am. Chem. Soc.*, **105**, 118 (1983).

45. C. Kruger, P. J. Roberts, Y. -H. Tsay, and J. B. Koster, *J. Organomet. Chem.*, **78**, 69 (1974).

46. G. A. Olah, J. S. Staral, and G. Liang, *J. Am. Chem. Soc.*, **96**, 6233 (1974); G. A. Olah, J. S. Staral, R. J. Spear, and G. Liang, *J. Am. Chem. Soc.* **97**, 5489 (1975).

47. G. A. Olah, G. Liang, and S. P. Jindal, *J. Org. Chem.*, **40**, 3259 (1975).

48. K. Schotz, T. Clark, and P. v. R. Schleyer, *J. Am. Chem. Soc.*, **110**, 1394 (1988).

49. H. D. Roth, *Topics Curr. Chem.*, **163**, 131 (1992).

50. G. A. Olah, G. Avensio, H. Mayer, and P. v. R. Schleyer, *J. Am. Chem. Soc.*, **100**, 4347 (1978).

51. G. R. Stevenson, Stabilization and Destabilization of Aromatic and Antiaromatic Compounds in J. F. Liebman and A. Greenberg (Eds.), *Molecular Structure and Energetics*, Vol. 3, VCH, New York, 1988, pp. 57–83.

52. M. Barfield, R. B. Bates, W. A. Beavers, I. R. Blacksburg, S. Brenner, B. I. Mayall, and C. S. McCulloh, *J. Am..Chem. Soc.*, **97**, 900 (1975).

53. J. M. Brown and J. L. Occolowitz, *J. Chem. Soc. Chem. Commun.*, 376 (1965); J. M. Brown and J. L. Occolowitz, *J. Chem. Soc.*, **B**, 441 (1968).

54. J. M. Brown, *J. Chem. Soc. Chem. Commun.*, 638 (1967); S. Winstein, M. Ogliaruso, M. Sakai, and J. M. Nicholson, *J. Am. Chem. Soc.*, **89**, 3653 (1967).

55. F. H. Kohler and N. Nertkorn, *Chem. Ber.*, **116**, 3274 (1983); M. Christl, H. Leininger, and D. Bruckner, *J. Am.. Chem. Soc.*, **105**, 4843 (1983).

56. W. N. Washburn, *J. Org. Chem.*, **48**, 4287 (1983).

57. J. B. Grutzner and W. L. Jorgensen, *J. Am. Chem. Soc.*, **103**, 1372 (1981).

58. E. Kaufmann, H. Mayer, J. Chandrasekhar, and P. v. R. Schleyer, *J. Am. Chem. Soc.*, **103**, 1375 (1981).

59. R. Lindh, B. O. Roos, G. Jonsaal, and P. Ahlberg, *J. Am. Chem. Soc.*, **108**, 6554 (1986).

60. J. M. Brown, R. J. Elliot, and W. G. Richards, *J. Chem. Soc. Perkin Trans II*, 485 (1982).

61. R. E. Lee and R. R. Squires, *J. Am. Chem. Soc.*, **108**, 5078 (1986).

62. N. Hertkorn, F. H. Kohler, G. Muller, and G. Reber, *Angew. Chem. Int. Ed. Engl.*, **25**, 468 (1986).

63. D. Cremer, B. Dick, and D. Christen, *J. Mol. Struct.* (THEOCHEM), **110**, 277 (1984).

64. K. N. Houk, R. W. Gandour, R. W. Strozier, N. G. Rondan, and L. A. Paquette, *J. Am. Chem. Soc.*, **101**, 6797 (1979).

65. R. D. Bach, G. J. Wolber, and H. B. Schlegel, *J. Am. Chem. Soc.*, **107**, 2837 (1985).

66. J. F. Liebman, L. A. Paquette, J. R. Peterson, and D. W. Rogers, *J. Am. Chem. Soc.*, **108**, 8267 (1986).

67. M. A. Miller, J. M. Schulman, and R. L. Disch, *J. Am. Chem. Soc.*, **110**, 7681 (1988).

68. J. M. Schulman, M. A. Miller, and R.L. Disch, *J. Mol. Struct.* (THEOCHEM), **169**, 563 (1988).

69. M. J. S. Dewar and A. J. Holder, *J. Am. Chem. Soc.*, **111**, 5384 (1989).

70. J. W. Storer and K. N. Houk, *J. Am. Chem. Soc.*, **114**, 1165 (1992).

71. D. Cremer and B. Dick, *Angew. Chem. Int. Ed. Engl.*, **21**, 865 (1992).

72. L. T. Scott, G. J. De Cicco, J. L. Hyun, and G. Reinhardt, *J. Am. Chem. Soc.*, **107**, 6546 (1985).

73. M. N. Paddon-Row and K. D. Jordan, Ch. 3, in J. F. Liebman and A. Greenberg (Eds.), *Molecular Structure and Energetics*, Vol. 6, VCH, New York, 1988, p. 115.

74. L. T. Scott, M. J. Cooney, D. W. Rogers, and K. Dejroongruang, *J. Am. Chem. Soc.*, **110**, 7244 (1988).

75. L. J. Schaad, B. A. Hess, and L. T. Scott, *J. Phys. Org. Chem.*, **6**, 316 (1993).

7

σ-AROMATICITY

7.1 σ-DELOCALIZATION AND σ-CONJUGATION

In the late 1970s and early 1980s the concept of σ-conjugation went through a certain renaissance due primarily to the work of Dewar [1–3]. This concept, whose origin can be traced back to Dewar and Pettit [4] and Sandorfy [5], had not been invoked for some time in studying the electronic and molecular structure of organic molecules to such an extent as was the case with the concepts of π-electronic delocalization and π-conjugation applied to unsaturated compounds. The concepts of σ-electronic delocalization and σ-conjugation were thus overshadowed by π-electron theory, which gained wide acceptance in organic chemistry. It was argued that the delocalization and all manifestations of nonadditivity (e.g., the aromaticity or antiaromaticity of cyclic structures) were typical of a π-electron system only, whereas a σ-electron system might be represented by a set of well-localized two-centered orbitals.

However, the real reasons for the observed additivity of bond energies in the saturated compounds, such as alkanes, are by no means connected with the absence of the σ-delocalization [3]. All the valence electrons in a molecule are delocalized. In a certain sense, the delocalization is characteristic also of inner shell electrons.

The notion of the σ-conjugated rests on the nonzero value of the resonance integrals β' between various hybrid orbitals of the same atom ("interatomic" or geminal integrals β' [5–7]) even if these AOs are orthogonal.

A simple analysis of the σ-conjugation can be made by means of the C-approximation of Sandorfy [5, 6], which, in its essence, is similar to Hückel's MO method that was extensively employed in calculations of π-electron molecules. In terms of the C-approximation $\beta' = m\beta$, where β is the resonance integral for the hybrid AOs of neighboring atoms ("vicinal" β) with $m < 1$. Hence the methylene fragment CH_2 in alkanes may be regarded as being iso-conjugated with the fragment $=CH-CH=$ in linear polyenes (Scheme 7.1).

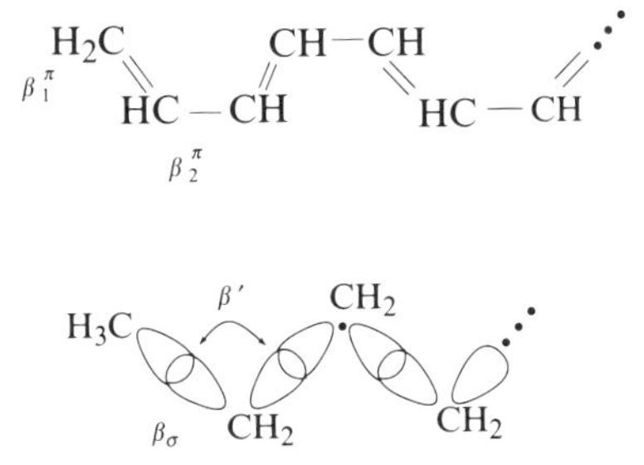

$$H_2C \quad CH-CH$$
$$\beta_1^\pi \quad HC-CH \qquad HC-CH$$
$$\beta_2^\pi$$

$$H_3C \qquad \beta' \qquad CH_2$$
$$\beta_\sigma \qquad CH_2 \qquad CH_2$$

Scheme 7.1

A distinguishing feature of the π-system is the fact that the overlap integral of the p-AOs of the neighboring atoms does not equal zero.

Thus, in examining various views as to σ-electron delocalization and σ-conjugation a comparison may be useful with the well-developed concepts of π-electron delocalization and π-conjugation. Since the values of the geminal integrals are fairly large : $\beta'\,(sp, sp) = \frac{1}{2}\,(I(s) - I(p)) = 5\text{ eV}$, $\beta'\,(sp^2, sp^2) = \frac{1}{3}\,(I(s) - I(p)) = 3.3\text{ eV}$, and $\beta'\,(sp^3, sp^3) = \frac{1}{4}(I(s)) - I(p)) = 2.5\text{ eV}$ [3], with $I(s)$ and $I(p)$ being the corresponding ionization potentials for the valence state, the degree of manifestation of σ-conjugation must be no smaller than in the case of π-conjugation.

There is ample evidence, both theoretical and experimental, confirming the existence of the σ-delocalization and σ-conjugation, in particular, the experimentally registered effects of the orbital interactions through bonds [8–11], as in **1** and **2** [8, 9].

An analysis of the PE spectra of **1** and **2** has shown that the orbital $n_- = (n_a - n_b)$ lies lower than the orbital $n_+ = (n_a + n_b)$ [9]. The ESR hyperfine splitting constants indicate the presence of delocalization of an unpaired σ-electron in the cation radicals of cyclopropane **3**, cyclobutane, and cyclopentane [12, 13]. The σ-delocalization has been confirmed by ESR studies also in the case of cation radicals of alkanes [14–16] (a typical SOMO has been shown for the cation radical of propane (**4**)). For linear alkanes, the σ-electron delocalization is evidenced moreover by the diminution of the ionization potentials (IPs) as the length of the chain is increased [17]. The σ-delocalization can be taken into account by means of a simple modification of the Hückel method: this approach allows one to describe satisfactorily the inductive effects [5, 6], the IPs [6, 18], and a number of properties of saturated compounds [6, 19, 20]. Analysis of the IP values of silanes [21, 22], germanes, and stannanes [23] shows that the σ-delocalization is characteristic of these compounds as well; in other words, it is not restricted to the chemistry of carbon. Certain properties of cyclic permethyl polysilanes $(Me_2Si)_n$ can be interpreted in terms of the σ-electron delocalization [24].

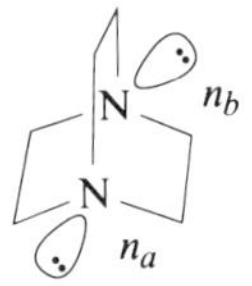

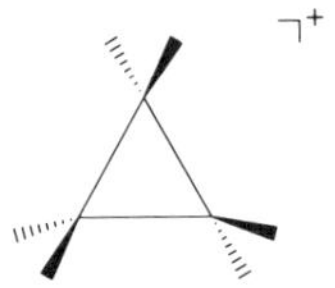

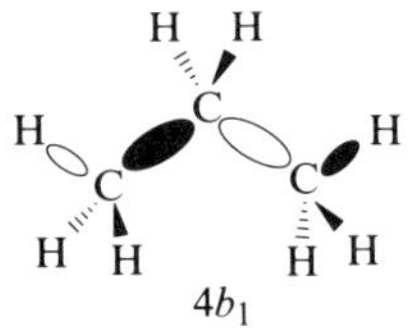

The structures are labeled **1**, **2**, **3**, and **4**. Structure 2 shows n_b and n_a labels on nitrogen atoms; structure 4 is labeled $4b_1$.

Here it is important to note (as has already been done in Chapter 1) that the conjugation is reflected in the one-electron properties (e.g., the above-mentioned IPs or the ESP hyperfine spin-coupling constants); on the other hand, it is a necessary but not sufficient condition for the existence of electron (bond) delocalization, which leads to nonadditivity of the collectivity properties, such as the heats of formation or diamagnetic susceptibility, and is indicated by the impossibility of their interpretation in terms of a model of localized bonds [25].

The presence of the σ-delocalization and σ-conjugation, the similarity of the systems with the π- and σ-conjugation (Scheme 7.1), and the isoconjugation of the fragments — $HC{=}CH$— and —CH_2—, all these facts suggest that there is a possibility of effects, which, using the concept of π-aromaticity as a paradigm, may be regarded as stemming from the σ-aromaticity. But does this possibility actually materialize? Next, we examine this question, which in A. Liberies' article "Delocalization" [26] has been formulated in the following expressive form: "Do the six π-electrons in benzene really account for its stability, or are the 36 σ-electrons actually responsible?"

7.2 σ-AROMATICITY

The introduction of various ideas regarding this or that type of aromaticity is necessitated by the inadequacy of the model of localized bonds for describing collective properties of structures of different topological types. Therefore the justification of such an introduction depends, in the first place, on whether there exists a nonadditive component of the total energy that might be regarded as the energy of the aromatic stabilization (antiaromatic destabilization). Bearing this in mind, we start the discussion of the σ-aromaticity with the problem of the determination of the above-mentioned energy component, which is central to this discussion. Afterwards we consider those specific features of the electronic structure that may be viewed as characteristic of this type of aromaticity. Furthermore, the points of analogy between the aromaticity under consideration and the π-aromaticity will be shown, which, as a rule, serves as a reference paradigm.

7.2.1 Cyclopropane: The Energy of Its σ-Aromatic Stabilization

Based on the isoconjugation of the fragments $-CH_2-$ and $-HC=CH-$ and, consequently, on the analogy between the systems of conjugation in cyclopropane (**5**) and benzene (**6**), Dewar suggested that certain attributes of cyclopropane might be explained as manifestations of the σ-aromaticity [1–3].

    5   **6**   **7**   **8**

Among the specific features characteristic of cyclopropane, one may single out the following:

1. Unexpectedly low conventional strain energy (CSE[1]) nearly equal to that of cyclobutane (**7**) (27.5 and 26.5 kcal/mol, respectively [27]), even though the latter is characterized by an appreciably lower Bayer strain [3]. The calculation of this strain using the CCC bending force constant and the values of the geometrical angles has yielded the values of 173 and 36 kcal/mol for **5** and **7**, respectively [30].

2. The ring strain should apparently weaken the CC bonds and accordingly, lengthen them. In fact, however, the lengths of these bonds are smaller in cyclopropane (1.51 Å) than in normal alkanes (1.53 Å) and cyclobutadiene (1.55 Å) [1, 3].

3. In the case of three-membered rings, the value of the spin–spin coupling constant ^1J (^{13}C, ^1H) exceeds those for analogous systems with cycles of larger size [31, 32]. For cyclopropane this value is 161 Hz (cf. 134 Hz for cyclobutadiene and 123 Hz for cyclohexane [32]) and the relationship shows that the s-character of the carbon orbital participating in the making of the C—H bond amounts to 32% [31]. This percentage may lead one to expect the protons to shift downfield as compared to the signals of the CH_2 protons in unstrained paraffins (σ 1.25). Actually, the ^1H NMR signal of cyclopropane protons is shifted upfield by ~1 ppm (σ 0.22) [31]. This may be accounted for by a considerable magnetic anisotropy, due to the ring current, which, similar to benzene, characterizes the cyclopropane ring. Since the arrangement of the cyclopropane protons relative to the ring plane is basically different from that of the benzene protons, they come to lie in the shield region (Fig. 7.1) [3].

Also in the ^{13}C NMR spectrum of cyclopropane, a 20 ppm upfield shift of the CH_2 carbon signal is observed relative to the ^{13}C shifts for alkanes [33]. The presence of the aromatic ring current in the cyclopropane ring plane is con-

---

[1]The CSE is the difference between ΔH_f°, the heat of formation calculated by an additive scheme making use of bond energies and group increments, and the experimental ΔH_f° [27–29].

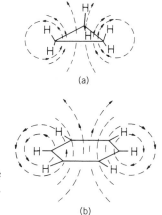

Figure 7.1 Magnetic lines of force in (a) cyclopropane (b) benzene. Reprinted with permission from W. J. S. Dewar, *J. Am. Chem. Soc.*, **106**, 669 (1984); 166.

firmed in this case by the value of the upfield component σ_C, perpendicular to this plane, of the ^{13}C shielding tensor (– 36 ppm from Me$_4$Si [33], – 40 ppm according to an *ab initio* calculation using the IGLO method [34]).

4. The orbitals of the cyclopropane fragment are capable of conjugation with multiple bond orbitals of the carbanion, radical, or carbenium center [35–38]. So the electron-donating capacities of the cyclopropyl group show up the bisected conformation (**9**), stabilized by the σ-conjugation [38], of the cyclopropylcarbinyl cation in which the methylene group lies perpendicular to the ring plane [35, 38–40].

9   **10**

According to *ab initio* calculations with the electron correlation taken into account (MP2/6-31G**), this structure corresponds to a minimum of the PES [41]. The stabilization of the bisected configuration over the perpendicular, according to *ab initio* calculations, amounts to over 35 kcal/mol [42]. Analogously, the bisected configuration (**10**) of the cyclopropylcyclopropenium cation proves stable. In **10**, the interaction occurs between the HOMO of the cyclopropane fragment and the LUMO (π*) of the cyclopropenium ion [43].

The interaction between the vacant 2p (C$^+$) orbital and the highest occupied Walsh orbital of the fused cyclopropane fragment leads to the stabilization of structure **11**, which is 23 kcal/mol stable (3-21G*) than the phenyl cation C$_6$H$_5^+$ relative to the corresponding hydrocarbons [44]. This value is in good agreement with the experimental estimate (27.6 kcal/mol) based on the energy of stabilization represented as the difference in the appearance energies of the cations **11** and $C_6H_5^+$ from their respective precursors [45]. Other examples of this type of conjugation are given by spiro[2,4]hepta-4,6-diene (**12**), cyclopropanecarbonitrile (**13**), and isocyanocyclopropane (**14**) [35, 38, 46].

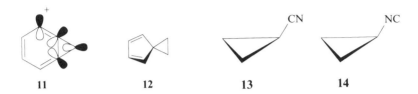

| **11** | **12** | **13** | **14** |

5. The cyclopropyl fragment is apt to form homoaromatic systems, such as **15** [47, 48] (for details see Chapter 6), whereas cyclopolyenes with a saturated four-membered ring do not exhibit homoconjugation (homoaromaticity) [48–50].

15

Dewar has shown that all the above-described anomalies of cyclopropane can be explained from a unified position represented by the notion of σ-aromaticity [1–3]. In estimating the energy of the σ-aromatic stabilization of **5**, Dewar assumed the C—C—C bending force constants to be equal in **5** and **7** and the energy of the σ-antiaromatic destabilization of cyclobutane (**7**) to be insignificant. Since in **5** the bond angles deviate from the standard strain-free CCC angle (109.5°) by 49.5° and in **7** by a mere 19.5°, the angle strain in **5** per CH_2 group should be $(49.5/19.5)^2 = 6.44$ times that of **7** and the total angle strain in **5** should be $(3/4)(6.44) = 4.83$ times that of **7**. Assuming that the CSE is equal to the sum of the angle strain energy R and the eclipsing strain energy, taken to be 12 kcal/mol [1], one obtains

$$CSE(\mathbf{5}) = 4.83R + 9 - A \qquad (7.1)$$

$$CSE(\mathbf{7}) = R + 12 \qquad (7.2)$$

where A is the energy of the σ-aromatic stabilization. Taking for CSE(**5**) and

---

[2]An interpath angle is the angle between two bond paths. In its turn, a bond path representing an image of the chemical bond is the line of maximum electron density (MED) linking the nuclei [47, 48, 51].

CSE(**7**) the values of 28.3 and 27.4 kcal/mol, respectively [29], Dewar found that in the case of **5** $A = 55.1$ kcal/mol [1].

Cremer and Kraka, using for paraffins the value of the CCC bending force constant of 1.07 mdyn·Å/rad^2 and the calculated interpath angles^2 (78.8°) rather than the geometrical ones (69°), estimated the total strain energy of **5** to be 75 kcal/mol [48]. Taking propane as an appropriate reference compound, one may see from the value of the theoretical strain energy (TSE) of the homodesmotic (see Chapter 2) reaction (7.3),

$$\mathbf{5} + 3CH_3CH_3 \rightarrow 3CH_3CH_2CH_3 \tag{7.3}$$

that **5** is destabilized by only 27 kcal/mol (using experimental values of heats of formation) [48]. Hence the total strain energy is compensated by the energy of the σ-aromatic stabilization of $75 - 27 = 48$ kcal/mol.

Thus, even though different schemes were used in [1, 3, 48] for determining the energies of the σ-aromatic stabilization, the values obtained were fairly close. However, these works were subjected to criticism [52, 53], with the closeness of those values regarded as merely an accidental ("magic") coincidence. As has been noted by Schleyer [52], the value of the strain energy (SE) is not necessarily evidence for the stabilization of cyclopropane: this value may be related to other causes. For example, the value of the SE of cyclobutane (**7**) may be anomalously high in consequence of the destabilizing Dunitz–Schomaker strain (1,3 CC interactions) [27, 52], while the value of the SE of **5** is normal, or else the SE values of both **5** and **7** may be anomalous due to the stabilization of **7**. Note also that in the last case the destabilization of cyclobutane is not associated with the σ-antiaromaticity, but rather with the above-mentioned 1,3-non bonded repulsions whose energy amounts to roughly 10 kcal/mol [52], and the stabilization of cyclopropane should not necessarily be related with σ-aromaticity since there is an alternative explanation, namely, the strengthening of the C—H bonds due to the rehybridization of carbon atoms [27, 52]. The conclusion drawn by Schleyer [52] is as follows: "there appears to be no need to invoke sigma aromaticity to explain the thermochemistry of cyclopropane."

The second critical point given by Grev and Schaefer [53] of Dewar's [1–3] and Cremer's [48] works relates to the problem of the choice of the value of the C—C—C bending force constant. Dewar [1, 3] assumed these values were equal for both **5** and **7**. However, by applying this approach to the calculation of the strain energy of cyclobutane, based on the value of the above constant for cyclopentane, one arrives at the conclusion in favor of the σ-aromaticity of cyclobutane [53]. The value of the force constant for paraffins was used by Cremer and Kraka [48] in calculating the strain energy of cyclopropane with the torsional strain neglected. This scheme as applied to cyclotrisilane points to σ-antiaromaticity of this molecule [53].

Cremer and Gauss [30] returned to the problem of determining the energy of the σ-aromatic stabilization of cyclopropane, where they presented a detailed analysis of the singling out of various contributions to the total strain energy of

5 and **7**. In particular, they showed that the stabilization of **5** on account of the strengthning of CH bonds comes to a more 6.4 kcal/mol (2.8 kcal/mol for **7**), while the Dunitz–Schomaker strain energy in **7** amounts to 12.0 kcal/mol. The adding up of all destabilizing contributions yields the strain energy of 50.8 kcal/mol for **5** and 29.9 kcal/mol for **7**. Since the TSE for **5** equals 28.0 kcal/mol and the stabilizing energy of the CH bond strengthening is 6.4 kcal/mol, the resulting value of the energy of the σ-aromatic stabilization turns out to be 16.4 kcal/mol [30] (with the refined value of the strain energy of propane taken into account [25]). The analysis of possible errors indicates that this value should be regarded as the lower estimate.

7.2.2 Cyclopropane: The Surface σ-Delocalization

The σ-aromatic stabilization of **5**, may be treated as a phenomenon caused by the σ-electronic delocalization, which is one of the modes of electronic delocalization along with the ribbon and volume types (see Chapter 3) [25, 30, 48, 54, 55]. In cyclopropane, the CC bonds are characterized by substantial ellipticites $\varepsilon = \lambda_1/\lambda_2 -1 = 0.11$ (see Chapter 2), while for cyclobutane $\varepsilon = 0.02$ [54].

The substantial in-plane ellipticities bear witness to the "π-like" charge distribution in the ring. Note that cyclopropane possesses an even greater nucleophilicity ($N = 6.4$) than ethene ($N = 4.7$) or ethyne ($N = 5.1$). This has been shown by quantitative evaluation of the gas-phase nucleophilicity by means of the empirical relationship $k\sigma = cEN$, where $k\sigma$ is the intermolecular stretching force constant for the hydrogen bond in the dimers $B\cdots HX$, E is the electrophilicity of the H end of HX, and N is the nucleophilicity of the acceptor region of B, and $c = 0.25$ N/m [56]. For cyclopropane (**5**) the bond and ring critical points are in close proximity and the concentration of the electronic charge is found not only in CC bonding regions, as in the case of **7**, but also inside the ring [48, 54]. Cyclopropane is characterized by an electronic density that amounts in the critical point of a ring to 82% of its value in the critical points of the CC bonds, while for cyclobutane and benzene the corresponding percentages are 33% and 7%, respectively [48]. As has been shown by Pan et al. [57], the total electron density in the center of a cyclopropane ring exceeds by 0.16 $e/Å^3$ that in the promolecule, which consists of three spherical free carbon atoms.

The experiment X–X difference electron densities in the ring centers of [3] rotane (**16**) are 0.05–0.10 $e/Å^3$ for the central ring and 0.10–0.15 $e/Å^3$ for the

16

other rings [58]. This supports the assumption of surface σ-electron delocalization in cyclopropane derivatives.

If sp^2 and p orbitals of the methylene fragments CH_2 (the radial and the tangential orbitals, respectively) are used as the basis orbitals constituting the valence MOs of skeletal CC bonds of cyclopropane, then the lowest valence MO a_1' will be constructed of radial orbitals only (see Fig. 3.3). A set of the latter orbitals composes a Hückel-type system, while the tangential orbitals make up a Möbius system of levels in odd-membered cycles and a Hückel system in the even-membered ones [59]. A characteristic feature of the MO a_1' is the considerable overlapping of three sp^2 hybrid orbitals in the center of the ring, allowing it to be regarded as a "surface" orbital [25, 30]. Since such overlapping diminishes exponentially as the size of the ring grows, the "surface" delocalization is transformed in this case to a ribbon-type electronic delocalization analogous to that of the π-type in cyclopolyenes. The filling of the surface orbital a_1' in cyclopropane leads to the formation of a two-electron three-center (2e-3c) bond, as, for example, in the H_3^+ ion.

In the skeletal MOs e', the contribution by tangential orbitals is predominant. If the orbitals that make an angle $90° \geq \tau > 45°$ with the straight line connecting nuclei of carbon atoms are classified under the π-type, and those characterized by the angles $45° > \tau \geq 0°$ are classified under the σ-type, then the MOs e' may be assigned to the π-type [25, 30]. As the size of the ring grows, a MO formed predominantly by the tangential orbitals changes its type from π to σ, but for a MO composed of the radial ones, this changes takes a reverse course—from σ to π (see Fig. 3.8). Thus, whereas for a three-membered ring the overlapping of the radial σ-orbitals is more preferable, with the increase in the ring size this preference falls to the tangential orbitals. This is keeping with the above relationship: the orbitals of the radial set will now correspond to the ribbon delocalization of the π-type, while the tangential orbitals will correspond to that of the σ-type (see Fig. 3.6) [25, 30]. The conclusion may be drawn that the surface delocalization is conceivable for small cycles only. More or less significant effects associated with the surface σ-delocalization may indeed be expected only in the case of the three-membered cycles [25, 30, 54, 55, 60].

7.2.3 Cyclopropane and Benzene as Aromatic Analogs

Thanks to its topology and the geometry of the ring, cyclopropane has a unique position among other cycloparaffins. The skeletal bonding in it is realized through a central two-electron three-center bond ("super σ-bond") and two peripheral four-electron three-center bonds ("π-bonds") [25, 30]. The representation of the electronic structure of cyclopropane by invoking the classical structure with localized bonds proves unsuitable for the description of the above-enumerated peculiarities of this molecule. But is the notion of "σ-aromaticity" more suitable for this purpose? The problems we are faced with in discussing this question are analogous to those that arise in arguments over whether the term aromaticity should be used at all. The chief reason against its use is that all those peculiarities can be described individually as a manifestation of different effects (see Chapter 1).

As to the legitimacy of the term σ-aromaticity, all pros and cons connected with its use were carefully weighed by Cremer [25]. His principal conclusion was that there is a possibility of interpretation of all specific features of cyclopropane not in terms of σ-aromaticity but rather on the basis of σ-electronic delocalization. The latter interpretation has certain advantages, the most important of which is the following: the term surface σ-delocalization corresponds to a specific type of distribution of electron density; that is, it is connected with an observable quantity and can consequently be verified by its analysis (e.g., see [30, 48]). The notion of the surface delocalization permits a simple description of the stabilization effect in the cyclopropane system. Indeed, the delocalization of the electrons lying in the lowest valence skeletal ("surface") orbital diminishes their kinetic energy, which, in turn, enhances the AO contraction on carbon atoms. As a result, the virial theorem is satisfied (for details see [60–62]), the total energy is decreased, and the length of CC bonds is reduced.[3] Thus the σ-electron delocalization is seen to dictate such an essential feature of the ring geometry as the length of the C—C bonds.

On the other hand, there is an important argument, valid also in the case of the π-aromaticity, in favor of the notion of σ-aromaticity, namely, the fact that, by applying it, all the molecular characteristics in question may be described from a unified position. In all probability, these characteristics could be interpreted separately on the basis of distinct molecules, but this approach may obscure the interrelation among them or, indeed, leave it out altogether. It should be noted that the presence of the surface delocalization does not necessarily involve the existence of the σ-aromaticity [25].

Cremer has revealed an important aspect in the analogy between the σ-aromaticity of cyclopropane and the π-aromaticity of benzene [25]. In describing the electronic structure of benzene, a resonance hybrid of two Kekule structures may be used [63]; however, in the case of cyclopropane a resonance hybrid of structures of three equivalent π-complexes [25] can be applied. Since for a π-complex, such as ethylene with a halogen cation, the representation in the form of the separate structure $\| \rightarrow X^+$ is consistent with the description in terms of the bond parts, such structures may be regarded as classical [25]. We may thus draw the conclusion that both the nonclassical structure of π-aromatic benzene and the structure of σ-aromatic cyclopropane can be represented by means of resonance hybrids of classical-type structures [25, 60]:

$$(7.4)$$

---

[3]Note, however, that in cyclopropane (**5**) the bond path length R_b exceeds the equilibrium value R_e of the internuclear separation so that the value of $R_b = 1.53$ Å turns out to be close to R_e (1.54 Å) in ethane [54].

$$\underset{\substack{| \\ H_2C-CH_2}}{\overset{CH_2}{}} \longleftrightarrow \underset{\substack{CH_2 \\ CH_2}}{\overset{CH_2}{}} \longleftrightarrow \underset{\substack{H_2C \quad CH_2}}{\overset{CH_2}{}} \equiv \underset{\substack{H_2C-CH_2}}{\overset{CH_2}{}} \qquad (7.5)$$

According to the energy criterion of aromaticity, the value of the topological resonance energy (TRE) (see Chapter 2) for benzene is positive, equaling 0.273 in β units. For cyclopropane, a calculation by the Sandorfy scheme of C-approximation [5], analogous to the MO method of Hückel, also yields a positive value of TRE = 0.019 [64]. This value is in β units with $m = 0.3$ and $\beta' = m\beta$, where β' is the interatomic (geminal) resonance integral and β is the vicinal resonance integral between the sp^3 hybrid orbitals of neighboring carbon atoms; for $m = 0.7$ [3], TRE = 0.170 [64].

7.2.4 σ-Aromaticity of Five-Membered Ring Systems

By exclusively associating the notion of the σ-aromaticity with the surface σ-electron delocalization, we restrict its application to the following cases: the three-membered rings formed by carbon atoms or the atoms of Si (17) and Ge (in the last two cases the σ-aromatic stabilization is manifested to a much lesser degree than in (5) [65]), the structures of organometallic compounds, such as 18 [66], and the bicyclic molecular structures composed of three-membered rings (e.g. [1.1.1]propellane (19)) [67]. The effects of σ-aromaticity grow weaker with the increase in the size of rings, and this weakening proceeds more quickly than in the case of the π-aromaticity [1, 64]. This is explained as follows: the levels of a skeletal σ-orbitals of a ring composed of ($2k + 1$) atoms correspond to a system of π-levels of [$4k + 2$]annulene, more precisely, to the annulene structures of the Kekule type characterized by alternation of bond lengths due to the alternation of the geminal and vicinal resonance integrals β. In other words, the systems of the levels of the skeletal orbitals of cyclopropane does not correspond to a system of π-levels of the high-symmetry structure of benzene (D_{6h}) but rather to a Kekule-type structure of D_{3h} symmetry [64] (Fig. 7.2). This is seen from the calculated value of TRE for cyclopentane equaling a mere 0.0012β ($m = 0.3$) [64].

L = CO, CNR

17   18   19

At the same time, while manifestations of the σ-aromaticity in a five-membered ring may be regarded as residual [3], the concept of the σ-electronic

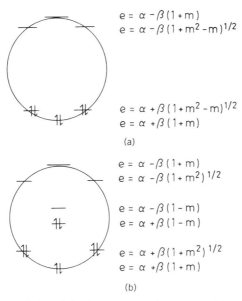

$$e = \alpha - \beta(1 + m)$$
$$e = \alpha - \beta(1 + m^2 - m)^{1/2}$$

$$e = \alpha + \beta(1 + m^2 - m)^{1/2}$$
$$e = \alpha + \beta(1 + m)$$

(a)

$$e = \alpha - \beta(1 + m)$$
$$e = \alpha - \beta(1 + m^2)^{1/2}$$

$$e = \alpha - \beta(1 - m)$$
$$e = \alpha + \beta(1 - m)$$

$$e = \alpha + \beta(1 + m^2)^{1/2}$$
$$e = \alpha + \beta(1 + m)$$

(b)

Figure 7.2 The level of the skeletal σ-MOs cyclopropane (a) and cyclobutane (b) obtained by the C-approximation method.

delocalization and the corresponding stabilization effects has been applied successfully not only to cyclopropane but to cyclopentane as well [68]. Using this approach, it has, for example, been found that the enthalpies of activation characterizing reactions of formation of three- and five-membered rings are lower than in the case of the corresponding even-membered ($n = 4,6$) less strained rings [69]. This fact may be regarded as being due to the formation, upon cyclization to three- or five-membered systems, of Hückel-type six or ten electron σ-aromatic transition states, whereas in the case of a four-membered ring the transition state is antiaromatic [69]. The stabilization of configuration **20** of dimethylcarbene is thought to be associated with the formation of a σ-aromatic ten-electron five-center system (**20a**) [70].

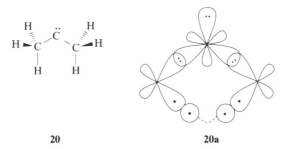

20   **20a**

264 REFERENCES

REFERENCES

1. M. J. S. Dewar, *Bull. Soc. Chim. Belg.*, **88**, 957 (1979).
2. M. J. S. Dewar and M. L. McKee, *Pure Appl. Chem.*, **52**, 1431 (1980).
3. M. J. S. Dewar, *J. Am. Chem. Soc..*, **106**, 669 (1984).
4. M. J. S. Dewar and R. Pettit, *J. Chem. Soc.*, 1625 (1954).
5. C. Sandorfy, *Can. J. Chem.*, **33**, 1337 (1955).
6. W. C. Herndon, *Prog. Phys. Org. Chem.*, **9**, 99 (1972).
7. O. Sinanoglu, in R. B. King (Ed.), *Chemical Applications of Topology and Graph Theory*, Elsevier, Amsterdam, 1983, Ch. 4 pp. 57–74.
8. R. Hoffmann, *Acc. Chem. Res.*, **4**, 1 (1971).
9. R. Gleiter, *Angew. Chem. Int. Ed. Engl.*, **13**, 696 (1974).
10. M. N. Paddon-Row, *Acc. Chem. Res.*, **15**, 245 (1982).
11. H.-D. Martin and B. Mayer, *Angew. Chem. Int. Ed. Engl.*, **22**, 283 (1983).
12. K. Ohta, H. Nakatsuji, H. Kubodera, and T. Shida, *Chem. Phys.*, **76**, 271 (1983).
13. M. Tabada and A. Lund, *Chem. Phys.*, **75**, 379 (1983).
14. M. Iwasaki, K. Toriyama, and K. Numone, *J. Am. Chem. Soc.*, **103**, 3591 (1981).
15. K. Toriyama, K. Numone, and M. Iwasaki, *J. Chem. Phys.*, **77**, 5891 (1982).
16. K. Numone, K. Toriyama, and M. Iwasaki, *J. Chem. Phys.*, **79**, 2499 (1983).
17. J. L. Holmes and F. P. Lossing, *Can. J. Chem.*, **60**, 2365 (1982).
18. K. Fukui, H. Kato, and T. Yonezawa, *Bull. Chem. Soc. Jpn.*, **33**, 1197 (1960).
19. M. V. Bazilevsky, *MO Method and Reactivity of Organic Molecules*, Khimiya, Moscow, 1969.
20. G. Klopman, *Tetrahedron*, **19**, 111 (1963).
21. H. Bock, W. Ensslin, F. Feher, and R. Freund, *J. Am. Chem. Soc.*, **98**, 668 (1976).
22. A. Herman, B. Dreczewski, and W. Wojnowki, *Chem. Phys.*, **98**, 475 (1985).
23. C. G. Pitt, M. M. Bursey, and P. F. Rogerson, *J. Am. Chem. Soc.*, **98**, 519 (1976).
24. R. West, *Pure Appl. Chem.*, **54**, 1041 (1982).
25. D. Cremer, *Tetrahedron*, **44**, 7421 (1988).
26. A. Liberies, *J. Chem. Educ.*, **54**, 479 (1977).
27. K. W. Wiberg, *Angew. Chem. Int. Ed. Engl.*, **25**, 312 (1986).
28. S. W. Benson, *Thermochemical Kinetics*, Wiley, New York, 1976.
29. A. Greenberg and J. Liebman, *Strained Organic Molecules*, Academic, New York, 1978.
30. D. Cremer and J. Gauss, *J. Am. Chem. Soc.*, **108**, 7467 (1986).
31. H. Günter, *NMR Spectroscopy*, Wiley, New York, 1980.
32. G. C. Levy and G. L. Nelson, *Carbon-13 Nuclear Magnetic Resonance for Organic Chemists*, Wiley, New York, 1972.
33. K. W. Zilm, A. J. Beeler, D. M. Grant, J. Michl, T. C. Chou, and E. L. Allfred, *J. Am. Chem. Soc.*, **103**, 2119 (1981).
34. A. M. Orendt, J. C. Facelli, D. M. Grant, J. Michl, F. H. Walker, W. P. Dailey, S. T. Waddell, K. B. Wiberg, M. Schindler, and W. Kutzelnigg, *Theor. Chim. Acta*, **68**, 421 (1985).

35. A. de Meijere, *Angew. Chem. Int. Ed. Engl.*, **18**, 809 (1979).

36. H. D. Roth, M. L. M. Schilling, and F. C. Schilling, *J. Am. Chem. Soc.*, **107**, 4152 (1985).

37. A. I. Ioffe, V. A. Svyatkin, and O. M. Nefedov, *The Structure of the Cyclopropane Derivatives*, Nauka, Moscow, 1986.

38. D. Cremer and E. Kraka, *J. Am. Chem. Soc.*, **107**, 3811 (1985).

39. H. C. Brown, E. N. Peters, and M. Ravindranathan, *J. Am. Chem. Soc.*, **99**, 505 (1977).

40. W. J. Hehre and P. C. Hiberty, *J. Am. Chem. Soc.*, **96**, 302 (1974).

41. W. Koch, B. Liu, and D. J. de Fress, *J. Am. Chem. Soc.*, **110**, 7325 (1988).

42. M. Saunders, K. E. Laiding, K. B. Wiberg, and P. v. R. Schleyer, *J. Am. Chem. Soc.*, **110**, 7652 (1988).

43. R. A. Moss, S. Shen, K. Krogh-Jespersen, J. A. Potenza, H. J. Schugar, and R. C. Munjal, *J. Am. Chem. Soc.*, **108**, 134 (1986).

44. Y. Apeloig and D. Arad, *J. Am. Chem. Soc.*, **107**, 5285 (1985).

45. E. Uggerud, D. Arad, Y. Apeloig, and H. Schwarz, *J. Chem. Soc. Chem. Commun.*, 1015 (1989).

46. P. Reinders and G. Schrumpf, *J. Mol. Struct.*, (THEOCHEM), **150**, 297 (1987).

47. D. Cremer, E. Kraka, T. S. Slee, R. F. W. Bader, C. D. H. Lau, T. T. Nguen-Dang, and P. J. MacDougall, *J. Am. Chem. Soc.*, **105**, 5069 (1983).

48. D. Cremer and E. Kraka, *J. Am. Chem. Soc.*, **107**, 3800 (1985).

49. R. C. Haddon, *Tetrahedron Lett.*, 4303 (1974).

50. W. L. Jorgensen, *J. Am. Chem. Soc.*, **98**, 6784 (1976).

51. R. F. W. Bader, *Atoms in Molecules*, Claredon Press, Oxford, 1990.

52. P. v. R. Schleyer, The Contrasting Strain Energies of Small Ring Carbon and Silicon Rings, The Relationship with Free Radical Energies in H. G. Viehe, R. Janoschek, and R. Merenyl (Eds.), *NATO Advanced Research Workshop on Substituent Effects in Radical Chemistry*, Reidel Publishing Co., Dordrecht, 1986, pp. 69–81.

53. R. S. Grev and H. F. Schaefer, *J. Am. Chem. Soc.*, **109**, 6569 (1987).

54. R. F. W. Bader, T. S. Slee, D. Cremer, and E, Kraka, *J. Am. Chem. Soc.*, **105**, 5061 (1983).

55. D. Cremer and E. Kraka, The Concept of Molecular Strain in A. Greenberg and J. F. Liebman (Eds.), *Structure and Reactivity. Molecular Structure and Energies*, Vol. 7, VCH, New York, 1988, p. 65.

56. A. C. Legon and D. J. Millen, *J. Chem. Soc. Chem. Commun.*, 986 (1987).

57. D. K. Pan, J. N. Gao, H. L. Huang, and W. H. E. Schwarz, *Int. J. Quant. Chem.*, **29**, 1147 (1986).

58. R. Boese, T. Miebach, and A. de Meijere, *J. Am. Chem. Soc.*, **113**, 1743 (1991).

59. V. I. Minkin, R. M. Minyaev, and Yu. A. Zhdanov, *Nonclassical Structures of Organic Compounds*, Mir Publishes, Moscow, 1987.

60. E. Kraka and D. Cremer, Chemical Implications of Local Features of the Electron Density Distribution in Z. B. Maksić (Ed.), *Theoretical Models of Chemical Bonding*, Part 2, Springer, Berlin, 1990, pp. 453–542.

61. K. Ruedenberg, in O. Chalvet, R. Daudel, S. D. Finer, and J. P. Malrieu (Eds.),

Localization and Delocalization in Quantum Chemistry, Vol. 1, Reidel, Dordrecht, 1975, pp. 223–245.

62. S. Nordholm, *J. Chem. Educ.*, **65**, 581 (1988).

63. T. E. Peacock, *Electronic Properties of Aromatic and Heterocyclic Molecules*, Academic, London, 1965.

64. V. I. Minkin, M. N. Glukhovtsev, and B. Ya. Simkin, *J. Mol. Struct.*, (THEO-CHEM), **181**, 93 (1988).

65. D. Cremer, J. Gauss, and E. Cremer, *J. Mol. Struct.*, (THEOCHEM), **169**, 531 (1988).

66. C. Mealli, *God. Jugosl. Cent. Kristallofr.*, **21**, 29 (1986).

67. I. Lee, K. Yang, and H. S. Kim, *Tetrahedron*, **41**, 5007 (1985).

68. W. W. Schoeller, *Tetrahedron*, **29**, 929 (1973).

69. S. M. van der Kerk, J. W. Verhoeven, and C. J. M. Stirling, *J. Chem. Soc. Perkin Trans. II*, 1355 (1985).

70. D. Cremer, J. S. Binkley, J. A. Pople, and W. J. Hehre, *J. Am. Chem. Soc.*, **96**, 6900 (1974).

8

IN-PLANE AND RADIAL AROMATICITY

The in-plane aromaticity may formally be regarded as a particular case of the σ-aromaticity, it may be termed the "homo-σ-aromaticity." The stabilization of the molecule is achieved through the surface σ-delocalization of the electrons occupying radial orbitals (sp_{in}-orbitals, see Fig. 8.1, or p_σ-orbitals) located in the plane of a cyclic arrangement of non-neighboring atoms. An example of the in-plane aromaticity is given by the $C_6H_3^+$ cation (1) in which the three largely p atomic orbitals at alternating bare carbons of the six-membered ring form the orbital system 1a possessing the same symmetry properties as the p_π-orbitals of the cyclopropenium ion [1].

1   1a

Since in the σ-system of cation 1 there are only two electrons (apart from those that form the two-center σ-bonds C—H and C—C), they occupy the bonding MO σ_1 of a_1' symmetry, whereas both antibonding σ-MOs of e'-type stay vacant (Fig. 8.1). Such occupation results in a trishomo-σ-aromatic $6\pi2\sigma$-electron system.

The stability of 1 has been assessed from calculations of the energy of the isodesmic reaction (8.1) [1, 2]:

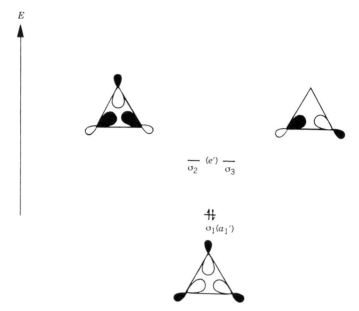

Figure 8.1 The Hückel system of the σ-MOs formed by three radial σ-AOs of p_σ or sp^n types.

$$1 \; + \quad \text{(benzene)} \quad \longrightarrow \quad \text{(benzene cation)} \quad + \quad \text{(biradical)} \tag{8.1}$$

$\Delta E = 38.5$ kcal/mol (MINDO/3), 14.4 kcal/mol (4-31G) [1], 35.2 kcal/mol (MP4SDTQ/6-31G**//MP2/6-31G*) [2].

The $C_6H_3^+$ ions generated from mercaptobenzene and isomeric dihalobenzenes were mass-spectroscopy detected. Although the structure of these could not be firmly established, the presence of ion **1** may be assumed since it should be one of the more stable $C_6H_3^+$ isomers [1].

The 1,2,3- and 1,2,4-cyclic isomers are 10.7 and 23.9 kcal/mol higher in energy than **1** at MP4SDTQ/6-31G**//MP2/6-31G* [2]. At the same level open-chain $HCC_4CH_2^+$ and $HCCCHCCCH^+$ structures also have 14.1 and 33.0 kcal/mol higher energies than that of **1**.

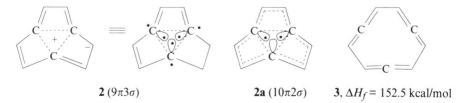

2 $(9\pi3\sigma)$ **2a** $(10\pi2\sigma)$ **3**, $\Delta H_f = 152.5$ kcal/mol

The possibility of the occurrence of an analogous three-center, two-electron bond incorporated into another perimetrical π-system was examined by Fukunaga and co-workers [3]. Clearly, the stabilization of such a bond, referred to by the authors as the "trefoil bonding," by lowering the energy level of the a'_1 MO, will be more effective as the overlap of the three interacting σ-orbitals increases. To achieve this, the carbon centers have to be brought closer together. Such a geometry is realized in [5.5.5]trefoilene C_9H_6 (**2**) and may conceptually be visualized as being formed from triquinacene by removing its central CH group and the three ring hydrogens at the positions to which this group is attached. By shifting one σ-electron in **2** to the π-system, a doubly aromatic $10\pi2\sigma$ electron structure (**2a**) with the trefoil bond is achieved. A number of the other structures possessing the trefoil aromaticity have been thought of where a three-center two-electron bond is embedded in the plane of the annulene perimeter containing $(4n + 2)$ out-of-plane π-electrons.

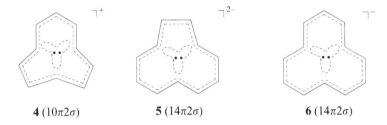

4 $(10\pi2\sigma)$ **5** $(14\pi2\sigma)$ **6** $(14\pi2\sigma)$

The stability of the **2**, **4**, **5**, and **6**, structures to isomerization and decompositions is determined by the degree to which the energy of the aromatic stabilization, due to the formation of the trefoil bond, can offset the great strain. The MINDO/3 and MNDO calculations [4] have shown that although the electronic structure of **2a** conforms to the double aromaticity of this system and the bond alternation in the perimeter is, accordingly, small, structure **2a** is highly unstable and does not correspond to the minimum on the PES of C_9H_6. The valence isomer of **2a**, cyclonona-1,2,4,5,7,7-hexaene (**3**), possessing nonplanar D_3 structure, has a heat of formation reduced by nearly 100 kcal/mol from that of [5.5.5] trefoilene whose symmetry is D_{3h}. By contrast, the energy gap between [3.6.6.] trefoilene (**7**) and cyclonona-1,2,4,8-tetraen-6-yne, which is the valence isomer of **7**, is much smaller. Trefoilene (**7**), possessing a nearly planar structure, represents, according to MNDO calculations [4], a local minimum on the C_9H_6 PES. It is assumed that this structure, similar to **8**, may be trapped in reactions that generate carbene (**9**), for example, in the decomposition of 4-diazoindene. Like **3**, **7**, and **8**, carbene (**9**) and the corresponding allene (**10**) represent the minima on the PES of C_9H_6.

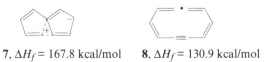

7, $\Delta H_f = 167.8$ kcal/mol   **8**, $\Delta H_f = 130.9$ kcal/mol

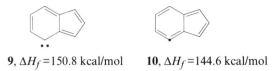

9, $\Delta H_f = 150.8$ kcal/mol   **10**, $\Delta H_f = 144.6$ kcal/mol

An extensive computational search for structures exhibiting the in-plane aromaticity effect due to the homoconjugative overlap of radially oriented $p_\sigma(sp^n)$-orbitals was undertaken by McEwen and Schleyer [5]. It has been shown that neither the D_{4h} structure of hypothetical hydrocarbon C_8H_4 (**11**) nor $C_{10}H_5^-$ (D_{5h}) (**12**) is stabilized, although both of them have the potential to form the double π and σ arrangement of p-orbitals:

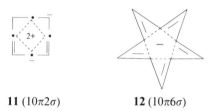

11 $(10\pi2\sigma)$ **12** $(10\pi6\sigma)$

No more promising are the [n]pericyclynes, which possess a topology allowing the in-plane homoaromaticity:

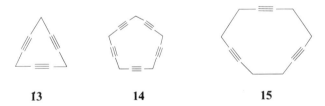

13   **14**   **15**

In photoelectron spectra these compounds manifest a substantial cross-ring in-plane π_σ-interaction [6]. It has, however, a repulsive destabilizing character [5, 7, 8]. This may be understood on the basis of MINDO/3 [9] and *ab initio* [10] calculations of the symmetric trimerization of acetylene to benzene. In transition-state structure, when carbons forming the new bond draw together to a distance of 2.2 Å, the MP3/6-31G* energy barrier amounts to 62 kcal/mol. At larger distances, which is the case with compounds **13**–**15**, a repulsive part of the PES is realized.

A special case of the in-plane aromaticity is presented by [1.1.1.1] and [2.2.1.1]pagodane dications (**16** and **17**), whose central four-membered ring may be viewed as a four-center two-electron "bishomoaromatic" system [11, 12]. These dications and some of their derivatives, for example, bis-lactone (**18**) [12, 13], are unprecedentedly stable, especially **16**, which can be stored in the SbF$_5$/SO$_2$ClF solution at ambient temperature for hours.

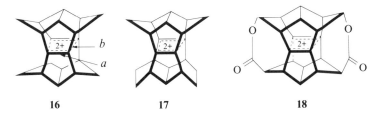

16   17   18

Three synthetic routes leading to dication **16** are conceivable, namely, the two-electron oxidation of pagodane (**19**) under stable ion-solution conditions, the oxidation of its valence isomer, the diene **20**, and the ionization of dibromide (**21**) (Scheme 8.1).

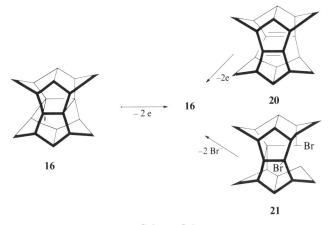

Scheme 8.1

Charge-stripping mass spectroscopy has also allowed one to observe the formation of dications from [1.1.1.1] and [2.2.1.1]pagodanes $C_{20}H_{20}^{2+}$ and $C_{20}H_{24}^{2+}$, respectively [14] (see also [15]).

Clear evidence in favor of the formation of a dicationic species, with the positive charges distributed mainly over the central cyclobutane ring, retaining the initial D_{2h} symmetry of **19** and **20**, has been presented by the ^{13}C NMR spectra of **16**.[1] The ^{13}C chemical shift additivity analysis [16, 17] ruled out the possibility of describing the [1.1.1.1]pagodane dication as an averaged structure of several classical dications (**22**) (only the central cyclobutane fragment is shown in the formulas):

22a   **22b**   **22c**

[1]The similarity of the cyclobutanoid ^{13}C shifts for **16** (δ 251.0) and tetramethylcyclobutadiene dication (δ 209.7) is remarkable.

22d   **22e**   **22f**

Since the D_{4h} form (**23**) is also excluded on the grounds of the symmetry of the NMR spectrum observed, the only alternative forms of the dication are non-classical structures **24** and **25** (again, only the part of the molecule containing the cyclobutane fragment is given).

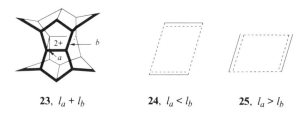

23, $l_a + l_b$ **24**, $l_a < l_b$ **25**, $l_a > l_b$

Both the chemical behavior of the dication in quenching experiments and the fact that it has been derived from diene **20**, while another isomeric diene with double bonds in positions b is not obtainable, show that the nonclassical structure **24** ≡ **16** is strongly favored. As the ^{13}C NMR spectra of the [1.1.1.1] pagodane dication are temperature-independent down to $-130°$ C, it has been concluded [11] that the dication possesses a static D_{2h} structure and cannot be represented by a rapidly equilibrating pair of classical forms **22d** ⇌ **22d′**, **22e** ⇌ **22e′**, or **22f** ⇌ **22f′**.

The dications **16–18** are obviously topologically equivalent to the transition state **26** for the orbital symmetry-allowed cycloaddition of ethylene to ethylene dication.

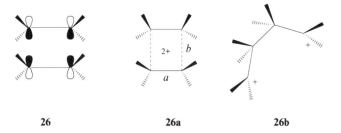

26   **26a**   **26b**

The D_{2h} structure **26a** and its C_{2v} open-chain isomer (classical dication **26b**) are interesting as models for the pagodane and norbornadiene dications [11, 12]. While at the SCF level **26b** is more stable than its "in-plane aromatic" counterpart **26a**, at MP2/6-31G* **26b** transforms into **26a** upon geometry optimization [12]. However, neither **26a** nor **26b** are minima. Stabilization may be achieved with the involvement of **26a** into a rigid molecular framework.

This may be exemplified by the pagodane dication. MNDO and 3-21G calculations on the [1.1.1.1]pagodane dication indicated that there were three local minima related to structures $16 \equiv 24$, $22a$, and $22d$ (structure 25 was not a minimum at this level of approximation), the l_a and l_h distances in 16 being equal to 1.461 and 2.687 Å, respectively. Whereas at the MNDO level 16 is 4 kcal/mol less energy favored compared to $22d$, no minima for the latter and the $22a$ structures exist at the MNDO/3 and AM1 levels as well as at 3-21G//3-21G, with 16 being the true minimum at all levels [11, 12]. This confirms that, in spite of the large strain imposed by the tight cyclobutane moiety embedded in 16, there occurs a strong in-plane aromatic stabilization allowing one to qualify dications 16–18 as "bishomoaromatic"[2] 2π-species [11, 12].

In-plane aromaticity is only one possible topological possibility for the $p_\sigma - p_\sigma$ overlap to stabilize separate cyclic systems. Further possibilities are seen in various manifestations of the so-called radial aromaticity [5] whose distinctive feature is the formation of a three-dimensional array of p (sp^n)-orbitals, their lobes being directed inward and overlapped inside a given molecular cage. The stabilization of such a molecular framework occurs where the number of electrons in the thus formed orbital subsystem does not exceed that needed to fill in all stabilized energy levels. An aesthetically appealing example of the radial aromaticity is given by a recent discovery [18] of the nonclassical tetratrishomoaromatic 1,3-dehydro-5, 7-adamantanediyl cation $C_{10}H_{12}^{2+}$ (27):

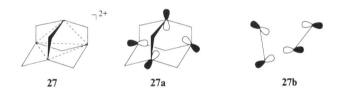

27   27a   27b

Four orbitals in the bridgehead positions extend towards the center of the adamantane framework ($27a$). Obviously, this orbital arrangement may be derived from 26 by a turn of one pair of p-orbitals to the orthogonal plane. Figure 8.2 shows the well-known orbital interaction diagram for four σ-orbitals placed at the apexes of the tetrahedron. The only bonding level is that of a_1-symmetry resulting from an in-phase combination of orbitals. The occupation of this MO in the two-electron dication (27) system should lead to its stabilization, provided the overlap is of sufficient magnitude. Indeed, that was found to be the case when Schleyer and co-workers [18–20] succeeded in preparing the dication 27 as a stable species in superacid media.

The non-classical structure of 27 is strongly supported by the ^{13}C NMR spectra data, revealing strong shielding of the bridgehead carbons ($\delta = 6.6$ at $-71°$) despite the presence of the two positive charges. Shielded values are characteristic features of hypercoordinate carbocation centers [17]. To estimate

[2] l_a is rather shorter than l_h.

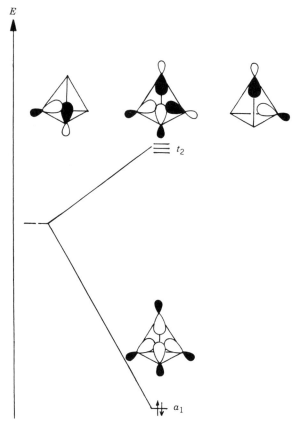

Figure 8.2   Orbital interaction diagram for four tetrahedrally oriented σ-orbitals.

(8.2)

28   **29**

30   **31**   **32**   **33**
$\Delta\Delta H_f = -25.4$ kcal/mol (MINDO/3)

qualitatively the degree of stabilization of **27**, the energies of the isodesmic reactions (8.2) and (8.3) are to be compared.

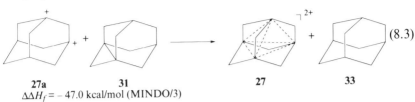

$$\text{27a} \qquad\qquad \text{31} \qquad\qquad\qquad\qquad\qquad \text{27} \qquad\qquad \text{33}$$

$$\Delta\Delta H_f = -47.0 \text{ kcal/mol (MINDO/3)}$$

It follows from the MINDO/3 calculations [18] that the extension of the two-electron delocalization from three centers in the trishomoaromatic 1,3-dehydro-5-adamantyl cation (**32**) to four bridgehead centers in **27** results in virtually doubling stabilization. This explains the striking stability of the non-classical dication **27**, which may also be viewed [18, 21] as a peculiar example of the molecular system with three-dimensional aromaticity (see Chapter 9).

The stability of the two-electron system **27** may therefore be considered as a simple consequence of the $(4n + 2)$ magic electron-count rule extended to three dimensions [22]. It has been shown that neutral polyolefinic systems containing radial p-orbitals arranged in cyclic conjugated arrays also obey this rule, revealing the effects peculiar to the in-plane and radial aromaticity. Modest but definite π-stabilization energies were indicated by MMP2 calculations [5] of cyclic conjugated [m]peristylapolyenes (**34**) ($m = 3 - 5$), p-[4^2. 5^8]hexadecahedra-polyene, dodecahedrapolyenes (**35**), and tetraquinapolyenes. An excerpt of the results of the calculations [5] is given in Table 8.1. It is seen from the data presented that resonance energies found for compounds with conjugated radial p-orbitals systems have sufficiently large positive values only in the case of alternating cyclic systems containing $4n + 2$ ($n = 1, 2$) p-electrons that is, Numbers 1, 3, and 4 in Table 8.1. Antiaromatic $4n$-electron systems (compound Numbers 2 and 5) possess negative values of resonance energy, whereas for nonalternative conjugated systems (compounds Numbers 6 and 7) resonance energies are small, their values being virtually independent of the number of radial p-electrons. It may therefore be predicted that these compounds will not display additional stabilization and are to be assigned to non-aromatic systems. Other criteria employed for evaluating aromaticity in these compounds, such as heats of hydrogenation corrected for strain energy differences (see Table 8.1), resonance integrals, bond orders, and the degree of bond alternation, have been found to be in line with the general conclusion as to the applicability of the $(4n + 2)$ rule for the description of the in-plane and radial aromaticity.

Can four unsupported atomic centers be held together by two-electrons? The simplest possibility is the H_4^{2+} tetrahedral structure (**36**). However, unlike the 3c-2e prototype, H_3^+ (D_{3h}), **36** is not a minimum [23]. Since p-orbitals in tetrahedral species like **27** can overlap centrally, p-contributions to the bonding a_1 MO formed by these orbitals may be essential for 4c-2e bonding. The simplest species affording the possibility of significant p-orbital involvement is the tetralithium dication, Li_4^{2+} (T_d, **37**). Structure **37** has a positive binding energy

TABLE 8.1   Evaluation of Aromaticity in the Alternating (Numbers 1–5) and Nonalternating (Numbers 6 and 7) Hydrocarbons According to MMP2 Calculations [5]

Number	Compound	Number of p-electrons	RE^a, kcal mol	REPE, kcal/mol	ΔH^b_{shy}, kcal/mol
1	**34**, $m = 3$	6	14.84	2.47	—
2	**34**, $m = 4$	8	−2.02	−0.26	−23.6
3	**34**, $m = 5$	10	5.22	0.52	−15.8
4	**35**	10	8.10	0.81	−16.0
5	**35**	12	−4.36	−0.36	−24.4
6	**34**, $m = 5$	6	1.92	0.32	−19.2
7	**34**, $m = 5$	8	−0.63	−0.08	−20.8

^{a}Resonance energy is defined as $RE = E - E_{\text{ref}}$, where E is the MMP2 calculated bonding energy due to the p-AOs in a cyclic conjugated system and E_{ref} is the value calculated from group increments obtained by using the noncyclic conjugated polyenes. Positive sign of RE corresponds to stabilization.

^{b}Strainless heat of hydrogenation: $\Delta H_{\text{shy}} = \Delta H_{\text{hyd}} - \Delta H_{\text{strain}}$.

(18.48 kcal/mol at QCISD(T)(full)/6-311G(2d)//MP2(full)/6-311G(2d)) and corresponds to a minimum at MP2/6-31G* [24]. Thus, despite the large Coulomb repulsion in **37**, unsupported 4c-2e bonding is capable of stabilizing this doubly charged metal cluster.

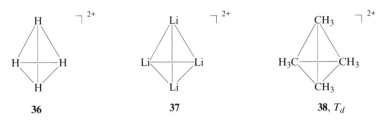

However, in contrast to **37**, unsupported 4c-2e bonding is not generally able to provide stability. For instance, 4e-2e bonding does not lead to structural stability of the $(CH_3)_4^{2+}$ T_d configuration (**38**) comprised of four methyl fragments. This loose complex was calculated to be a fourth order saddle point at HF/6-31G* [24].

REFERENCES

1. J. Chandrasekhar, E. D. Jemmis, and P. v. R. Schleyer, *Tetrahedron Lett*, 3707 (1979).

2. M. N. Glukhovtsev and P. v. R. Schleyer, unpublished results.

3. T. Fukunaga, H. E. Simmons, J. J. Wendoloski, and M. D. Gordon, *J. Am. Chem. Soc.*, **105**, 2729 (1983).

4. R. W. Alder, J. C. Petts, and T. Clark, *Tetrahedron Lett.*, **26**, 1585 (1985).

5. A. B. McEwen and P. v. R. Schleyer, *J. Org. Chem.*, **51**, 4357 (1986).

6. T. L. Scott, *Pure Appl. Chem.*, **58**, 105 (1986).

7. M. J. S. Dewar and M. K. Holloway, *J. Chem. Soc. Chem. Commun.*, 1188 (1984).

8. K. N. Houk, R. W. Strotzier, C. Santiago, R. W. Gandour, and K. P. C. Volhardt, *J. Am. Chem. Soc.*, **101**, 5183 (1979).

9. K. N. Houk, R. W. Gandour, R. W. Strotzier, N. G. Rondan and L. A. Paquette, *J. Am. Chem. Soc.*, **101**, 6797 (1979).

10. R. D. Bach, G. J. Wolber, and H. B. Schlegel, *J. Am. Chem. Soc.*, **107**, 2837 (1985).

11. G. K. S. Prakash, V. V. Krishnamurthy, R. Herges, R. Ban, H. Yuan, G. A. Olah, W.-D. Fessner, and H. Prinzbach, *J. Am. Chem. Soc.*, **110**, 7764 (1988).

12. R. Herges, P. v. R. Schleyer, M. Schindler, and W. -D. Fessner, *J. Am. Chem. Soc.*, **113**, 3649 (1991).

13. R. Pinkos, J. P. Melder, H. Fritz, and H. Prinzbach, *Angew. Chem. Int. Ed. Engl.*, **28**, 310 (1989).

14. T. Drewello, W.-D. Fessner, A. J. Kos, C. B. Lebrilla, H. Prinzbach, P. v. R. Schleyer, and H. Schwarz, *Chem. Ber.*, **121**, 187 (1988).

15. M. Saunders and H. A. Jiménez -Vázquez, *Chem. Rev.*, **91**, 375 (1991).

16. P. v. R. Schleyer, D. Lenoir, P. Mison, G. Liang, G. K. S. Prakash, and G. A. Olah, *J Am. Chem. Soc.*, **102**, 683 (1980).

17. G. A. Olah, G. K. S. Prakash, R. E. Williams, L. D. Field, and K. Wade, *Hypercarbon Chemistry*, Wiley, New York, 1987, Chapter 5.

18. M. Bremer, P. v. R. Schleyer, K. Scholz, M. Kausch, and M. Schindler, *Angew. Chem. Int. Ed. Engl.*, **26**, 761 (1987).

19. P. v. R. Schleyer, in G. A. Olah (Ed.), *Cage Hydrocarbons*, Wiley, New York, 1987, Chapter 1, pp. 1–38.

20. P. Buzek, P. v. R. Schleyer, and S. Sieber, *Chem. unser. Zeit*, **26**, 116 (1992).

21. K. Lammertsma, P. v. R. Schleyer, and H. Schwarz, *Angew. Chem. Int., Ed. Engl.*, **28**, 1321 (1989).

22. E. D. Jemmis and P. v. R. Schleyer, *J. Am. Chem. Soc.*, **104**, 4781 (1982).

23. M. N. Glukhovtsev, P. v. R. Schleyer, N. J. R. v. E. Hommes., J. W. de M. Carneiro, and W. Koch, *J. Comput. Chem.*, **14**, 285 (1993).

24. M. N. Glukhovtsev, A. Stein, and P. v. R. Schleyer, *J. Phys. Chem.*, **97**, 5541 (1993).

9

THREE-DIMENSIONAL AROMATICITY

Among the distinctive features of structures and reactivity contributing to the general definition of aromaticity, the following two rank over many others: (a) enhanced stability of a given structural type providing favorable conditions for the cyclic electronic delocalization and (b) persistence of the molecular framework against destruction in chemical reactions. The above features are not restricted to the planar conjugated molecular systems considered in Chapters 4 and 5. It has been recognized that a large set of polyhedral main-group and transition-metal compounds, such as boranes [1, 2], pyramidal, and sandwich-like π-complexes [3, 4], reveal a marked thermodynamic stability and propensity to substitution reactions characteristic of aromatic systems, provided their molecules possess certain valence electron shells.

9.1 RESONANCE ENERGIES OF POLYHEDRAL ORGANOMETALLICS

Ferrocene (1) is perhaps the most famous example of organometallic compounds showing strong covalent bonding between a metal and organic moieties, that is, cyclopentadienyl rings, the latter being prone to virtually every kind of electrophilic substitution.

To estimate the aromaticity of ferrocene, Aihara [5] employed the graph theory, allowing one to derive the relevant TRE value (see Section 2.2.5). A Hückel-type MO model, in which five equivalent hybridized $3d$ orbitals of iron, each treated formally as five heteroatoms, has been used. For the sake of simplicity the eclipsed configuration (2) has been considered instead of the slightly more energy-preferable staggered form (**1**). This is of no importance for analyzing the degree of aromaticity of ferrocene since no detectable energy barrier exists for the **1** ⇌ **2** interconversion [4]. Graph **3** with all orbitals forming the three-dimensional conjugated system represents the bonding scheme of ferrocene.

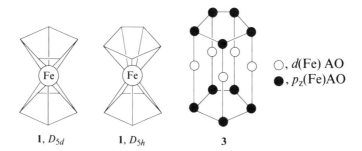

○, d(Fe) AO
●, p_z(Fe)AO

1, D_{5d} **1**, D_{5h} **3**

By deriving the characteristic polynomial of **3** and the corresponding reference polynomial for a hypothetical acyclic polyene-like structure in which the presence of cycles in **2** is ignored within the graph formalism, the resonance energy of ferrocene was calculated to be equal to 0.253 β_{C-C} (with the Coulomb integral for iron d-orbitals set equal to that of carbon $2p_z$ orbitals). This value is comparable to the TREs of benzene, 0.273 β_{C-C} and the cyclopentadienide ion, 0.371 β_{C-C} [6, 7], even though considering the presence of two carbon rings in ferrocene, its aromaticity should be estimated as about half that of benzene.

That the degree of aromaticity of ferrocene is relatively high follows from a comparison of its TRE magnitude with that of so-called open ferrocene, that is, bis(pentadienyl)iron (**4**), calculated in accordance with the same procedure. The

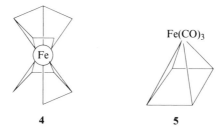

Fe(CO)₃

4   **5**

negative value of TRE = – 0.160 β_{C-C} found for (**4**) indicates its antiaromatic character. Another polyhedral iron π-complex for which the value of TRE is positive is the pyramidal tricarbonyliron cyclobutadiene (**5**) [8]. Along with *closo*-boranes and some other π-complexes of olefins, compounds **1**, **5** constitute a broad category of organometallic polyhedral compounds possessing positive resonance energies due to cyclic conjugation. So the calculations of SRTRE yield the value of 1.80 eV for **5**, whereas for the cyclobutadiene ligand this value is – 0.65 eV [9]. Thus the results of calculations of the TRE and

SRTRE values for organometallic three-dimensional systems may serve as a basis for the conclusion that the concept of aromaticity may be extended to such three-dimensional conjugated systems as well as the recently discovered non-classical organic ions and molecules with polyhedral molecular frameworks [8, 10]. The term three-dimensional aromaticity has been coined to describe the properties of relevant compounds that were considered as characteristic of aromatic systems only. In the following sections of this chapter we shall consider the most important topological types of three-dimensional aromatic system with an emphasis on organic molecules and ions. A straightforward connection between the electronic structure of polyhedral organic, organo-metallic, and metal–carbon cluster compounds will be elucidated through an analysis of electron-count rules common to a given topology regardless of the origin of the atoms forming the molecular framework.

9.2   PYRAMIDAL ORGANIC MOLECULES AND IONS

Pyramidal organic structures can be formed through populating apexes of a trig-onal, square, pentagonal, and so on pyramid with the C, CH groups, or those isoelectronic to them.

The first compound of this series is tetrahedrane (**6**), a valence isomer of cyclobutadiene. The problem of synthesizing tetrahedrane had been formulated more than half a century ago and stimulated intensive search for methods of its preparation from proper precursors [11, 12], eventually resulting in the isolation and complete structural characterization of tetra-*tert*-butyltetrahedrane (**7a**) [12, 13].

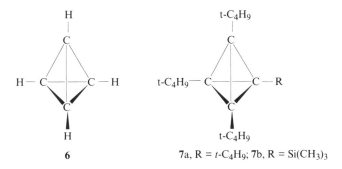

6

7a, R = *t*-C$_4$H$_9$; **7b**, R = Si(CH$_3$)$_3$

Although in view of the enormous angular strain (E_{str} = 140.0 kcal/mol, 6-31G* [14]) the energy required for breaking the C—C bond in **6** is a mere 10 kcal/mol (DZ + P basis set with the electron correlation taken into account using the CEPA scheme) [15]; numerous *ab initio* and semiempirical calculations (see [12, 16, 17] for references) consistently indicate sufficiently high kinetic stabil-ity of tetrahedrane in the absence of other reagents. In **7a** and another recently prepared astonishingly stable derivative of tetrahedrane, **7b** [18], the just mentioned condition is circumvented via a spatial shielding of the tetrahedrane framework with four bulky groups, the so-called corset effect [12, 18, 19].

The next member of the series of pyramidal hydrocarbons, the cation $(CH)_5^+$ (**8**), represents a structure that is not only strained but also nonclassical with a pentacoordinate carbon atom at the apex of a square pyramid.

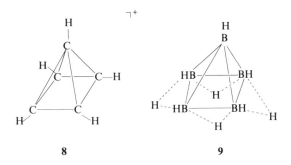

The possibility of the existence of such a cation was first assumed by Williams [20] who stressed that it should be isoelectronic to pyramidal pentaborane B_5H_9 (**9**). Stohrer and Hoffmann [21] made the first calculations on **8** (EHT) and suggested a basis orbital interaction scheme that offered a clear explanation of the nature of its stability. Soon after this a number of derivatives of **8**, for example, 1,2-dimethyl (**10**) [22], 1,2,4-trimethyl (**11**) [23], and bishomo $(CH)_5^+$ cation (**12**) [24] (see [16, 25–27] for comprehensive reviews), were prepared under the stable ion condition and their structures were unambiguously characterized by ^1H and ^{13}C NMR spectroscopy.

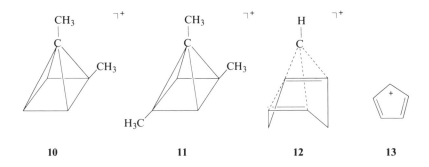

The stability of the pyramidal C_{4v} structure (**8**) has been confirmed in a series of both semiempirical [28–30] and *ab initio* [10, 31, 32] calculations. At MP2/6-31G**//HF/6-31G* the square pyramidal structure (**8**) is only 3.7 kcal/mol less energy favorable than the singlet cyclic isomer (**13**) [32].

In 1973 Hogeveen and Kwant reported the synthesis [33] and *ab initio* calculation [34] of a new pyramidal C_{5v} hydrocarbon structure, the hexamethyl derivative of the $(CH)_6^{2+}$ dication (**14**):

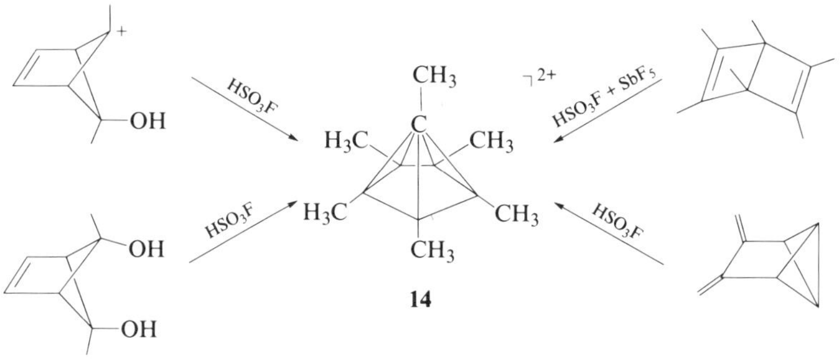

(9.1)

14

The ^{1}H and ^{13}C and NMR spectral data proved the pyramidal structure of the dication **14**. Both spectra contains only two groups of signals in the 5:1 ratio, with the lesser intensity signal in the ^{13}C NMR spectrum lying in the stronger field (–2.0 ppm) than the signal of tetramethylsilane. This result, together with recent studies [35] where the isotopic perturbation technique was employed, rules out the explanation of the NMR spectra in terms of signal averaging stemming from a fast exchange (Eq. (8.2)) by the mechanism of 1,2-shifts in a dication with the classical structure (**14a**):

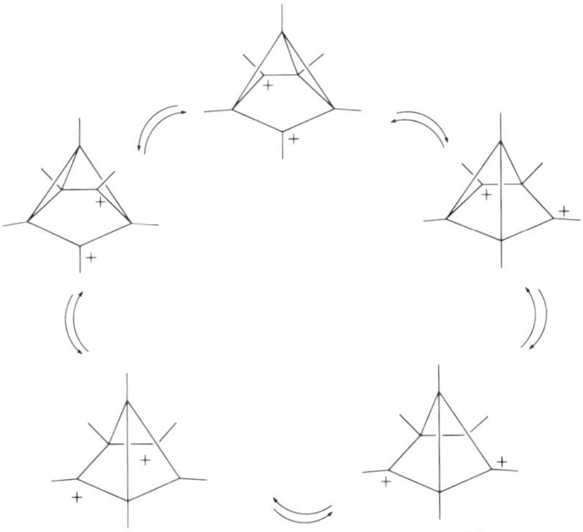

(9.2)

14a

The stability of the pyramidal structure (**14**) has been confirmed by *ab initio* [36] as well as semiempirical [37] calculations on $(CH)_6^{2+}$. According to Krogh-Jespersen [36], the most plausible candidate for the structure corresponding to the global minimum on the PES of $(CH)_6^{2+}$ would be C_{5v} pyramidal structure (**15**), which is lower in energy than the fulvene dication (**16**). Structures of the D_{6h} triplet (**15a**) (since the HOMO benzene is double generated, **15a** has the triplet ground state) and of the fulvene dication (**16**) have higher energies than that of **15** [36] (see Chapter 4).

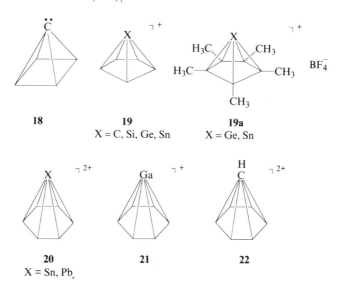

15	**15a**	**16**	**17**

A number of other stable pyramidal structures composed of a basal conjugated carbocycle had a first-row cap have been revealed by MO calculations or identified experimentally. Both semiempirical (MINDO/3 [38], MNDO [39]) and *ab initio* (3-21G [10], 4-31G [40], and MP2/6-31G* [39]) calculations have led to the conclusion that the C_{4v} structure of nonclassical carbene (**18**) (X = C) (named pyramidane) belongs to a rather deep local minimum on the PES of C_5H_4. Also, the C_{5v} carbene cation (X = C) (**19**), a conjugated base to the dication **14**, was found to be a local minimum on the $C_6H_5^+$ PES [41–43].

Analogous C_{5v} structures (**19**) were suggested on the basis of semiempirical calculations [44–46] for the ions $C_5H_5X^+$ detected by means of mass spectroscopy [47]. Jutzi and co-workers reported the synthesis of stable pentamethyl derivatives of cations **19**. They proved their C_{5v} structure (**19a**) by direct X-ray diffraction studies (see [48] for a comprehensive review).

X-ray diffraction studies (see [48, 49] for a summary) have shown that the tin and the lead dications (**20**) as well as the gallium cation (**21**) possess a C_{6v} symmetry in the crystalline state. On the other hand, the carbocation $(CH)_7^+$ formed from norbornadiene precursors, for which Winstein and Ordonneau [50] suggested the nonclassical structure **22**, has actually a less symmetrical structure. The calculations [51] have shown that structure **22** does not correspond to the minimum on the PES of $(CH)_7^+$.

18	**19** X = C, Si, Ge, Sn	**19a** X = Ge, Sn

20 X = Sn, Pb	**21**	**22**

Semiempirical and *ab initio* (STO-3G, 4-31G) calculations [38, 52] indicate that the square pyramidal thiophene structure (**23**) is a hill top on the C_4H_4S PES, whereas for its dication (**23a**) the C_{4v} form is a minimum on the $C_4H_4S^{2+}$ PES.

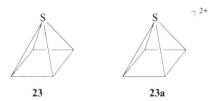

23   23a

The foregoing facts evidence the possibility of stabilization for a fairly large number of pyramidal organic and nontransition-metal organic compounds and point to a clearly pronounced dependence of their thermodynamic stability on the number of valence electrons participating in the formation of the three-dimensional molecular framework. This last relationship implies straight-forward association with aromatic and antiaromatic systems, the former ones obeying the paradigmatic $(4n + 2)$ π-electron-count rule.

Understandably, there has been a temptation to derive an analogous electron-count scheme for pyramidal delocalized systems so as to assign those that are stable to three-dimensional aromatic structures. The necessary first step in the development of such a scheme, usually based on a one-electron (orbital) approximation, is to single out a certain subsystem of orbitals governing the essential bonding in a given structural type (e.g., π-orbitals in conjugated cyclic polyenes). It is assumed that the order and electron population in this orbital subsystem formed as a rule by frontier MOs are very sensitive to structural variations, whereas the rest of the orbitals are not much affected by these.

To select the crucial orbital subsystem in the above-considered pyramidal conjugated molecules and ions, they should be viewed as π-complexes formed by an [*n*]annulene conjugated fragment and an apical atom X or bond X—R [10, 16, 21, 25, 26, 48].

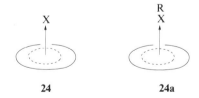

24   24a

For example, the structures of tetrahedrane, the $(CH)_5^+$ cation, and pyramidane may be composed of the cyclopropenyl ring and the CH group, the cyclobutadiene fragment and CH^+, and the cyclobutadiene fragment and a carbon atom, respectively.

The electrons of the bonds C—H and C—C are localized at the corresponding 2c-2e bonds; therefore these two units are bound together through delocali-

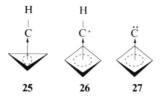

zation of the π-electrons of the ring to the cap and the back donation. The orbital interaction diagram portraying this type of multicenter bonding is readily derived [10, 21, 25, 53]. It is represented in Fig. 9.1 in the most simplified form. Mixing the lowest lying π-MO a_1 with a pair of sp_z hybridized AOs of an apical atom yields three orbitals: strongly bonding, nearly nonbonding, and strongly antibonding a_1 MOs of the C_{nv} pyramidal structure. The stabilization

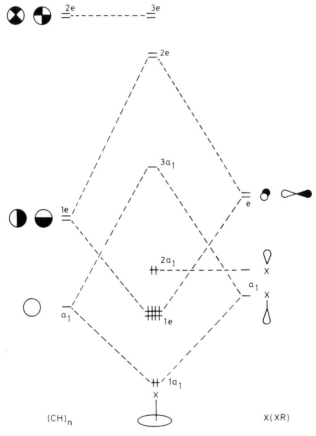

Figure 9.1 Orbital interaction diagram for the C_{nv} pyramidal structures **24** (X = bare atom). In the case of compounds **24a**, the energy level of the $2a_1$ MO falls and lies usually between the $1e$ levels.

originates mainly from mixing of the lowest [n]annulene $1e$-MOs and p-AOs of a cap, resulting in the formation of the bonding $1e$ MOs of C_{nv} pyramidal structures. Therefore three bonding MOs exist in the pyramidal systems **24** and **24a**, which may be populated by no more than six "interstitial" electrons. An additional two electrons can be placed into the nonbonding $2a_1$-MO (in compounds **24a** this MO is localized in the X—R bond and does not contribute to the bonding of the basal ring to an apex). By counting π-electrons of the basal ring and all valence electrons of the cap in the pyramidal structures **24** and **24a**, the eight electron rule (8e rule) for determining their stability has been formulated [16, 25, 54, 55]. In a way, this rule may be understood as a tendency of the main-group element center X to fill up its electron shell so as to obtain the shell of the respective noble gas.

One may easily show by a direct electron count that for the stable C_{nv} pyramidal species **6**, **8**, **9**, **12**, **14**, **18**–**21**, and **23a**, the number of interstitial electrons is indeed equal to eight, while structures **22** and **23** are ten-electron species. For adapting the 8e rule, the latter molecules should lower their symmetry to switch off the surplus π-electrons. The case of the norbornadienyl cation (**22**) distorted to a fluctuating C_s structure obeying the 8e rule is illustrative [56]:

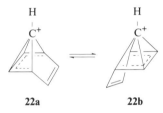

22a	**22b**

Bearing the 8e rule in mind, one may speculate about some unusual C_{nv} pyramidal molecules and ions (**28**–**31**) that meet this requirement:

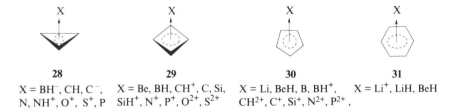

28	**29**	**30**	**31**
X = BH^-, CH, C^-, N, NH^+, O^+, S^+, P	X = Be, BH, CH^+, C, Si, SiH^+, N^+, P^+, O^{2+}, S^{2+}	X = Li, BeH, B, BH^+, CH^{2+}, C^+, Si^+, N^{2+}, P^{2+},	X = Li^+, LiH, BeH

Semiempirical and *ab initio* calculations have been carried out for most of the ring-cap combinations cited; see [16] for a review. They show that in all cases the 8e rule correctly predicts the correspondence of given structures to a local minimum on the corresponding PES.

An alternative formation of the 8e rule has been proposed [10], especially attractive in the context of the concept of three-dimensional aromaticity. As noted earlier, only six electrons actually contribute to the ring-cap bonding in the pyramidal structures **28–31**, while one electron pair of the apical center is employed either in the formation of the X—R bond or in the populating of the nonbonding (lone pair) orbital. In some cases (Li, Be) this orbital is not populated at all. The magic number six of the remaining interstitial electrons and the structures of the respective electronic shell $(a_1)^2(e)^4$ (Fig. 9.1) make it possible to formulate the $(4n + 2)$ interstitial electron rule [10, 57, 58] and thus emphasize the analogy between aromatic annulenes and their pyramidal counterparts.

The $(4n + 2)$ interstitial electron rule may easily be extended to other organic polyhedral structures, such as in sandwich (**32**) or bipyramidal (**33**) molecules and ions formed by conjugated rings with cap(s) X defined as in **19**:

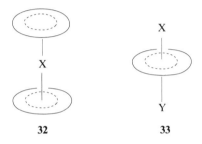

32	**33**

From the fragment MO analysis, it follows that there are only three bonding MOs in both structures drawing the above fragments together. Taking into account the nonbonding orbitals in each group X, eight-electron and ten-electron rules of stability (three-dimensional aromaticity) were formulated [16, 54] for the sandwich (**32**) and the bipyramidal (**33**) structures, respectively. These rules explain, for example, why polyhedral cations, $(CH)_5^+$ and $(CH)_6^{2+}$, prefer the pyramidal, **8** and **19** (X = CH⁺), rather than the bipyramidal structures, **34** and **35**:

34	**35**

However, the substitution of some CH vertices in **34** and **35** by BH groups gives rise to well-documented stable *closo*-carboranes $C_2B_{n-2}H_2$ [1, 2], for instance **36–39**, which conform to the 10e rule:

$$C_6H_6 + Li^+ \longrightarrow C_6H_6\,Li^+ \quad (31) \quad -84.3\ \text{kcal/mol}$$
$$C_6H_6 + LiH \longrightarrow C_6H_6LiH \quad (31) \quad -46.1\ \text{kcal/mol}$$

CH
HB———BH
BH
CH

36

BH
BH
HB———CH
CH
BH

37

CH
BH
HB———BH
BH
CH

38

BH
HB—BH
HC———CH
BH
BH

39

Although these equations are not isodesmic and the values of the heats of reaction may contain errors inherent in the method and stemming from the neglect of electron correlation, the high exothermicity of the reactions definitely points to a substantial ring–cage bonding in the pyramidal structures with the three-dimensional aromaticity.

The concept of isolobal analogy [59–62] between the main-group-centered fragment and the transition-metal-centered fragment defines the relationship between the type of valence orbitals and their number. Its application allows straightforward extension to be made of the above-considered electron rules

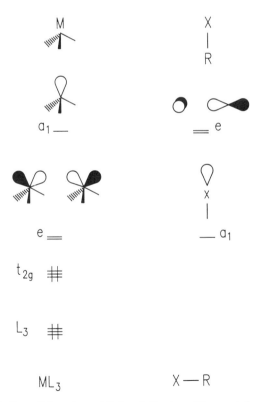

Figure 9.2 Isolobality of frontier orbitals of the transition-metal-centered fragment, ML_3, and the first-row element group X—R.

TABLE 9.1   Isolobal Correlations

First-Row Group	Number of Valence Electrons	Number of Interstitial (skeletal) electrons	Organometallic Group
CH, N, O$^+$	5	3	Co(CO)$_3$, Ni(η^5-Cp), W(CO)$_2$(η^5-Cp), Re(CO)$_4$, Rh(η^6-C$_6$H$_6$)
CH$^+$, C, BH	4	2	Fe(CO)$_3$, Co(η^5-Cp)
CH^{2+}, C$^+$, BeH	3	1	Mn(CO)$_3$, Fe(η^5-Cp)

describing the conditions for the three-dimensional aromaticity to the transition-metal complexes. Figure 9.2 features a similarity in the number, symmetry, and spatial characteristics of the valence orbitals of the first-row groups and those of transition-metal fragments ML$_3$, where L is the two-electron ligand. In Table 9.1 the most important transition-metal fragments isolobal to the first-row groups are listed.

By substituting the groups CH, CH$^+$, and others in the polyhedral structures by isolobal organoelement fragment or, conversely, proceeding from organoelement transition-metal compounds, straightforward correlations are obtained between the three-dimensional aromatic compounds discussed (Fig. 9.3).

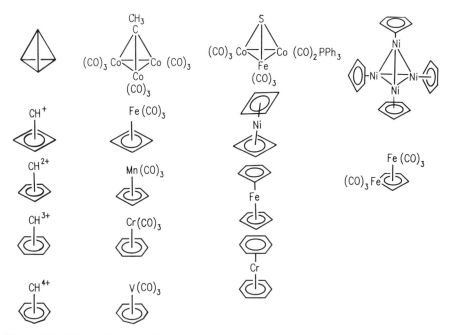

Figure 9.3   Three-dimensional aromatic organic and organometallic pyramidal structures.

Interestingly, the structure of ferrocene was ascertained in 1952 and the first (tetraphenyl) derivative of the iron tricarbonyl complex of cyclobutadiene (**5**) was isolated in 1959 (see [3, 4]). However, two more decades had passed before the conclusion was drawn that organic nonclassical compounds of analogous type can possess pyramidal nonclassical structure. This conclusion was arrived at independently, and the analogy in question did not become clear immediately, but rather in the late stages. A detailed analysis of a broad variety of transition-metal complexes can be found elsewhere [61–63].

9.3   THREE-DIMENSIONAL AROMATICITY AND ELECTRON COUNTING IN CLUSTERS

The stability of the above-considered polyhedral organic and organometallic compounds is the result of the specific structure of their valence electron shells inherent in the aromatic systems. The described electron-count techniques (8e, 10e or $(4n + 2)$ interstitial electron rules) are largely based on these specificities. It should be kept in mind, however, that rules ought to be considered in a broader context of the general electron-count rules worked out for cluster structures of boron hydrides, carbo- and heteroboranes, and organometallic and bare metal clusters [63–68].

The most fundamental approach to the problem in question is represented by the polyhedral skeletal electron pair theory (PSEPT) developed by Wade [63, 64] and Mingos [67]. It assumes that each transition metal in the vertex of a molecular polyhedron uses six AOs for M—L bonding and occupation by non-bonding electrons leaving six AOs available for skeletal bonding. These three orbitals are d_{sp}-hybrids shown in Fig. 9.2. In the case of main-group elements, isolobal p- and sp-AOs correspond to them. Only these three orbitals participate in the skeletal bonding responsible for the stability of a three-dimensional structure. By symmetry arguments and direct calculations, it has been found that in deltahedral structures, that is, those in which all faces are triangles, there are $(m + 1)$ bonding or nearly bonding MOs (m is the number of vertices). They can be populated by $(2n + 2)$ skeletal electrons k, which are counted as follows:

$$\begin{aligned} \text{Main-group elements} \quad & k = v + x - 2 \\ \text{transition metals} \quad & k = v + x - 12 \end{aligned} \qquad (9.2)$$

Here v is the number of valence electrons of the central atom or group occupying one of the apexes, and x is the number of one-electron ligands (the lone pair serves as a phantom ligand). Clearly, the 10 electron difference is, in fact, the difference in the requirements of atoms of the nontransition and transition elements for the 8 and 18 electron shells, respectively.

The polyhedral structures realizable for boron hydrides, carbo- and heteroboranes, and organometallic compounds characterized by triangular faces (deltahedra) can be derived from *closo* (closed) structures of bipyramidal type for 5–7 vertices, from the triangular dodecahedron for 8 apexes, and so on. The *nido* (nest-like) structures are obtained from *closo* forms through truncation of one apex (e.g., pyramids are obtained from bipyramids). In the same way, *arachno* (web-like) structures are derived from *nido* structures. The *closo–nido–arachno* transformation is illustrated by the general scheme shown in Fig. 9.4.

For *nido* structures there are $(m + 2)$ bonding MOs, while for *arachno* structures there are $(m + 3)$ bonding MOs. Thus count rules for skeletal electrons corresponding to complete filling of the bonding MOs of polyhedral structures of transition and nontransition elements alike can be formulated as follows:

$$\begin{aligned} closo: \quad & 2m + 2 \\ nido: \quad & 2m + 4 \end{aligned} \tag{9.4}$$

where m is the number of vertices.

It is easy to verify that, for example, the pyramidal cation $(CH)_5^+$ (**8**) possesses a shell consisting of 14 skeletal electrons ($n = 5$): each of the four CH groups con-

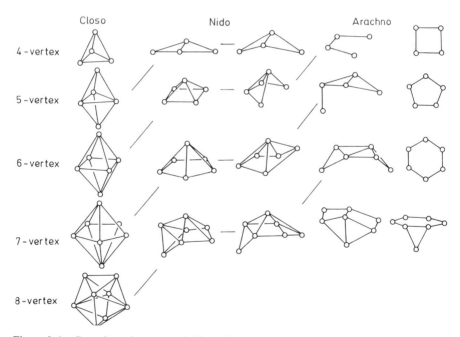

Figure 9.4   Genesis and structural hierarchy for *closo*, *nido*, and *arachno* structures. Tilted bars designate truncations; horizontal bars denote conformational transitions. Stable *closo-*, *nido-*, *arachno-*boranes are $B_nH_n^{2-}$, $B_nH_n^{4+}$, and $B_nH_n^{6-}$ ions, respectively. (Adapted from [65].)

tributes 3 while the CH^+ group contributes 2 skeletal electrons. Since the group $Fe(CO)_3$ is isolobal to CH^+, the pyramidal structure of the π-complex (**5**) equally corresponds to the rule of 14 skeletal electrons: for $Fe(CO)_3$, $k = (8 + 3) + 3 - 12 = 2$. In the case of *arachno* hydrocarbons (i.e., cyclic polyenes) derived from *closo* structures by truncating two apexes, the dianion $(CH)_4^{2-}$, the anion $(CH)_5^-$, and the cation $(CH)_7^+$ must be stable, in accordance with the $(2m + 6)$ rule. Evidently, this situation is in complete agreement with the $(4n + 2)$ Hückel rule.

It is not difficult to show also that the rule of $(4n + 2)$ interstitial electrons is fully equivalent to the count rules for skeletal electrons. For example, the 8e and 10e rules are analogous to the $(2m + 4)$ and $(2m + 2)$ rules for skeletal electrons, respectively. Indeed, for a pyramidal *nido* structure comprising m apexes, out of the total $(2m + 4)$ skeletal electrons, $2(m - 1)$ electrons form endocyclic σ-bonds of the basal cycle, while 2 electrons participate in the exocyclic bond (in the electron pair) of the apical group. Thus the number of electrons bonding the fragments in question comes out to $(2m + 4) - 2(m - 1) + 2 = 8$ (8e rule). For a bipyramidal (*closo*) structure, one obtains in the same manner $(2m + 2) - 2(m - 2) + 2 + 2 = 10$ (10e rule).

Some other general rules of electrons counting have been suggested that take into account not only the skeletal but also all the valence electrons of the framework and cluster structures. They may be applied to a great variety of metal and organometallic structures having from 5 to 15 apexes; for reviews see [67–69]. All these formulations depend on a given structural type (bipyramids, pyramids, cycles, etc.), irrespective of the nature of the vertex groups, and predetermine a definite number of valence, skeletal and interstitial orbitals of a cluster, thus setting a limit on the number of the electrons populating these. Going beyond this limit leads to instability and, as a result, rearrangement into a stable structure. This conclusion is illustrated by the redox relationship known for the series of boron hydrides and carboranes [65, 70]:

$$closo \underset{-2e}{\overset{+2e}{\rightleftarrows}} nido \underset{-2e}{\overset{+2e}{\rightleftarrows}} arachno \tag{9.5}$$

which may easily be extrapolated to polyhedral organic structures in the form [16] of

$$(n - 2)\text{-gonal bipyramid (10e)} \underset{-2e}{\overset{+2e}{\rightleftarrows}} (n - 1)\text{-gonal bipyramid (8e)}$$

$$\underset{-2e}{\overset{+2e}{\rightleftarrows}} n\text{-gonal cycle (6e)} \tag{9.6}$$

As an example, one may point to the oxidation of Dewar benzene (*arachno* form) to the pyramidal dication **14** (*nido* form); see Eq. (9.1). Further oxidation could lead to the electron-precise bipyramidal tetracation **40a**; however, such a multicharged species would be too unstable for electrostatic reasons.

The foregoing implies that the electron-counting rules defining the stability of polyhedral structures (i.e., their three-dimensional aromaticity) are topological in nature. It has indeed been shown [71] that the PSEPT scheme in various modifications is on the whole equivalent to the topological electron-counting (TEC) scheme based on Euler's theorem: $E = V + F - 2$, where E, V, and F are the numbers of edges, vertices, and faces of a polyhedron, respectively. TEC may be related to PSEPT through some simple algebraic manipulations [72].

For polyhedral organic structures, such as cations $(CH)_n^{c+}$, the following relationship has been established between the number of CH apexes and the charge by means of the graph theory [73]:

$$n = 2h + c + 2 \tag{9.7}$$

where h is the number of "holes," which indicates all faces of a polyhedron containing more than three edges. For *closo* structures of type **40**, $h = 0$ (all faces are triangular). The octahedral structure $(CH)_6$ may, according to Eq. (9.7), be stable only at $c = +4$. The same prediction may equally be inferred from the 10e or $(2n + 2)$ rules.

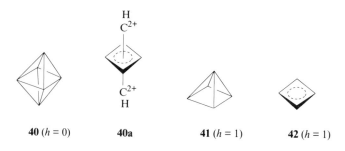

40 ($h = 0$)   **40a**   **41** ($h = 1$)   **42** ($h = 1$)

The pyramidal structure of $(CH)_5$ (**41**) has one nontriangular face and, according to Eq. (9.7), the charge $c = +1$. The *arachno* form $(CH)_4$, **42** obtained through truncation of the apex in **41** and corresponding to cyclobutadiene, must be noncharged.

Thus the combining of conjugated cycles and the inclusion of certain main-group and transition-metal vertices give rise to polyhedral structures, which, provided that their electronic shells correspond to relevant filling rules, exhibit enhanced stability. Considered from this angle, the electron-count rules may be regarded as a development of the concept of aromaticity. In going from planar polyene cyclic structures to the three-dimensional ones, this concept acquires an additional dimension so that we speak of the three-dimensional aromaticity.

REFERENCES

1. R. N. Grimes, *Carboranes*, Academic, New York, 1975.
2. G. A. Olah, G. K. Prakash, R. E. Williams, L. D. Field, and K. Weid, *Hypercarbon Chemistry*, Wiley, New York, 1987.
3. R.C. Mehrotra and A. Singh, *Organometallic Chemistry*, Wiley, New Delhi, 1991.

4. A. Yamoto, *Organotransition Metal Chemistry*, Wiley, New York, 1986.

5. J. Aihara, *Bull. Chem. Soc.* Jpn., **58**, 266 (1985).

6. J. Aihara, *J. Am. Chem. Soc.*, **98**, 2750 (1976).

7. J. Gutman, M. Milun, and N. Trinaistić, *J. Am. Chem. Soc.*, **99**, 1692 (1977).

8. J. Aihara, *J. Am. Chem. Soc.*, **100**, 3339 (1978).

9. W. C. Herndon, *J. Am. Chem. Soc.*, **102**, 1538 (1980).

10. E. D. Jemmis and P. v. R. Schleyer, *J. Am. Chem. Soc.*, **104**, 1538 (1982).

11. A. Greenberg and J. F. Liebman, *Strained Organic Molecules*, Academic, New York, 1978.

12. G. Maier, *Angew. Chem. Int. Ed. Engl.*, **27**, 309 (1988).

13. G. Maier, S. Pfriem, U. Schafer, and R. Matusch, *Angew. Chem.*, **90**, 552 (1978).

14. K. B. Wiberg, R. F. W. Bader, and C. D. H. Lau, *J. Am. Chem. Soc.*, **109**, 985 (1987).

15. H. Kollmar, *J. Am. Chem. Soc.*, **102**, 2617 (1980).

16. V. I. Minkin, R. M. Minyaev, and Yu. A. Zhdanov, *Nonclassical Structures of Organic Compounds*, Mir, Moscow, 1987.

17. B. A. Hess and L. J. Schaad, *J. Am. Chem. Soc.*, **107**, 865 (1985).

18. G. Maier and D. Born, *Angew. Chem. Int. Ed. Engl.*, **28**, 100 (1989).

19. G. Maier, S. Pfriem, U. Schafer, K. D. Malsch, and R. Matusch, *Chem. Ber.*, **114**, 3965 (1981).

20. R. E. Williams, *Inorg. Chem.*, **10**, 210 (1971).

21. W. D. Stohrer and R. Hoffmann, *J. Am. Chem. Soc.*, **94**, 1661 (1972).

22. S. Masamune, M. Sakai, H. Ona, and A. J. Jones, *J. Am. Chem. Soc.*, **94**, 8956 (1972).

23. V. I. Minkin, N. S. Zefirov, M. S. Korobov, N. A. Averina, and A. M. Boganov, *Zh. Org. Khim.*, **17**, 2616 (1981).

24. S. Masamune, *Pure Appl. Chem.*, **44**, 661 (1975).

25. V. I. Minkin and R. M. Minyaev, in I. Csizmadia (Ed.), *Progress in Theoretical Organic Chemistry*, Vol. 3, Elsevier, Amsterdam, 1982, Chapter 2 p. 65.

26. G. A. Olah, G. K. Prakash, and J. Sommer, *Superacids*, Wiley, New York, 1985.

27. H. Schwarz, H. Thies, and W. Franke, in M. A. A. Ferreira (Ed.), *Ionic Processes in the Gas Phase*, Reidel, New York, 1984, p. 267.

28. M. J. S. Dewar and R. E. Haddon, *J. Am. Chem. Soc.*, **95**, 5836 (1973).

29. H. Kollmar, H. O. Smith, and P. v. R. Schleyer, *J. Am. Chem. Soc.*, **95**, 5834 (1973).

30. G. W. Jefford, J. Mareda, H. Perlberger, and U. Burger, *J. Am. Chem. Soc.*, **101**, 1371 (1979).

31. W. J. Hehre and P. v. R. Schleyer, *J. Am. Chem. Soc.*, **95**, 5837 (1973).

32. J. Feng, J. Leszczynsky, B. Weiner, and M. C. Zerner, *J. Am. Chem. Soc.*, **111**, 4648 (1989).

33. H. Hogeveen and P. W. Kwant, *Tetrahedron Lett.*, 1665 (1973).

34. H. Hogeveen and P. W. Kwant, J. Postma, and P. T. Duynen, *Tetrahedron Lett.*, 4351 (1974).

35. H. Hogeveen and E. M. Krutchen, *J. Org. Chem.*, **46**, 1350 (1981).

36. K. Krogh-Jespersen, *J. Am. Chem. Soc.*, **113**, 417 (1991).

37. M. J. S. Dewar and K. Holloway, *J. Am. Chem. Soc.*, **106**, 6619 (1984).

38. V. I. Minkin, R. M. Minyaev, I. I. Zakharov, and V. I. Avdeev, *Zh. Org. Khim.*, **14**, 3 (1978).

39. V. Bataji and J. Michl, *Pure Appl. Chem.*, **60**, 189 (1988).

40. V. I. Minkin, R. M. Minyaev, and G. V. Orlova, *J. Mol. Struct.* (THEOCHEM), **110**, 241 (1984).

41. W. A. M. Gastenmiller and H. M. Buch, *Rec. J. R. Neth. Chem. Soc.*, **96**, 207 (1977).

42. K. Krogh-Jespersen, J. Chandrasekhar, and P. v. R. Schleyer, *J. Org. Chem.*, **45**, 1608 (1980).

43. M. Tasaka, M. Ogaba, H. Ichikawa, *J. Am. Chem. Soc.*, **103**, 1885 (1981).

44. Yu. A. Borisov and Yu. S. Nekrasov, *Izv. Akad. Nauk SSSR, Ser. Khim.*, 1693 (1980).

45. Yu. A. Borisov and Yu. S. Nekrasov, and V. F. Sizoi, *Izv. Akad. Nauk SSSR, Ser. Khim.*, 494(1982).

46. J. Chandrasekhar, P. v. R. Schleyer, R. O. W. Baumgartner, and M. T. Reetz, *J. Org. Chem.*, **48**, 3453 (1983).

47. Yu. S. Nekrasov, V. F. Sizoi, D. V. Zagorevsl, and Yu. A. Borisov, *J. Organomet. Chem.*, **205**, 157 (1981).

48. P. Jutzi, *Adv. Organomet. Chem.*, **26**, 217 (1986).

49. H. Schmidbaur, *Angew. Chem.*, **97**, 893 (1985).

50. S. Winstein and C. Ordonneau, *J. Am. Chem. Soc.*, **82**, 2084 (1960).

51. C. Cone, M. J. S. Dewar, and D. Landmann, *J. Am. Chem. Soc.*, **99**, 372 (1977).

52. R. M. Minyaev and V. I. Minkin, *Zh. Org. Khim.*, **18**, 2008 (1982).

53. H. Hogeveen and P. W. Kwant, *Acc. Chem. Res.*, **8**, 413 (1975).

54. J. B. Collins and P. v. R. Schleyer, *Inorg. Chem.*, **16**, 152 (1977).

55. V. I. Minkin, R. M. Minyaev, *Zh. Org. Khim.*, **15**, 225 (1979).

56. G. A. Olah, G. K. Prakash, T. N. Rawdah, D. Wittaker, and J. S. Rees, *J. Am. Chem. Soc.*, **101**, 3935 (1979).

57. E. D. Jemmis and P. v. R. Schleyer, *J. Am. Chem. Soc.*, **104**, 7017 (1982).

58. P. Buzek, P. v. R. Schleyer, and S. Sieber, *Chem. unser. Zeit*, **26**, 116 (1992).

59. M. Elian and R. Hoffmann, *Inorg. Chem.*, **14**, 1058 (1975).

60. J. W. Lanker, M. Elian, R. M. Summerville, and R. Hoffmann, *J. Am. Chem. Soc.*, **98**, 3219 (1976).

61. R. Hoffmann, *Angew. Chem.*, **94**, 725 (1982); Science, **211**, 995 (1981).

62. T. A. Albright, J. K. Burdett, and A. H. Whangbo, *Orbital Interactions in Chemistry*, Wiley, New York, 1985.

63. K. Wade, in B. F. G. Johnson (Ed.), *Transition Metal Clusters*, Wiley, New York, 1985, pp. 193–264.

64. K. Wade, Adv. *Inorg. Chem. Radiochem.*, **18**, 1 (1976).

65. R. W. Rudolph, *Acc. Chem. Res.*, **9**, 46 (1976).

66. D. M. P. Mingos, *Adv. Organomet. Chem.*, **15**, 1 (1977).

67. D. M. P. Mingos, *Acc. Chem. Res.*, **17**, 311(1984).

68. Yu. L. Slovokhotov and Yu. T. Struchkov, *Usp. Khim.*, **54**, 556 (1985).

69. S. M. Owen, *Polyhedron*, **7**, 253 (1988).

70. J. P. Fritch and K. P. C. Vollnardt, *Angew. Chem.*, **92**, 570 (1980).

71. B. K. Teo, *Inorg. Chem.*, **23**, 1251, 1627 (1984); *Inorg. Chem.*, **24**, 115, 4209 (1985).

72. D. M. P. Mingos, *Inorg. Chem.*, **24,** 114 (1985).

73. A. T. Balaban and D. H. Rouvray, *Tetrahedron*, **36**, 1851 (1980).

10

SPHERICAL AROMATICITY

As has already been mentioned in Chapter 3, the concept of separability may be extended to cover the three-dimensional molecules. With this aim, based on the POAV1 and POAV2 schemes, a unified definition for the π-orbital was suggested applicable in all dimensions (see Fig. 3.7 and [1]). This approach which permits the HMO theory to be extended into three dimensions (3D-HMO method, see Chapter 3) proves a convenient tool for analyzing the electronic structure of conjugated nonplanar molecules, including the spheroidal carbon clusters C_n. In the last five years these clusters have been targets of numerous studies, both experimental and theoretical (for reviews see [2–7], initiated by Kroto and Smalley's experiments [8]. They found that in the mass spectrum of products of the solid graphite laser vaporization C_{60}^+ mass peak was largest and, under certain clustering conditions, the C_{60} molecules completely dominate in the resulting cluster distribution amounting to over 50% of the total large cluster abundance [8].

There is high stability in the truncated icosahedron 1 with 32 faces, of which 20 are hexagons and 12 are pentagons (I_h symmetry). The spheroidal structure was dubbed by Kroto et al. [8] as "buckminsterfullerene" in honor of Buckminster Fuller, who studied the so-called geodesic polyhedra (constructed on the surface of a sphere by a network of intersecting geodesic lines [9]).[1] The possibility of its existence was suggested in the original experimental work [8] on the basis of the notion of aromaticity, since in 1 "all valence are satisfied and the molecule appears to be aromatic."

---

[1]This structure may remind one of a football, for which reason it has the alternative name "footballene."

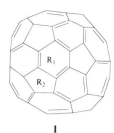

1

The possibility of the existence of stable large carbon clusters, in particular, of structure **1**, was predicted back in the early 1970s by turning to the notion of aromaticity first investigated by Joshida and Osawa [10] and independently a little later by Botchvar and Gal'pern [11]. The latter authors have shown by calculations in the ρ-approximation (an analog of the π-approximation) that structure **1**, which they designated as "carbo-s-icosahedron," has a closed electronic shell, rather wide HOMO–LUMO energy gap, and a resonance energy that is fairly high (see also their subsequent EHT calculations [12]).

Noteworthy also are the works [13–15] in which structure **1** was equally predicted. Haymet's prediction [14, 15] was based on calculations of the HRE—equal to 0.5527β per carbon atom for **1**—which exceeds the corresponding value for benzene (Table 10.1) but is less than that for graphite (0.576β).

Intensive studies of the last years have shown (for reviews see [2, 3, 6]) that other closed cages containing five- and six-membered rings (whose family name is "fullerene") are also stable. Even an aromaticity rule has been proposed [19, 20] for fullerenes, analogous to the Hückel ($4n + 2$) rule, according to which

TABLE 10.1   Results of 3D-HMO Calculations of Buckminsterfullerene C_{60} [16, 17]

Property	C_{60} (**1**)		Benzene	Graphite[a]
	Planar[b]	3D[c]		
HRE/C, β[d]	0.553	0.365	0.033	0.576
λ, β^{-1}	0.893	1.016	1.0	[e]
L^{+}, β	2.767	2.432	2.536	—
L, β	2.629	2.311	2.536	—
L^{-}, β	2.490	2.189	2.536	—
F[f]	0.179	0.367	0.399	0.156
ΔE (HOMO–LUMO), β	0.757	0.665	2.0	0.0[g]

[a]Infinite sheet, C_{∞}.

[b]Hypothetical planar case.

[c]POAV1/3D-HMO Calculations with $\rho = 0.879$ [17].

[d]HRE per carbon atom.

[e]Stable to distortion.

[f]Free valence (unitless).

[g]See [18].

the fullerenes with the number of electrons equal to $6n + 60$ ($n = 0,1,2,...$) should be regarded as stable. Studies into large carbon clusters are continuing. Considerable attention is attracted, for example, by the ability of some clusters to form structures of the metallofullerene type [21] (calculations reported in [22]). Incidentally, during the Chernobyl disaster structures of the $C_{60}X$ type could have acted as dangerously mobile airborne transporters of radioactive elements, such as ^{90}Sr [3].

Buckminsterfullerene has been doped with combinations of alkali and other metals to produce high-temperature superconductors [7]. For alkali metal doped C_{60} phases (Me_xC_{60}), superconductivity has been discovered in the case of $M = K$ ($T_c = 18$ K) and $M = Rb$ ($T_c = 28$K) dopants [7].

But how justified is the description of fullerene-type structures in terms of aromaticity and will the idea of aromaticity in the later stages prove as helpful as at the start?

Let us consider the legitimacy of the assignment of buckminsterfullerene (**1**) to the aromatic class. Its relative aromaticity compared to benzene and graphite (for details on graphite as an aromatic system, see [18]) can be assessed from the values of indices devised in terms of the energetic, structural, and magnetic criteria (see Chapter 2).

According to 3D-HMO calculations [16, 17] with corrections of resonance integrals ($\rho\beta$) for nonplanarity ($\rho = 0.879$, see also Chapter 3), for structure **1** the values of such indices as the HRE per carbon atom, the highest eigenvalue of the bond–bond polarizability matrix (β^{-1} units) λ, the localization energy for electrophilic (L^+), radical ($L^{\cdot}$), and nucleophilic (L^-) attacks (see Section 2.5.1.2.), and the HOMO–LUMO energy gap ΔE turn out to be close to the corresponding values for benzene (Table 10.1).

The rehybridization in **1** is quite considerable[2] ($s^{0.0928}p$ and $s^{0.0877}p$ for the π-orbital, see Fig. 3.6 where $s^m p$ is explained) according to POAV1 and POAV2 schemes [23]. But, as may be seen from Table 10.1, the inclusion of nonplanarity of the structure does not change qualitatively the ratio between the values of the aromaticity indices for **1** and benzene. As a result, the resonance stabilization of **1** comes to about 90% of the corresponding value for a size-consistent planar fragment [17, 23].

However, as was noted in Chapter 2 and emphasized especially for the case of C_{60} **1**, the HRE (HREPE) values cannot serve to predict the potential stability, as is convincingly exemplified by polyacenes.

The values of REPE listed in Table 10.2, calculated for **1** by means of different schemes, show that buckminsterfullerene is inferior to benzene in aromatic stability and is by no means a superaromatic system. Structure **1** has a lower value of the per cent TRE[3] (%TRE = 1.79, the %TRE of a given molecule

---

[2] Note that for 9,9′,10,10′-tetradehydrodianthracene the degree of pyramidalization is even greater than in **1** and for the ethylene-type carbon atom in it m (POAV1) = 0.1201 [23].

[3] In [27], a reference polynomial was constructed for calculating the TRE of **1** which to date is record-breaking in complexity.

TABLE 10.2 Resonance Energies for Buckminsterfullerene and Benzene

		Resonance Energy			
		Buckminsterfullerene		Benzene	
Scheme of Resonance		REPE			
Energy Calculation	RE	Uncorrected	Corrected[a]	RE	REPE
HRE, β, Eq. (2.5)	33.16 [14, 15]	0.553	0.365	2	0.333
HSRE, β, Eq. (2.9)	1.87 [24]	0.031	0.027	0.39	0.065
CCMRE, eV Eq. (2.26) { Randic's parameters[b] [25]	6.96	0.116	0.097	0.869	0.145
{ Herdon's parameters[c] [25]	7.20	0.120	0.101	0.841	0.140
LM, eV, Eq. (2.30)	11.18 [26]	0.186	0.157	0.821	0.137
AS, β, Eq. (2.48)	1.96 [26]	0.033	0.029	0.44	0.073
TRE, β, Eq. (2.25)	1.643 [27]	0.027	0.024	0.276	0.046

[a]According to 3D-HMO model, $\rho = 0.879$ [17]; the correction for nonplanarity in CCMRE and LM calculations is 0.84 [28].
[b]$R_1 = 0.869$ eV, $R_2 = 0.246$ eV, $Q_1 = -1.60$ eV.
[c]$R_1 = 0.841$ eV, $R_2 = 0.336$ ev, $Q_1 = -0.65$ eV (see [25]).

is defined as 100 times the TRE, divided by the total π-electron energy of its olefinic-type reference structure [27]) than the unstable hexacene (%TRE = 1.99) and heptacene (%TRE = 1.91, β units) [27].

As for the structural criteria of aromaticity, it is to be noted that, according to MNDO [22, 29], PRDDO [30], AM1 [31], and *ab initio* [32–35] calculations, the benzene-type six-membered rings in **1** have a Kekule-type structure with bond length alternation ($R_1 = 1.370$ and $R_2 = 1.448$ Å (TZP-type basis set) [35])). The experimental values are 1.450 and 1.387 Å [36]. As has been shown by Glukhovtsev et al. [37], for the first member of the series of Archimedean solids to which C_{60} (**1**) belongs, namely, for the truncated tetrahedron C_{12} (**2**), the allowance made for the bond length alternation may qualitatively alter the degree of aromaticity.

2

The positive value of the TRE pointing to insignificant aromaticity of **2** was obtained only when the bond length alternation was taken into account. This alternation in **2** arises because of the π-electron system—an analogous structure of cycloalkane $(CH)_{12}$ has equal CC bond lengths (*ab initio*) [38].

In the case of C_{60} (**1**), the inclusion of the alternation increases the value of the HRE [39]. As has been confirmed by calculation of the π-electron ring-current magnetic susceptibility of C_{60} (**1**) [40, 41], this specificity in the geometry ought not to be neglected when discussing stability (earlier it was often assumed not to affect the qualitative conclusion, see [12, 13]). With the bond lengths taken to be equal, buckminsterfullerene is predicted to be weakly paramagnetic (i.e., an antiaromatic species according to the magnetic criterion, see Section 2.4). By contrast, when the bond length alternation is taken into account, C_{60} (**1**) is found to be diamagnetic. Thus *ab initio* calculations [34] with $R_1 = 1.376$ Å and $R_2 = 1.465$ Å (STO-3G geometry [23]) indicate the strong diamagnetism of **1** typical of aromatic molecules.

Thus there are weighty reasons for the assertion that structure **1** does not possess any superior aromaticity; moreover, the aromaticity is apparently not the factor determining the remarkable stability of buckminsterfullerene. An important stabilizing factor may be the absence in **1** of peripheral reactive carbons and of hydrogens on its surface [26, 27]. The last circumstance makes substitution reactions for C_{60} (**1**) impossible without cage rupture [42, 43].

Whatever the role of aromaticity for the case under discussion, the notions developed within the framework of the aromaticity concept can be used for ascertaining qualitative regularities in the relative stability of isomeric C_n clusters and interpreting the specificity of their electronic structure. For example, the RE values calculated by means of various schemes indicate that structure **1**

TABLE 10.3   REPE for Carbon Clusters

Carbon Cages	HSRE, β [46][a]	RE Calculated by Method of Moments, β [47]	CCMRE, eV [46][b]
C_{60}	0.031	0.032	0.120
C_{80}	0.022	—[c]	0.080
C_{140}	0.038	—[c]	0.130
C_{180}	0.045	0.045	0.166
C_{240}	0.047	0.046	0.155
C_{420}	—	0.047	—
C_{720}	—	0.048	—
C_{1980}	—	0.049	—
Graphite	0.053	0.049	0.168

[a]For HSRE calculations for C_n clusters, Eq. (2.6) reduces to simple expression HSRE(C_n) = $E_\pi(HMO) - 1.52\ln\beta$ [46].
[b]Herndon's parameters.
[c]0.040 (C_{80}^6) and 0.047 (C_{320}^6) [47].

is the most stable among the isomers [25, 28, 29, 44, 45], so it is reasonable to expect increased stability for giant fullerenes (Table 10.3).

There is a clairvoyant passage in the review by Kroto [3] that goes like this: "it could be that we are entering a new age for just as the pre-Columbian assumption that the earth was flat made way for a round world-view, it may be that, post buckminsterfullerene, the traditional assumption that polyaromatic organic chemistry is essentially a flat field may also make for a bright, nonplanar future."

The concept of aromaticity, whose services both in the prognostication of the structure of a stable carbon cluster [10, 11, 14, 15] and in the interpretation of experimental data on C_{60} [8]—"the first example of a spherical aromatic molecule" [16]—has been quite valuable in the "pre-Columbian" age and will definitely be of use in the epoch to come.

REFERENCES

1. R. C. Haddon, *Acc. Chem. Res.*, **21**, 243 (1988).

2. W. Weltner and R. J. van Zee, *Chem. Rev.*, **89**, 1713 (1989).

3. H. W. Kroto, *Pure Appl. Chem.*, **62**, 407 (1990).

4. H. W. Kroto and D. R. M. Walton, in E. Osawa and O. Yonemitsu (Eds.), *Carbocyclic Cage Compounds: Chemistry and Applications*, VCH, New York, 1992, pp. 91–100.

5. H. W. Kroto, *Angew. Chem.*, **104**, 113 (1992).

6. F. Diederich and R. L. Whetten, *Acc. Chem. Res.*, **25**, 119 (1992).

7. R. C. Haddon, *Acc. Chem. Res.*, **25**, 127 (1992).

8. H. W. Kroto, J. R. Heath, S. C. O'Brien, R. F. Curl, and R. E. Smalley, *Nature*, **318**, 162 (1985).

9. R. Buckminster Fuller, *Synergetics: Explorations in the Geometry of Thinking*, Macmillan, London, 1982.

10. Z. Yoshida and E. Osawa, *Aromaticity*, Kagakudojin, Kyito, 1971 (in Japanese).

11. D. A. Botchvar and E. G. Gal'pern, *Dokl. Akad. Nauk SSSR*, **209**, 610 (1973).

12. D. A. Botchvar and E. G. Gal'pern, and I. V. Stankevich, *Zh. Strukt. Khim.*, **30**(3), 38 (1989).

13. R. A. Davidson, *Theor. Chim. Acta*, **58**, 193 (1981).

14. A. D. J. Haymet, *Chem. Phys. Lett.*, **122**, 421 (1985).

15. A. D. J. Haymet, *J. Am. Chem. Phys. Soc.*, **108**, 319 (1986).

16. R. C. Haddon, L.E. Brus, and K. Raghavachari, *Chem. Phys. Lett.*, **124**, 459 (1986).

17. R. C. Haddon, L. E. Brus, and K. Raghavachari, *Chem. Phys. Lett.*, **131**, 165 (1986).

18. C. Minot, *J. Phys. Chem.*, **91**, 6380 (1987).

19. R. W. Fowler, *Chem. Phys. Lett.*, **131**, 444 (1986).

20. R. W. Fowler and J. I. Street, *J. Chem. Soc. Chem. Commun.*, 1403 (1987).

304   REFERENCES

21. P. J. Fagen, J. C. Calabrese, and B. Malone, *Acc. Chem. Res.*, **25**, 134 (1992).

22. M. L. McKee and W. C. Herndon, *J. Mol. Struct.* (THEOCHEM), **153**, 75 (1987).

23. R. C. Haddon, *J. Am. Chem. Soc.*, **112**, 3385 (1990).

24. B. A. Hess and L. J. Schaad, *J. Org. Chem.*, **51**, 3902 (1986).

25. T. G. Schmalz, W. A. Seitz, D. J. Klein, and G. E. Hite, *Chem. Phys. Lett.*, **130**, 203 (1986).

26. M. Randic, S. Nicolic, and N. Trinaistić, *Croat. Chim. Acta*, **60**, 595 (1987).

27. J. I. Aihara and H. Hosoya, *Bull. Chem. Soc. Jpn.*, **61**, 2657 (1988).

28. D. J. Klein, T. G. Schmalz, G. E. Hite, and W. A. Seitz, *J. Am. Chem. Soc.*, **108**, 1301 (1986).

29. M. D. Newton and R. E. Stanton, *J. Am. Chem. Soc.*, **108**, 2461 (1986).

30. D. S. Marynick and S. Estreicher, *Chem. Phys. Lett.*, **132**, 383 (1986).

31. J. M. Schulman, R. L. Disch, M. A. Miller, and R. C. Peck, *Chem. Phys. Lett.*, **141**, 45 (1987).

32. R. L. Disch and J. M. Schulman, *Chem. Phys. Lett.*, **125**, 465 (1986).

33. H. P. Lüthi and J. Almlöf, *Chem. Phys. Lett.*, **135**, 357 (1987).

34. P. W. Fowler, P. Lazzeretti, and R. Zanasi, *Chem. Phys. Lett.*, **165**, 79 (1990).

35. G. E. Sauseria, *Chem. Phys. Lett.*, **176**, 423 (1991).

36. H.-B. Bürgi, E. Blanc, D. Schwarzenbach, S. Liu, Y. Lu, M. M. Kappes, and J. A. Ibers, *Angew. Chem.*, **104**, 667 (1992).

37. M. N. Glukhovtsev, B. Ya. Simkin, and I. A. Yudilevich, *Teor. Eksp. Khim.*, 229 (1990).

38. J. M. Schulman, R. L. Disch, and M. L. Sabio, *J. Am. Chem. Soc.*, **108**, 3258 (1986).

39. K. Balasubramanian and X. Liu, *J. Comput. Chem.*, **9**, 406 (1988).

40. V. Elser and R. C. Haddon, *Nature*, **325**, 792 (1987).

41. R. B. Mallion, *Nature*, **325**, 760 (1987).

42. S. Nicolic and N. Trinaistić, *Kem. Ind.*, **36**, 107 (1987).

43. J. R. Dias, *J. Chem. Educ.*, **66**, 1012 (1989).

44. A. J. Stone and D. J. Wales, *Chem. Phys. Lett.*, **128**, 501 (1986).

45. T. G. Schmalz, W. A. Seitz, D. J. Klein, and G. E. Hite, *J. Am. Chem. Soc.*, **110**, 1113 (1988).

46. D. J. Klein, W. A. Seitz, and T. G. Schmalz, *Nature*, **323**, 703 (1986).

47. Y. Jiang and H. Zhang, *Pure Apppl. Chem.*, **62**, 451 (1990).

11

IS THE PHYSICAL NATURE OF AROMATICITY KNOWN?

We set out to achieve two main goals: first, to carry out an in-depth critical analysis of the very concept of aromaticity, and, second, to demonstrate its merits and limitations in studying and predicting electronic and geometrical aspects of molecular structure. The reader will have judged by now to what extent we have been able to realize those goals.

The concept of aromaticity is one of the key notions that constitute the rationale of theoretical organic chemistry; it can be applied to a wide variety of often quite dissimilar compounds see (Scheme 1.1). But there is something of a paradox about this notion since nobody knows for sure what the physical nature of aromatic stabilization actually is.

It has already been pointed out (see Chapter 1) that the concept of aromaticity did not originate by way of inductive reasoning from physical experiment. Thus we are confronted here with the habitual problem of theoretical chemistry debated again and again when putting such concepts as hybridization, bond orders, and bond dipoles to use. In essence, this problem comes down to the following question. Can we, by proceeding from first principles, provide a physical interpretation of these valuable and commonly accepted notions, or are we, of necessity, to be content with blindly utilizing these without trying to understand the underlying physical mechanisms, if indeed they exist at all?

Several attempts are known for ascertaining the nature of aromatic stabilization. The most accurate gauge for assessing the aromaticity is, in our view, the energetic criterion, which is based on the determination of aromatic stabilization by making use of various schemes for calculating the resonance energy. So we turn, in the first place, to the question of whether there is any physical reality behind the term "resonance energy" of an aromatic or antiaromatic compound.

As has been noted in Section 2.2.1, the calculation of the RE is based on the determination of the difference between the energy of a cyclic conjugated molecule and that of a corresponding reference structure calculated by means of the model of bond-energy additivity. Thus the RE is not a quantity determinable by direct physical experiment; rather, it is a formal hypothetical quantity, albeit rigorously defined with the logical framework of the aromaticity concept.

Given the fact that such widespread schemes for the determination of the RE as the TRE and HSRE are devised in terms of the HMO method [1], the following questions are relevant. What are the HMO energies, and do they have any physical meaning?

For the total HMO energy of acyclic polyenes the additivity relationship is fulfilled. *Ab initio* calculations bear witness to the additivity of the total HF energy rather than of its π-portion or of the HF electronic energy. The energy by the π-electron MO theory corresponds to the negative value of the kinetic energy of π-electrons (π-KE) in the *ab initio* theory [1]. Since for the total kinetic energies of acyclic polyenes ($CH_2=CH(CH=CH)_nCH=CH_2$) the relationship (11.1) holds, the kinetic resonance energy (KRE) of the π-electron may be defined as in Eq. (11.2):

$$\pi\text{-}KE = 2.51224n + 5.04275 \text{ a.u. hartree} \tag{11.1}$$

$$KRE = E_\pi^{\text{total}} \left[\left(\begin{array}{c} \end{array} \right) \right]_n - 2.51224(n + 1) \text{ a.u.} \tag{11.2}$$

The correlation between HRE and KRE allows one to expect that the aromaticity and antiaromaticity phenomena could be interpreted in terms of the π-KE [1]. An analysis of the behavior of the kinetic, potential, and total energies in the π-bond formation in ethylene ion and ethylene [2] indicates that the initial driving force in forming the π-bond is the release of the kinetic energy pressure. In other words, the role of kinetic energy in the formation of the π-bond in these species turns out analogous to the significance of initial lowering of the kinetic pressure in the process of σ-bond formation in the H_2 molecule [3]. A similar analysis for the molecule of vinylamine has confirmed that the stabilization caused by conjugation is determined by a decrease in the kinetic energy [4].

Since the aromaticity is due to effects of stabilization brought about by the π-electron cyclic conjugation, one may assume that aromatic stability is caused by the relaxation of the kinetic energy pressure of π-electrons in isolated double bonds. Indeed, in benzene the π-electrons are uniformly delocalized between atoms; in antiaromatic molecules, such as cyclobutadiene, π-electrons are characterized by localization. This is reflected in the values of the π-kinetic energy per electron (KEPE) for diagonal terms ($\Sigma_A E_{\pi A}$) and nondiagonal terms ($\Sigma_{A>B} E_{\pi AB}^T$) of π-KE, see [5]. For benzene the KEPE of the diagonal term is less

(1.1624 a.u.) than for cyclobutadiene (1.29173 a.u., STO-3G [1]). Correspondingly, for benzene the value of the nondiagonal KEPE is greater (0.07962 a.u.), than in the case of cyclobutadiene (0.04483 a.u.). Note that in acyclic polyenes the diagonal and nondiagonal parts of the KEPE have practically constant values of 1.25874 and 0.08079 a.u., respectively [1]. Thus one may conclude that the "aromatic stability is caused by releasing the kinetic-energy pressure on the atom by allowing delocalization between atoms" [1]. Note that the treatment of the aromatic stabilization as being determined by lowered kinetic energy of π-electrons is consistent with the interpretation of the forbiddenness of the concerted reactions, developing via an antiaromatic transition state, as stemming from kinetic energy of the π-electrons [6].

The results of a study reported by Ichikava et al [7] illustrate the fruitfulness of such understanding of the physical meaning of aromaticity, where this approach has revealed an intrinsic difference between the resonance stabilizations of the aromatic cyclic molecules and the acyclic conjugated ones assigned to a class of compounds with the "polymethinic" resonance stabilization, such as **1–3**.

Analysis of the changes of the kinetic and potential energies that occur in consequence of the interruption of π-conjugation effected by protonation of benzene, naphthalene, and **1–3**, as well as of hexatriene taken as the reference structure, has shown the following. For the first two molecules it is the kinetic energy that increases more substantially than in the case of the reference structure, while for the last three molecules the potential energy grows to a greater degree. These results highlight the fundamental difference between the resonance stabilization of the cyclic conjugated molecules ("topological aromatics") and the polymethinic resonance stabilization of the acyclic conjugated molecules, such as **1–3** [7]. The stabilization of the former type of molecules arises from the lowering of the kinetic energy of electrons compared to the acyclic molecules, where it is due to the lowering of the potential energy.

This seems a good model. But is it sufficient enough to account for all the multifarious manifestations of aromaticity, in particular, for the specificity (if there is one) of the aromatic stabilization? Further studies may eventually give an answer to this question.

A different interpretation of the aromatic stabilization is suggested in a series of papers [8–12], where the authors analyzed the structure of benzene and several conjugated molecules with the aid of the spin-coupled theory, which is based on ideas of the VB method. According to that theory, the orbitals are no longer required to be purely atomic but can be expressed as linear combinations of basic functions drawn from all the atomic centers in the molecules. The N-electron system is described by N orbitals, all of which are allowed to be distinct and nonorthogonal. The authors claim that the π-electrons in benzene are almost certainly localized and the characteristic properties of such a system arise from the mode of spin-coupling and not from any supposed delocalization of the orbitals. In this manner the conclusion is drawn as to the illegitimacy of the customary description of the aromatic systems as the ones with delocalized π-orbitals.

A substantial argument against the ideas put forward in [8–12] is McWeeney's comment [13] on [8]. He states that it hardly makes sense to search for reasons for the aromatic stabilization in the context of a method (VB in this case) alternative to the MO methods seeing that the results of both approaches to the description of chemical bonding must be mathematically identical: both seek an approximation to the many-electron wave function and both use orbitals as "building blocks"; it is only in the simplest approximations of both types that disagreement is possible. Or, to put it briefly, for the issue in hand, the VB theory has no advantage over the MO description.

Even so, other attempts to invoke the VB scheme are known. For example, Schultz and Messmer [14], by using it and rejecting the σ, π-approximation, have found that the pattern of bonding in this molecule may be depicted in terms of Ω or double-bent bonds. Being interesting in itself, this fact does not, however, explain the reasons for the aromatic stabilization.

More promising are the approaches that extend beyond the framework of the one-electron approximation, for example, those resorting to pair orbitals [15] or highly correlated Cooper pairs [16]. The latter approach was originally applied to electrons near the Fermi surface in the BCS theory of superconductivity. The aromatic molecule is treated as a structure of positive carbon ions in a ring surrounded by a gas of free electrons. It should be noted that Squire's work [16] was subjected to serious criticism [17], the gist of which is as follows: "it seems most unfortunate to extend the use of highly correlated Cooper pairs toward another (rather ill-defined) concept like aromaticity, when the role of these pairs is now much in discussion in the field where they were originally introduced." (See also the answer in [18].)

Among other investigations dealing with the effect of electron correlation on aromaticity (antiaromaticity) manifestations for the $(4n + 2)$ and $(4n)$ π-electron monocycles [19, 20] (see also Chapter 2), the interesting results [21, 22] should be mentioned. The mean-square deviation of the electron charge at the ith π-center (charge fluctuation) has been used as a quantitative measure of the electron localization. Bond length alternation in $(4n)$ π-electron monocycles was found to be accompanied by an enhancement of the corresponding charge fluctuations.

Recently Schultz and Messmer using the bent bond model, have shown that the nature of resonance is untimely intertwined with the kinetic energy [23]. From the viewpoint based on this model, antiaromaticity can be associated with an unfavorable electron delocalization. Streitwieser, Vollhardt, Weinhold, and their co-workers [24] have found with the use of localized orbital analysis of benzene that π-delocalization strongly stabilizes the symmetric D_{6h} geometry of benzene. This supports the conclusion about the validity of the structural criteria of aromaticity (Section 2.3.1).

Thus, having spent a good deal of time trying to describe and rationalize to ourselves and the reader the intricacies relating to the effect of aromatic stabilization, we have to concede in the end that the commodity we have been dealing in is indeed somewhat illusory. At any rate, up to this time nobody has been able to produce a physically sound explanation to this effect. That is one fact.

The other fact is the undeniable and very real usefulness of the aromaticity concept challenging the investigative minds to persevere in the search for the physical reality in which it may be rooted.

We adhere to the view that there is intrinsic value in the idea of aromaticity and we hope that the present publication may invite a still greater attention to this interesting and enigmatic problem.

REFERENCES

1. H. Ichikava and Y. Ebisawa, *J. Am. Chem. Soc.*, **107**, 1161 (1985).
2. H. Ichikava, Y. Ebisawa, and A. Shigihara, *J. Phys. Chem.*, **92**, 1440 (1988).
3. K. Ruedenberg, in O. Chalvet, R. Daudel, S. Diner, and J. R. Malrieu (Eds.), *Localization and Delocalization in Quantum Chemistry*, Vol. 1, Reidel, Dordrecht, pp. 223–245, 1975.
4. H. Ichikava, Y. Ebisawa, and K. Sameshima, *Bull. Chem. Soc. Jpn.*, **61**, 659 (1988).
5. H. Ichikava, *J. Am. Chem. Soc.*, **105**, 7467 (1983).
6. L. Salem, *Electrons in Chemical Reactions: First Principles*, Wiley, New York, 1982.
7. H. Ichikava, J. I. Aichara, and S. Daehne, *Bull. Chem. Soc. Jpn.*, **62**, 2798 (1989).
8. D. L. Cooper, J. Gerratt, and M. Raimondi, *Nature*, **323**, 699 (1986).
9. D. L. Cooper, S. C. Wright, J. Gerratt, and M. Raimondi, *J. Chem. Soc. Perkin Trans. II*, 255, 263 (1989).
10. D. L. Cooper, S. C. Wright, J. Gerratt, P. A. Hyams, and M. Raimondi, *J. Chem. Soc. Perkin Trans. II*, 719 (1989).
11. D. L. Cooper, J. Gerratt, and M. Raimondi, *Topics in Current Chemistry*, Vol. 153, Springer-Verlag, Berlin, 1990, pp. 42–55.
12. J. Gerratt, *Chem. Br.*, 327 (1987).
13. R. McWeeney, Nature, **323**, 666 (1986).
14. P. A. Schultz and R. P. Messmer, *Phys. Rev. Lett.*, **58**, 2416 (1987).
15. P. E. Schipper, *Chem. Phys. Lett.*, **142**, 393 (1987).
16. R. H. Squire, *J. Phys. Chem.*, **91**, 5149 (1987).
17. G. van Hooydouk, *J. Phys. Chem.*, **92**, 1700 (1988).
18. R. H. Squire, *J. Phys. Chem.*, **92**, 1701 (1988).
19. D. J. Klein and N. Trinajstic, *J. Am. Chem. Soc.*, **106**, 8050 (1984).
20. S. Kuwajima and Z. G. Soos, *J. Am. Chem. Soc.*, **109**, 107 (1987).
21. J. Schütt and M. C. Böhm, *J. Phys. Chem.*, **96**, 604 (1992).
22. J. Schütt and M. C. Böhm, *J. Am. Chem. Soc.*, **114**, 7252 (1992).
23. P. A. Schultz and R. P. Messmer, *J. Am. Chem. Soc.*, **115**, 10943 (1993).
24. E. D. Glendening, R. Faust, A. Streitwieser, K. P. C. Vollhardt, and F. Weinhold, *J. Am. Chem. Soc.*, **115**, 10952 (1993).

INDEX